Digital Transformation of Identity in the Age of Artificial Intelligence

Kazuhiko Shibuya

Digital Transformation of Identity in the Age of Artificial Intelligence

Springer

Kazuhiko Shibuya
Faculty of System Design
Tokyo Metropolitan University
Hino, Tokyo, Japan

ISBN 978-981-15-2247-5 ISBN 978-981-15-2248-2 (eBook)
https://doi.org/10.1007/978-981-15-2248-2

This Springer imprint is published by the registered company Springer Nature Singapore Pte Ltd.
The registered company address is: 152 Beach Road, #21-01/04 Gateway East, Singapore 189721, Singapore

Preface

This book aims to edit an overview on digital social science for social scientists in the age of digital social transformation. Principally, it articulates the nature of *identity of the human-being,* which related to the cutting-edge knowledge on the AI and social sciences. The contents of this book are to widely argue on *identity* by comprehensive investigations in digital social sciences, and it interweaves with wider disciplines related to philosophy, ethics, sociology, STS (including ELSI), computer sciences, engineering, medical sciences except for theology. It firstly reviews contemporary conditions proliferated by advanced technological tendencies, and it further unveils social mechanisms on identity of the human in depth.

Let me reminisce own academic investigations. This book shows a part of research records by the author. The author's latest research process fortunately overlapped last three decades as *the Heisei Era* in Japan (from 8th January 1989 to 30th April 2019), and the author continually intends to commit next researches since the beginning of *the Reiwa Era* (from 1st May 2019). And then this book includes long-term research results by the author, and it arranges a set of collections on findings and thoughts of the author on *identity* sorted by the categories. Namely, before the advent of the AI and big-data fascinations, the author devoted to investigating relevant studies. Contemporarily, there are innumerable controversial issues on the AI, medical sciences, data sciences and other engineering, and then naturally those advancements have strong optimisms for the future in progress. However, many risk concerns have also occurred against those matters. Both differences are actually coined in the common reason, and each of both looks at a possibility on each side of positive or negative. Such dichotomy of a possibility should be critically verified in wider diversifications.

In addition, this book seems to be also a lecture note in university. The author has been undertaking an undergraduate course '*Risk Management*' at Tokyo Metropolitan University, Japan. This curriculum intentionally diversifies the risks from genetics, medical sciences, chemical safety, disaster, computer security, and social issues related to emerging risks in term of computational social science. Through teaching those topics, the author got an idea to condense the diversity on those risks to a concept '*identity*' relating with those different disciplines. Every risk of both social and natural phenomena usually intersects with a common root of the problems on the humankind, and then the author intended to explore hidden relevant linkages on identity interweaving with those fields.

The author has a disposition to eagerly review any social phenomena under the continuous dynamics and variating progress such as historical changes, moving patterns of any variables and complexity. Even though the research career of the author started from philosophy and social sciences, research experiences by computational sciences and practical solutions have enriched to deepen such activities. Conceptual thinking and thought experiments could be grounding in computable solutions by mathematical modeling, simulations and social designs by computational services. And conversely, possible technological solutions could be feed-backed to integrating meta-level foresights.

Of course, the author is not omnipotent, and this book has many spaces to be upgraded. But such contributions for the future studies will be continued by the author if possible. This book will become a first step for such pursuing the goals.

Tokyo, Japan

Kazuhiko Shibuya

Acknowledgement

A part of this work (Research Project Leader: Shibuya, K) was officially supported by JSPS KAKENHI Grant (Number 26590105: Grant-in-Aid for Challenging Exploratory Research: *An Exploring Study on Networked Market Disruption and Resilience*) and ISM Cooperative Researches (Number 19-0008: *A Study on the Research Evaluation of Science & Technology and the Rationality of Decision Making*, and Number 28-0017: *A Study on Mathematics for International Cooperation*).

Contents

List of Figures

List of Tables

Part I
Foreseeing for Digitized Society

Chapter 1
Digital Social Science

1.1 What Does the Advanced Technology Mean for Human-Being?

1.1.1 When AI Excesses the Leader of the Primates

The fascinated concerns with the AI and big data have been growing in these days. Over the past few years, a considerable number of industrial and business solutions have been already conducted in these areas. It is called either the AI or big-data revolution. In light of the technological paradigm shift since the beginning of the human history, occasionally, there were innumerable times being opened novel views to enlighten next steps for us in the past. Reminiscently, industrial revolution, personal computer boom, internet' tremendous upheaval, and smarter mobile phones' solutions were good examples. Such technologies had innovated on our life.

Now, we indeed face the next bifurcation, and namely the AI and big data vividly induce us to step forward further destination. Probably, the humankind cannot return to the backward. Since we obtained prototypes of those handful tools, we already sought to be eagerly climbing over the plateau of technological limitations. Until now, meanwhile traditional products made by technological paradigm shifts had renewed mostly exterior and physical environment of our daily life, current digital transformation progress which led by the AI and big data rather lay much weight on enhancing our unseen intellectual and rational functions.

Moreover, the autonomous mechanisms led by the AI-based services have the capabilities to exhaustively learn own experiences and assimilate necessary data through the large amount of big data, and consequently a part of the performance of such facility on intellectual tasks already exceeds more than the humankind.

Some theory further predicts that the singularity of the AI will sooner or later come in the near future around 2045. Of course, many of such forecasts have not enough evidences. It feels to be exaggerated by broadcasting news. The most important is to prepare latent risks against the brink of the future, and such criteria will be

K. Shibuya, *Digital Transformation of Identity in the Age of Artificial Intelligence*, https://doi.org/10.1007/978-981-15-2248-2_1

tipping point of the superiority loss of the human against the AI. When the AI exponentially excesses our intellectual level (Bostrom 2016), it must reconsider our existence itself and modality of identity (Copeland 1993). Since the humankind historically emerged on the earth, we will face at the first-contact to the something smarter existence than ours. Many of the people will pursue a meaning what an existential modus of them is. And it will further call to redefine a raison d'etre of each individual in own daily life.

1.1.2 Big Data Accumulates Everything

We cannot still transparently look over the whole landscape of the big data. It exponentially keeps growing to be bigger day by day. In such a paradigm shift, digital transformation since the beginning of our civilization means a progressive form for social revolution, which exceeds our intellectual limitations.

To date, big data and its statistical analysis can reveal the unique value of each individual, and it collectively devotes to exploring implicit meaning of unidentified patterns from multiple data pools such as either deindividualized or personalized data. It includes both physical existence itself and biological data of each person, while those data often omitted the important value on privacy information of each individual, and thus it can be universally plotted in mathematical space (Fogel and Kvedar 2018). Namely the tremendous extent of those big data such as genome, DNA sequences, health data, behavior logs in daily life, transaction data with interconnected communication, geospatial location data, intellectual abilities, skill levels, CV as life-long achievements, human relationships, hobby, historical evidences, and other specifications of each person has been widely accumulated, stored, and managed little by little. The AI driven system impressively runs the computing procedures which coded any programs to predict, explain, and control various phenomena for the pursuit of academic purposes and practices on the bases of those large amounts of big data.

Further, interconnected services by XaaS (Everything as a Service) extended from MaaS (Mobility as a Service) have been universally overlapping all over the world. Everything in the world has been already possibly interconnected with each person, mobile devices, wearable devices, IoT (internet of things), M2M (machine-to-machine), connected vehicles, AI driven machines, smart houses, fabrications, plants, smart grid network, smarter cities (Geng 2017), and virtual nations through wireless networking (e.g., 5G). As those big data have been growing its tremendous size over than our cognitive capacities of total controls, all of them can be operationalized by the AI driven mechanisms.

1.2 The Sign of Latent Risks

1.2.1 How Can We Depict a Good Future Opened by Advanced Technologies?

In the light of those contexts, we vigorously rush into the door of digitized society by investigating harder such themes.

However, as already known, many of the citizens had noticed disadvantageous discomforts on such trends (Papacharissi 2014; United Nations Security Council 2018; Berkich and D'Alfonso 2019; Congressional Research Service (USA) 2019). For example, there are warning about AI driven military forces and genome editing. Those trends have intuitive vulnerability against the human-being in the future, and its solution should be collaborated with stakeholders (e.g., nations, developers, medics). But the gap among their opinions is now far from each other.

Actually, on March 2019, it was too suggestive that two significant congresses related to the United Nations held in Geneva, Switzerland. The former was an expert committee organized by the WHO in which genome editing was to be discussed, and they intended to recommend global registration and monitoring system against illegal and unethical activities using genome editing. Secondly, the latter case continued to discuss controversial issues on advanced AI weapons, it is especially called LAWS (Lethal Autonomous Weapon Systems) and other military specifications (UNODA 2017; UNOG 2018).

Both technologies can be definitely recognized as one of our state-of-the-art. Of course, these technologies further include regenerative medicine (e.g., iPS cells), many fields on the AI, computational engineering, data sciences, and other progresses. Concretely, these technologies have been envisaging another possibility of the possible worlds for our next steps, and some of our dreams have been achieved by the efforts of the human.

However, those congresses share latent anxieties against such ongoing progress of technologies. Can we a good future truly open by advanced technologies? Rather, it should be restated more accurate how we can regulate such solutions by the humankind. What should it regulate by the humankind? Advanced AI? Military robots and weapons driven by the AI? Is it limitless lifespan by regenerative medicines? Or, is it other technologies in the future?

1.2.2 Asilomar Conference on Beneficial AI

Regarding latent threats and ethical issues on the AI, on January 2017, Asilomar Conference on Beneficial AI was held in the USA, and it had lastly proposed *the 23 Asilomar AI Principles*. Those principles consist of ethics and moral for the AI, and it enumerates noticeable articles as follows.

Research Issues

1. *Research Goal: The goal of AI research should be to create not undirected intelligence, but beneficial intelligence.*
2. *Research Funding: Investments in AI should be accompanied by funding for research on ensuring its beneficial use, including thorny questions in computer science, economics, law, ethics, and social studies, such as:*
   - *How can we make future AI systems highly robust, so that they do what we want without malfunctioning or getting hacked?*
   - *How can we grow our prosperity through automation while maintaining people's resources and purpose?*
   - *How can we update our legal systems to be more fair and efficient, to keep pace with AI, and to manage the risks associated with AI?*
   - *What set of values should AI be aligned with, and what legal and ethical status should it have?*
3. *Science-Policy Link: There should be constructive and healthy exchange between AI researchers and policy-makers.*
4. *Research Culture: A culture of cooperation, trust, and transparency should be fostered among researchers and developers of AI.*
5. *Race Avoidance: Teams developing AI systems should actively cooperate to avoid corner-cutting on safety standards.*

Ethics and Values

6. *Safety: AI systems should be safe and secure throughout their operational lifetime, and verifiably so where applicable and feasible.*
7. *Failure Transparency: If an AI system causes harm, it should be possible to ascertain why.*
8. *Judicial Transparency: Any involvement by an autonomous system in judicial decision-making should provide a satisfactory explanation auditable by a competent human authority.*
9. *Responsibility: Designers and builders of advanced AI systems are stakeholders in the moral implications of their use, misuse, and actions, with a responsibility and opportunity to shape those implications.*
10. *Value Alignment: Highly autonomous AI systems should be designed so that their goals and behaviors can be assured to align with human values throughout their operation.*
11. *Human Values: AI systems should be designed and operated so as to be compatible with ideals of human dignity, rights, freedoms, and cultural diversity.*
12. *Personal Privacy: People should have the right to access, manage and control the data they generate, given AI systems' power to analyze and utilize that data.*
13. *Liberty and Privacy: The application of AI to personal data must not unreasonably curtail people's real or perceived liberty.*
14. *Shared Benefit: AI technologies should benefit and empower as many people as possible.*

15. *Shared Prosperity: The economic prosperity created by AI should be shared broadly, to benefit all of humanity.*
16. *Human Control: Humans should choose how and whether to delegate decisions to AI systems, to accomplish human-chosen objectives.*
17. *Non-subversion: The power conferred by control of highly advanced AI systems should respect and improve, rather than subvert, the social and civic processes on which the health of society depends.*
18. *AI Arms Race: An arms race in lethal autonomous weapons should be avoided.*

Longer-Term Issues

19. *Capability Caution: There being no consensus, we should avoid strong assumptions regarding upper limits on future AI capabilities.*
20. *Importance: Advanced AI could represent a profound change in the history of life on Earth, and should be planned for and managed with commensurate care and resources.*
21. *Risks: Risks posed by AI systems, especially catastrophic or existential risks, must be subject to planning and mitigation efforts commensurate with their expected impact.*
22. *Recursive Self-Improvement: AI systems designed to recursively self-improve or self-replicate in a manner that could lead to rapidly increasing quality or quantity must be subject to strict safety and control measures.*
23. *Common Good: Superintelligence should only be developed in the service of widely shared ethical ideals, and for the benefit of all humanity rather than one state or organization.*

Above list has a strong ethical assertion for all of the human-being. In science and technology and ethical studies, "*Asilomar*" has another meaning of the Asilomar conference on bioethics in 1975. The coincidence between two "Asilomar" was too suggestive for us. It was certain that this "Asilomar" on the AI intended to explicitly organize the rules of conduct in AI researches and its doctrines. Commonly, many of the researchers hold and share serious concerns that those technologies have sustainable risks against the humankind. Unless it will not prepare any effective solutions, the human-being will confront such risks at the brink of the future for the future generations.

1.3 Grounding in the Actuality

1.3.1 Grounding Between Technologies and Thought

Recently, many publications and online materials related to AI, big data, and medical sciences have been popularly published in the world. Such stories were participating by renowned scholars such as Hawking (theoretical physics), Harari (history), and the other academic scholars in multi-disciplines. But the discourses on technology forecast for the future will be often gradually moving away from the truth and writer's original intentions, and consequently confusions about such knowledge among naive citizens will become extremely more vague and unexpected interpretations. Actually, some of them were often underlying in imaginary understanding. Why will the humankind be overthrown by advancement of the AI in line of popularized stories?

Recently, Cave and Dihal (2019) reviewed an intrigued report to categorize types of SF stories on the AI, and they concluded the taxonomy of either optimism or pessimism on the AI such as *the hope for much longer lives ("immortality"), the fear of losing one's identity ("inhumanity"), the hope for a life free of work ("ease"), the fear of becoming redundant ("obsolescence"), the hope that AI can fulfill one's desires ("gratification"), the fear that humans will become redundant to each other ("alienation"), the hope that AI offers power over others ("dominance")*, and *the fear that it will turn against us ("uprising")*. This taxonomy clearly teaches me an outline of discussion on the AI. In particular, the goal of the author is to discuss "*inhumanity*," and of course it delves into each of those viewpoints.

Certainly, military applications of the AI have important concerns for us ("*dominance*" and "*uprising*"), but there are also a lot of contaminations of fake stories and overexpression in daily publications. Such materials offer only trendy discussions for ordinary citizens rather than academic verifications. Then, it should be carefully verified. Such story telling can depict a lattice of the whole, and any evidences would denote a certainty turned from one of the possibilities. No story has described all of the possible bifurcation patterns. Future forecast based on the past and present conditions will be often constrained by statistical variations, emergent properties, and complexed interactions within co-occurring variables.

Technologically, popularized publications are apt to write one side on either optimism or pessimism such as singularity of the AI, regenerative medicine (and immortality), big-data analysis, and other engineering progress. In such contents, a part of them included a fiction, and it should be rather managed as a kind of fake news. It should verify each of the evidence which is often popularized. And it must be laid weight on multiple and compound factors rather than single factor, and each of them and emerging matters ought to be respectively reexamined in scientific manner.

- **Discourse statement**: "Advanced computation overcomes everything."
- **It is FALSE now**.
- **A reason**: For example, computer simulation on the whole human body is impossible now even if one of the best supercomputers is assigned to solve such tasks. Nevertheless earth weather simulations and many simulations on civil engineering can manage innumerable variables using supercomputers, our biological living system has impossibly more complex and dynamic interactions than such solutions. And quantum computing has not been completely built, even though many of quantum mechanisms in physics had been theorized and proved. It is now undergoing development, and it is necessary to be coding specific operations for quantum computations (Wallden and Kashefi 2019).

- **Discourse statement**: "the human-being understand perfectly own living phenomena."
- **It is FALSE now.**
- **A reason**: Human genome had been completely parsed in bioinformatics (International Human Genome Sequencing Consortium 2004). But, nobody knows completely the living mechanisms of our body. In the present, even if computer simulation on a cell eliminated from a physical body is possible, it is still too difficult to manage whole living body because each inner cell has dynamics of complexity on many interactions by many compounds and gene networks (Arabnia and Tran 2016). And then, Tomita et al. (1999) investigated a simple primitive cell (i.e., *E. coli*) for their simulation (E-Cell project). Moreover, the living complexity of gene networking and epigenetics have not been understood in all cases, and those mechanisms could not be unveiled clearly yet.
- **Discourse statement**: "Cancer can be perfectly cured by the state-of-the-art of the human-being."
- **It is FALSE now.**
- **A reason**: Cancer disease has been intensively investigated by medical sciences during several decades, but much effort will be required more over. Even though the extent of knowledge by medical science researches has been steadily increasing, it has not reached the horizon as final goal yet. And molecular dynamics among cells are ruled in specific physical principles (e.g., intermolecular force, van der Waals force), and those mechanisms often bring tough headaches to researchers.
- **Discourse statement**: "Regenerative medicine can cure for every patient."
- **It includes both FALSE and TRUE now.**
- **A reason**: Using iPS (induced pluripotent stem) cells and ES (embryonic stem) cells (Wert and Mummery 2003), regenerative medicines have been growing to apply for each clinical case. In the present, several clinical cases can be applied. For example, the retina transplantation case has already been approved officially by the Japan authority.[1] In the near future, skin, bone, blood, and other organs will be regenerated from iPS cells, and medics transplant it to the body of patients.
- **Discourse statement**: "Genome-editing can modify biologically each of the all livings."
- **It includes TRUE now.**
- **A reason**: Genome editing has much potential to redesign the living species (including our human-being) on the earth. But the consequences of its genetic operations further call our responsibilities for all living species (Martin et al. 2012). Both all tools and modified materials should be regulated in both laws and bioethics (Cohen 2019). For example, one of the most influential science journals *Nature* has published special topics on genome editing regulation and called for discussion on March, 2019 (Lander et al. 2019). Their augments stood on

---

[1] https://www.amed.go.jp/en/seika/fy2018-05.html

ethical conducts rather than scientific research progress, and international researchers critically claim against a false start case in China (Cyranoski 2019).

- **Discourse statement**: "the human-being gains reincarnation and immortality."
- **It includes FALSE now**.
- **A reason**: To date, there is a good promise that advancement of traditional and regenerative medical science and its engineering prolongs our lifespan which cared by effective ways (Surani 2012), and those efforts certainly provide the chances to fulfill each life goal for us. And the National Institute on Aging (USA)[2] has been intensively promoting aging researches, and further one of the largest ICT companies Google already invested the larger amount of financial supports to one of their child companies *Calico*[3] which has been investigating biological and medical sciences such as aging controls and *senolytics* (i.e., an academic study to explore and develop senolytic agents which are to delay, prevent, alleviate, or reverse age-related diseases. Its ultimate goals indeed include regenerative controls for cancer diseases, Alzheimer's diseases, near immortality, and prolonged lifespan) in particular. Health and genome big data can be matched for computational ways to explore the unknown gene functions and the secrets of the nature.
- However, regenerative medical science has been examining cellular senescence and its immortalization derived from mice body (Utikal et al. 2009; Odell et al. 2010), and then it cannot sooner apply for the human-being. If possible, when medical science success to gain both reincarnation and immortality, it will call rather another problem. For example, there will be overpopulation, food shortages, living environmental matters, and severe issues on social welfares at least. Such controversies are discussed later in detail.
- **Discourse statement**: "AI is (and will become) our threats and enemy."
- **It includes FALSE**.
- **A reason**: No AI could solve all models and theorems. AI driven performance has been owned by the large amount of somewhat big data. And ordinary performances of domain-specific AI have not been critically applied to the general purposes. AI has not autonomously produced own consciousness and metacognition yet. Whatever either domain-specific AI or general purpose AI will be built, such systems cannot be omnipotent (namely God-like machine: Deus Ex Machina). Higher order inference will not be managed by such AI, because Turing' theorem (it partly depends on Gödel's incompleteness theorem) had already proved the case of computational limitation (Turing 1950; Copeland and Shagrir 2019; Ben-David et al. 2019). Besides, symbol grounding problem (Harnad 1990) and frame problem has been also arguing in those contexts. Concluding concisely, both the AI and the human-being will not ultimately leap the border of the God (what is "God"?) in this condition. As mentioned later, our

[2] https://www.nia.nih.gov/

[3] https://www.calicolabs.com/

enemy will be rather ourselves (e.g., malicious AI developers, terrorists, and military nations).

- **Discourse statement**: "the AI driven system will rob our jobs."
- **It includes both FALSE and TRUE now**.
- **A reason**: It will be certain that a larger amount of the AI driven systems and working robots will be vigorously employed to work for our alternative, and then many of us will be gotten rid of works. But it should say releases from harder duty of working rather than the loss of jobs. Certainly, a part of jobs will be still further enduring in some working fields such as scientific investigations on AI studies, big-data sciences, mathematics, physics, medical sciences, and other necessary academics which will be undertaken by parts of human scholars, for example. The most significant is what many of the human should do in vacant time. And there is another problem how to earn the money for our daily living. Nevertheless the AI driven robots will serve all necessaries for our daily living, as providing basic income (BI) and somewhat rewarding systems for all citizens, and still then, many of ordinary citizens will suffer how to spend their free time. Entertainments such as game, sports, and other hobbies as well as further motivations to achieve something by self-learning will be inspired by enterprise businesses.

In this way, breaking pseudoscience (including fake news) is to distinguish between subjective belief and objective fact (Shibuya 2021, in press). The former is only belief in our cognitive system and it justifies our world view. The latter is a fact and evidence, and it shall stress on scientific verification and prioritized importance rather than public preferences for acceptances. Moreover, it definitely claims that it needs falsifiability (Popper 1963).

For example, medical and life sciences have own incompatible issues among their scientific reexamination on papers. As many clinical research findings could not be often reproduced, NIH (National Institute of Health, USA) had already recommended that the results of significant findings should be revalidated (Ioannidis 2005; Fanelli 2018). The reason is that some of them are consisted of the conditions under suspicious of forgeries or ambiguous results in examinations (Steen et al. 2013).

Similarly, social evolutionism (e.g., Marxism) had own vulnerabilities not to be grounding between theory (ideology) and actuality (Popper 1963). Many of their failures were caused by systematic lacks of objective verification and updating cycles. Their stories were only deduced from their belief and conclusion (their first hypothesis and conclusion are almost same meaning). Theoretical-based studies and story-based ideologies do not often infill the gaps. Causal relationships of such story often remain many unaware gaps and latent missing links. Nobody knows all necessary variables yet. If possible, somebody can already formalize a model using such variables. It should be cared about a term: *If it never exists, it is never observable*.

Thus, hypotheses, observations, analyzations, theories, models, and interpretations conducted by digital social scientists (and all of the academics) must devote to grounding in actual verifiable evidence and genuine scientific manners.

1.3.2 Brief Historical Perspective of Digital Progress

Next, the author roughly visualized a historical map of a part of the digital progresses from 1990s to around 2045. Historical time arrow directs from the top-left to bottom-right. Top-right side depicted a kind of hardware, and bottom-left side plotted something relational to software.

At first, ubiquitous computing was originally proposed by Weiser (1993). "*Ubiquitous*" in Latin suggests "God exists pervasively everywhere." Fruits ripped from his conceptualization for ambient computation have now been accepted by all of the citizens, and namely these are AI driven services, smartphones, and widely internet-connected technologies. Past sensor networking systems and devices (e.g., Active Badge (Want et al. 1992)) were partly origins for IoT (internet of things) (Geng 2017). In the past, Abowd and Mynatt (2000) summarized the three directions of ubiquitous computing such as *natural interfaces*, *context-aware applications,* and *automated capture and access* in cyber-physical environment. Their future predictions are now established by many of industrial products.

In 1995, the internet revolution has publicly reconstructed our daily life and global world. Undoubtedly, WWW (world-wide-web: Berners-Lee 1999) had been approved as one of the most innovated solutions in twentieth century. Afterwards, Silicon Valley ventures and innovative companies had achieved new trends such as internet search engines, browsing applications, Java technologies, social networking services, and user-friendly devices during 1990s to 2010s (Fig. 1.1).

Supercomputing power has been keeping grown up its performance. For example, the Earth Simulator (Japan) gave land-breaking shocks to the US researchers, and it sensationally called "*Computenik*[4]" shock. This maximum performance ranked in LINPACK benchmark as 122.4 TFLOPS in 2009. Next generation of the Earth Simulator was named "*K*" computer[5] (it finally achieved more than ten petaflops), and it was launched in 2011. Additionally, "*Fugaku*" as an exa-scale supercomputer in Japan has launched under development (10^{18}, a quintillion calculations per second). Next generation of such supercomputer over than exa-scale flops has been developing in each developed country (Lu et al. 2018b; Skordas 2019). Furthermore, if quantum computing appears in the future, it will overwhelm the traditional architecture of supercomputing, and it exceeds more extra performances

[4] "The Divergence Problem," Horst D. Simon, Director, NERSC Center Division, Lawrence Berkeley Lab, November 19, 2002.

[5] http://www.nsc.riken.jp/index_j.html; http://www.riken.jp/~/media/riken/pr/publications/anniv/riken100/part2/riken100-2-4-7.pdf (in Japanese).

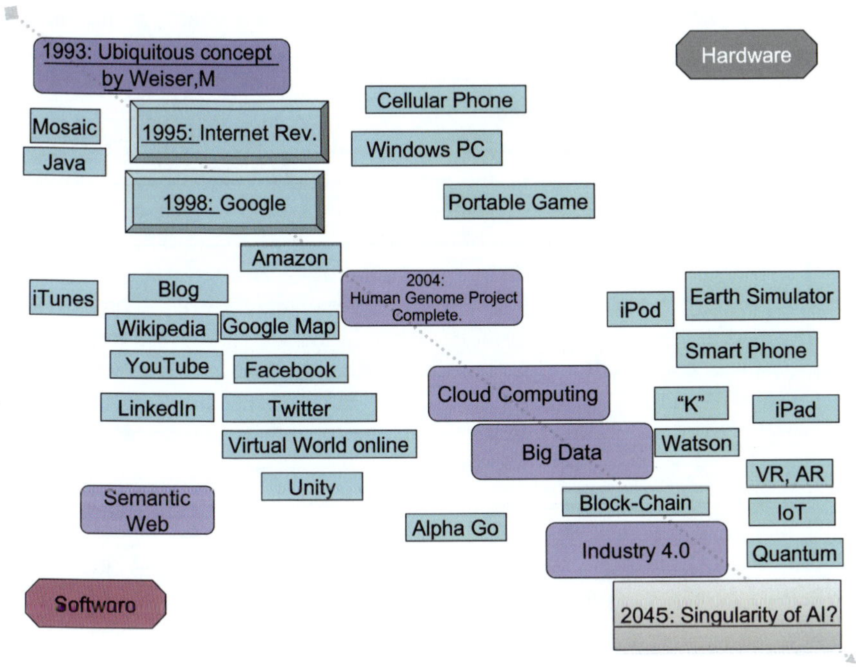

Fig. 1.1 Rough historical map on a part of information technology progresses

(Lu et al. 2018a). It will asymptotically approach the physical limitation to improve our requirements for digitized data analysis and simulations.

Further, AI technologies have been steadily growing during several decades. AI Watson made by IBM won champions of the TV quiz shows in 2011, and Alpha-Go made by Google beaten a champion player of game Go in 2017. Moreover, in 2019, Jaderberg et al. (2019) reported that multiple corporative agents produced by the DeepMind (one of the AI research companies of Google) won teams among human players in collaborative online game. Those cooperative mechanisms of the AI have been studying as one of the human-agents' interaction and multi-agents' coordination problems for adaptations in cyber-physical environment (Weiss 2000). To date, such AI driven system has lastly achieved their collaborative tasks through machine learning process. Generally, collective coordination among multiple AI driven systems is harder than single one. That is why there are greater needs to implement a set of intellectual mechanisms such as understanding oneself, belief for the worlds and other existences, inference for the others, acquiring sensual information, judgement what to do (and not to do), and adaptive behavior for coordination and cooperation among a large amount of agents (it is a kind of the AI driven software entity) at least. Namely, because collaboration among multiple AI driven agents and robots can be expanded for adapting in any interconnected environments such as XaaS and MaaS, it will be more valuable than winning by single AI. Connected autonomous cars and working robots can be enhanced by those coordination systems of equipped

AI in complicated daily environment, and collaborative assistances by the smarter robots for human citizens in disasters have been investigating (Kitano et al. 1999). So then, the AI driven system has been steadily evolving from the stage where singly become a strategic game winner to the higher stage where required collective coordination among numerous partners step by step. Therefore, those sensational results simultaneously announced a loss of one of the intellectual superiorities of the human. After that, AI driven system design and development have been opened the door to further step forward the advancements.

The AI frontier of machine learning progressed by deep-learning studies has been equipped by neural network models which are mathematically referred to neural circuits of the human brain (Aimone 2019). In this sense, both bio-inspired and neuro-inspired computation had been designed and examined for the purpose of next generation computing (Kim et al. 2019). Such computation has the greater needs to conquer the physical limitation of Moore's law for faster computer processing. Those neural networks within our brain have attractive benefits for computational engineering. Although many of those mechanisms are still in prototype stage, computer engineers are now encouraged to bridge between neural sciences and medical engineering, because those biological systems have useful specifications such as self-repairing, self-learning, and self-adaptiveness. Computational applications will become more resilient, evolutionary, redundant, and smarter design if such attempts success.

Additionally, after the middle of the twenty-first century and in the future, our society will certainly become more AI driven, big-data oriented and interconnected world. Social systems will be also underlying in information infrastructure of blockchain-based interconnections and distributed database. These data management and automatic analysis by AI have been already undertaking in various social conditions. Portable devices such as smartphone, wearable devices, and implanted microchips will approach human's behavior to cybernetics. Digitized information society aims at enhancing next industrial revolutions (industry 4.0 and more) crossing with different backgrounds such as academic researches, industries, business, and governments.

In economics, at least, some initials "S" are required such as *sharing*, *subscription*, *social*, *smart,* and *sustainable*. These factors improve our daily life from the paradigm of mass production, mass consuming, and mass disposal to the next paradigm of optimizing necessary plans on economy proposed by the AI systems. These features certainly originated the breakthrough of digitized services for daily life and working demands globally. And above all data has potentials to be earned as data economy. Consequently, it can decrease the total costs such as energy, wages of human resources, and other resources.

- *Sharing*: Time and space, common resources with partners.
- *Subscription*: It offers necessary services by fix amount of payment.
- *Social*: Social media and networking offers the platforms to achieve further possibilities.

- *Smart*: The AI supports various optimization, analyzation, simulation, and decision-making.
- *Sustainable*: Environment friendly consumptions and development are quite needed.

Therefore, AI and big data recently became one of the buzzwords. As historical process indicated above, these technologies have been investigated by various long-going efforts. Then, why does it become AI and big-data "revolution"? The author concisely summarizes below factors to become such revolutionary condition.

1. Computable solution becomes faster:
   Computer CPU (central processing unit) and clustering technologies can achieve faster calculations. Further quantum computing will break the limitation.
2. Thinkable complexity becomes deeper:
   As deep learning shows brilliant results to solve complexities on domain-specific problems and it takes advantages of dealing with pattern recognition and classification demands in computer architectures.
3. Manageable data become larger:
   Cloud computing and large sizes of data storages have been globally accumulating daily data, and the innumerable system runs to transact data with databases around the clock.
4. Tracking collective dynamics of human behaviors in real time:
   Using sensing devices and portable mobile tools with geospatial information, it can be analyzing collective dynamics of human behavior in real-time sequences (Shibuya 2004; Renso et al. 2013).
5. Inter-communicable faster:
   Global society launches 5G (the fifth generation) based wireless intercommunication, and it will globally enhance speeds and quantity of digitalized data transactions.
6. Inter-connectable wider:
   XaaS interconnected designs in whole society have been proposed. Using devices, human-being and computable entity (i.e., smart products, robots, IoT, sensors, connectable cars, etc.) have capacities to interconnect with each other. Those robust networking should be required in systems management and computer security.
7. Translatable accurately:
   The AI-based natural language process and machine translation systems have been exponentially evolved during last several years. It can proficiently offer multi-translations across various languages for global communication beyond the cultural and lexical differences.
8. Collaborating with different industries and partners:
   Possible combinations to collaborate with different industries and partners are increasing. For example, these are patterns such as medical science and engineering (i.e., patterns recognition for diagnosis, medical engineering), car industry and AI (i.e., connected car, auto-driving car, transportation scheduling and its

optimization), energy industry and AI (i.e., smart grid), municipal government and AI industry (i.e., city planning and digital government enhanced by AI and big data), sports industry and data science (i.e., sports analytics), and other combinations.

9. Multiplying above factors more comprehensively:

As listed above factors, recent *revolution* on AI and big data has already opened the gate to heighten our daily life and its possibilities.

In such backgrounds, developed countries eagerly invest a large amount of budgets for innovating and progressing both academic and industrial fields such as AI, big data, medical health, and information communication technologies (Executive Office of the President (USA) 2016). Especially, Japan government has approved a total budget of approximately 700 million US dollars for AI researches in 2018FY. But, the USA already prepares at least more than 4.5 billion US dollars for investment on AI, and it seems that China also invests more than 4.0 billion US dollars for investment on AI-related researches and developments in 2018FY. Both countries may excess more than six times of the extent of Japan's investment on AI.

In this concern, GAFA platformer companies (Google, Amazon, Facebook, and Apple) have been keeping their oligopoly in global markets by possessing greater part of global big data. In 2018FY, their total market capitalization could be estimated 3.4 trillion US dollars, and they still occupied 13.2% within S&P 500 companies (total market capitalizations were approximately 25.6 trillion US dollars). On the contrary, China plans to become the world's top-level AI innovation country by 2030 (China Institute for Science and Technology Policy at Tsinghua University 2018).[6] And China's unicorn companies (abbreviated as BATH (Baydu, Alibaba, Tencent, and Huawei)) have been increasing their presences in the world. Especially, China invests larger grants for intensification on the AI applications such as automatic driving cars, smart cities, supercomputing, FinTech, online shopping, social media, smart agriculture, robotics, biotechnology, military industry, and other R&D fields (Chen and Li 2018).

Then, new technology war between the USA and China has already provoked mutual serious tensions (McKinsey and Company 2017). Following with them, global competitions among other nations such as EU countries, Japan, and other nations have been investing large amount of budgets and supports for more intensive innovations (i.e., World Economic Forum: Artificial Intelligence and Robotics).

[6] Will China lead the world in AI by 2030? https://www.nature.com/articles/d41586-019-02360-7.

1.4 What Is a Role of Digital Social Science in Those Contexts?

1.4.1 Definition of Digital Social Science and Its Vector

Therefore, as those social surroundings based on above technological progresses already change all of our society, social science research faces the paradigm shift of own researches. It should be renewed in digital transformation and computational manners. Now, such academic study *by*, *for*, and *of* the social scientists is called "*digital social science*" ("*digital sociology*" (Orton-Johnson and Prior 2013; Marres 2017) and "*computational social science*" is sometimes used as synonym of digital social science).

According to a definition on "*digital social science*" by SSRC[7] (Social Science Research Council), they defined "*Digital Social Science: This working group will engage how social scientists use digital tools, methods, and data sources in their research. This includes "big data" (whether from the internet, social media, geo-spatial techniques, or more traditional quantitative and textual sources).*" In addition, academic courses on digital and computational social sciences have been globally launched in universities (e.g., University of Chicago, Stanford University, Oxford University, Santa Fe institute). Their curriculum suggests their motivations and directions for academic studies in this emerging field.

The AI and big-data age may eventually provoke rather many undesirable and uncomforted matters against the human. Impressively, in the twentieth century, Habermas (1976) had conceptualized the crises patterns between social systems and our identity for the integrations in the context of later capitalism, political, technologies, and other social factors ("*Legitimationsprobleme im Spätkapitalismus*").

The author thought as follows. Regarding digital data, during twenty-first century, traditional social sciences should be orchestrated for updating by computational knowledge, skills, and technologies, which are related to digital and computational social science, data science, digital humanities, and somewhat engineering. Regarding those backgrounds, as a pioneering role by digital social science, it should undertake to bridge between theorization and technological progresses. "*Social science*" in digitalization diversifies its own meanings. Here, the author should divide two parts such as sociological taxonomy and interdisciplinary one.

[7]Please see an item 'SSRC' of 'Further Researches' within 'References'.

1.4.2 Sociological Taxonomy

As borrowing ISA (international sociological association) RC (research committee) numbering categories (it often calls *Bindestrich-Soziologie*), the author has an idea to categorize topics in digital social science. Each RC[8] has own large amount of members, and total of RCs itself cover much fields in sociology (more than 50 RCs). But there is not a RC "digital sociology" (and "computational sociology") yet. Then, digital social science can be foreseen in various contexts by corresponding with each area to digitized social topics.

- RC01 (Armed forces and conflict resolution) [*Related to Chaps. 1, 2, and 13 in this book*]

- It includes issues on advanced AI driven military arms, robotics, sharp power, and other conflicts.
- RC02 (Economy and society) [*Related to Chaps.* 4, 5, 12, *and* 13 *in this book*]
- It includes issues on analysis of consumer behavior online, blockchain-based financial application, digital economy, and socioeconomic revolution in digital society.
- RC05 (Racism and nationalism), 31 (Migration) [Related to Chaps. 8, 9, *and* 10 *in this book*]
- It includes issues on hate speech online, fake news on racism in social media against antagonists, immigration crisis, and nationalism matters in digitized actuality.
- RC07 (Future research), 23 (Science and technology), 51 (Sociocybernetics) [*Related to Chaps.* 1, 2, 3, 4, 5, 12, *and* 13 *in this book*]
- It includes issues on directly forecasting for AI and big-data centered society.
- RC10 (Participation, organizational democracy, and self-management), 18 (Politics), 48 (Social Movement) [*Related to Chaps.* 8, 9, 10, *and* 13 *in this book*]
- It includes issues on social movement and participatory democracy depended on social media.
- RC11 (Aging), 38 (Biography), 41 (Population), 06 (Family Research) [*Related to Chaps.* 3, 4, 11, 12, *and* 13 *in this book*]
- It includes issues on demography, population statistics, aging, and social welfare matters.
- RC19 (Poverty, social welfare, and social policy), 28 (Social Stratification), 30 (Work), 47 (Social Classes) [*Related to Chaps.* 4, 11, 12, *and* 13 *in this book*]
- It includes issues on socioeconomic policy making for basic income, social welfare with AI-centered working and other working environment.
- RC33 (Logic and methodology), 55 (Social Indicators) [*Related to all Chaps. from* 1 *to* 13 *in this book*]
- It includes issues on advanced methodological investigations for computational and digital social sciences.

[8] https://www.isa-sociology.org/en/research-networks/research-committees

- RC39 (Disasters) [*Related to Chaps.* 10, 11, *and* 12 *in this book*]
- It includes issues on disaster management and sociological risks on hazards using ICT.
- RC40 (Agriculture) [*Related to Chap.* 13 *in this book*]
- It includes issues on smart agriculture, management in rural community using ICT, and food traceability for consumers using smartphone.
- RC42 (Social Psychology) [*Related to Chaps.* 7, 8, 9, 10, 11, 12, *and* 13 *in this book*]
- It includes issues on analyzing media usage and human factors of communications using social media and AI-centered interactions.
- RC43 (Housing and built environment) [*Related to Chaps.* 11 *and* 13 *in this book*]
- It includes issues on smart house, AI managed environment, smart grid system in community and auto-driving vehicles in community designs.
- RC45 (Rationality) [*Related to Chaps.* 4, 6, 7, *and* 13 *in this book*]
- It includes issues on rational choice theory based considerations for AI and big-data centered society.
- RC46 (Clinical sociology), 49 (Mental health and illness), [*Related to Chaps.* 4, 11, 12, *and* 13 *in this book*]
- It includes issues on technology stress, mental disorder caused by much usage of social networking, therapy by AI diagnosis, and health managements using ICT.
- RC54 (Body in sociology) [*Related to Chaps.* 2, 3, 4, 5, 12, *and* 13 *in this book*]
- It includes issues on contemporary dualism between cognition and body in digitization.
- RC56 (Historical sociology) [*Related to Chaps.* 1, 2, 8, 12 *and* 13 *in this book*]
- It includes issues on historical perspective of human civilization, history of technology, and those sociocultural backgrounds.
- RC57 (Visual sociology), 37 (Arts) [*Related to Chaps.* 3 *and* 12 *in this book*]
- It includes issues on archives, museum, art, image and video analysis based on big-data and AI pattern recognition in dynamics of society. And VR and AR will enhance understanding on the arts in various sociological contexts.

Digital transformation has been upgrading all such fields, and hence each sociological study can tackle to revise own research area.

1.4.3 Interdisciplinary Taxonomy

Other aspect is to stress on recent computational advancements, and there is a need of taxonomy which comprehensively recompiles them by coordinating and interweaving with each standpoint from meta-physics (i.e., philosophy, ethics), humanities, general social sciences, computational social science, behavior sciences, STS (science and technologies studies) (Latour 2002; Shibuya 2007), data sciences, computer sciences, other natural sciences and engineering (Borup et al. 2006).

Our digitized society shows the prisms of multifurcation which divides into multiple academic fields on *identity* such as:

- Metaphysical realm [*Related to Chaps.* 1, 2, 3, 4, 12, *and* 13 *in this book*]
  - Philosophy: As standing on computational and advanced technologies, it should argue Dasein, existence, the dualism, ontology, space, time, perception of the self, and meta-level value standard.
  - Ethics: As standing on computational and advanced technologies, it should argue *Summum bonum* (the supreme good), bioethics, way of life, and dignity of the human-being.
- Natural Realm: [*Related to Chaps.* 1, 2, 3, 4, 5, 11, 12, *and* 13 *in this book*]
  - Biology: As standing on computational and advanced technologies, it should argue and manage genome and DNA data for each person.
  - Medical Science: As standing on computational and advanced technologies, it should manage health, pharmaceutical, and medical data for each person.
  - Brain Science: As coordinating with computational and advanced technologies (e.g., AI, BMI (brain-machine interface)), it should argue the qualia for existence and recognize the nature of consciousness and neural system for intelligence of human-being.
- Artifactual Realm: [*Related to Chaps. from* 1 *to* 13 *in this book*]
  - Robotics: As developing computational and advanced technologies, it should achieve an enhancement for the human value on wearable robot-suits, robot-assisted exoskeleton, cyber-prosthesis, and others.
  - Computer science: As developing computational and advanced technologies, it should achieve an enhancement for the human value on AI, digitization, VR (virtual reality), AR (augmented reality), tele-existence, security management, secure by design, HPC (high performance computing) (e.g., supercomputer, quantum computer), and other engineered artifacts.
- Humanity Realm: [*Related to all Chaps. from* 1 *to* 13 *in this book*]
  - Personality Science: As grounding on above all of these issues, it should argue and recognize the potential meaning in depth on self, abnormal and ordinal personality.
  - Social Psychology: It mentioned above. As grounding on above all of these issues, it should argue and recognize the potential meaning in depth on self-esteem, social identification, groupness, justification, prejudice, and collective dynamics of human behavior.
  - Sociology: It mentioned above. As grounding on above all of these issues, it should argue and recognize the potential meaning in depth on socialization, social identification, nationalism, demography, immigration, social stratifications, human networking, social welfare (gerontology), and collective dynamics of human behavior.

- Economics: As grounding on above all of these issues, it should argue and recognize the potential meaning in depth on economic phenomena, basic income, signaling, game theory, medical economics, behavior economics, consumers, marketing, networking effect (i.e., network externality, tipping point), industrial relationships, innovations, neural economics, and others.
- International Affairs: As grounding on above all of these issues, it should argue and recognize the potential meaning in depth on nationality, race, ethnicity, and legitimacy by meta-level integration of the citizens.
- Law: As grounding on above all of these issues, it should argue and enact necessary laws on management for human rights, digital heritage of each person, the minimal entity for responsibility, holder of the rights and obligations in law.
- Computational Social Science: As grounding on above all of these issues, it should model, analyze, and simulate social phenomena above fields.
- Archive studies: As grounding on above all of these issues, it should manage digitized data in museum, library, data archives, and other archives.

Therefore, digital transformation and its revolution has been majorly upgrading above fields at least. Namely, it is time that vast amount of the entire knowledge holding among researchers should be revised by such perspectives.

References

Abowd, G. D., & Mynatt, E. D. (2000). Charting past, present, and future in ubiquitous computing. *ACM Transactions on Computer Human Interaction, 7*(1), 29–58.

Aimone, J. B. (2019). Neural algorithms and computing beyond Moore's law. *Communications of the ACM, 62*(4), 110–119.

Arabnia, H. R., & Tran, Q. N. (Eds.). (2016). *Emerging trends in applications and infrastructures for computational biology, bioinformatics, and systems biology: Systems and applications* (Emerging Trends in Computer Science and Applied Computing). Cambridge, MA: Morgan Kaufmann.

Ben-David, S., et al. (2019). Learnability can be undecidable. *Nature Machine Intelligence, 1*, 44–48.

Berkich, D., & D'Alfonso, M. V. (2019). *On the cognitive, ethical, and scientific dimensions of artificial intelligence: Themes from IACAP 2016* (Philosophical Studies Series). Cham: Springer.

Berners-Lee, T. (1999). *Weaving the web*. San Francisco: Harper.

Borup, M., Brown, N., Konrad, K., & Van Lente, H. (2006). The sociology of expectations in science and technology. *Technology Analysis & Strategic Management, 18*(3/4), 285–298.

Bostrom, N. (2016). *Superintelligence: Paths, dangers, strategies*. Oxford: Oxford University Press.

Cave, S., & Dihal, K. (2019). Hopes and fears for intelligent machines in fiction and reality. *Nature Machine Intelligence, 1*, 74–78.

Chen, W., & Li, X.-Y. (2018). Welcome to the china region special section. *Communication of the ACM, 61*(11), 38–87. (∗ multiple contents within special section).

China Institute for Science and Technology Policy at Tsinghua University. (2018). *China AI development report*. Retrieved from http://www.sppm.tsinghua.edu.cn/eWebEditor/UploadFile/China_AI_development_report_2018.pdf

Cohen, J. (2019). WHO panel proposes new global registry for all CRISPR human experiments. *Science*. https://doi.org/10.1126/science.aax3948.
Congressional Research Service, USA. (2019). *Artificial intelligence and national security*. Retrieved from https://fas.org/sgp/crs/natsec/R45178.pdf
Copeland, J. (1993). *Artificial intelligence: A philosophical introduction*. Oxford: Wiley-Blackwell.
Copeland, B. J., & Shagrir, O. (2019). The church-Turing thesis: Logical limit or Breachable barrier? *Communications of the ACM, 62*(1), 66–74.
Cyranoski, D. (2019). China to tighten rules on gene editing in humans. *Nature*. https://doi.org/10.1038/d41586-019-00773-y.
Executive Office of the President (USA). (2016). *Artificial intelligence, automation, and the economy*. Retrieved from https://www.whitehouse.gov/sites/whitehouse.gov/files/images/EMBARGOED%20AI%20Economy%20Report.pdf
Fanelli, D. (2018). Is science really facing a reproducibility crisis, and do we need it to? *PNAS, 115*(11), 2628–2631.
Fogel, A. L., & Kvedar, J. C. (2018). Artificial intelligence powers digital medicine. *NPJ Digital Medicine, 1, 5*. Retrieved from https://www.nature.com/articles/s41746-017-0012-2.pdf.
Geng, H. (2017). *Internet of things and data analytics handbook*. Hoboken: Wiley.
Habermas, J. (1976). *Legitimation crisis*. London: Heinemann.
Harnad, S. (1990). The symbol grounding problem. *Physica D, 42*, 335–346.
International Human Genome Sequencing Consortium. (2004). Finishing the euchromatic sequence of the human genome. *Nature, 431*, 931–945.
Ioannidis, J. P. A. (2005). Why most published research findings are false. *PLoS Medicine, 2*(8), 696–701.
Jaderberg, M., et al. (2019). Human-level performance in 3D multiplayer games with population-based reinforcement learning. *Science, 364*(6443), 859–865.
Kim, H., Bojor, D., & Fussenegger, M. (2019). A CRISPR/Cas9-based central processing unit to program complex logic computation in human cells. *PNAS, 15*, 7214–7219.
Kitano, H., et al. (1999). RoboCup rescue: Search and rescue in large-scale disasters as a domain for autonomous agents research. In *IEEE SMC'99 Conference Proceedings. 1999 IEEE International Conference on Systems, Man, and Cybernetics (Cat. No. 99CH37028)* (Vol. 6, pp. 739–743).
Lander, E., et al. (2019). Adopt a moratorium on heritable genome editing. *Nature, 567*, 165–168.
Latour, B. (2002). *Resembling the social: An introduction to actor-network theory*. Oxford: Oxford University Press.
Lu, C.-Y., Peng, C.-Z., & Pan, J.-W. (2018a). Quantum communication at 7,600km and beyond. *Communications of the ACM, 61*(11), 42–43.
Lu, Y., Qian, D., Fu, H., & Chen, W. (2018b). Will supercomputers be super-data and super-AI machines? *Communications of the ACM, 61*(11), 82–87.
Marres, N. (2017). *Digital sociology: The reinvention of social research*. Malden, MA: Polity.
Martin, J., et al. (2012). A programmable dual-RNA-guided DNA endonuclease in adaptive bacterial immunity. *Science, 337*(6096), 816–821.
McKinsey & Company. (2017). *Artificial intelligence: Implications for China*. Presented at the 2017 China Development Forum. Discussion Paper.
Odell, A., Askham, J., Whibley, C., & Hollstein, M. (2010). How to become immortal: Let MEFs count the ways. *Aging, 2*(3), 160–165.
Orton-Johnson, K., & Prior, N. (2013). *Digital sociology: Critical perspectives*. Basingstoke: Palgrave Macmillan.
Papacharissi, Z. (2014). *Affective publics: Sentiment, technology, and politics*. New York: Oxford University Press.
Popper, K. (1963). *Conjectures and refutations*. London: Routledge.
Renso, C., Spaccapietra, S., & Zimányi, E. (2013). *Mobility data*. Cambridge: Cambridge University Press.

Shibuya, K. (2004). A framework of multi-agent based modeling, simulation and computational assistance in an ubiquitous environment. *Simulation, 80*(7–8), 367–380.

Shibuya, K. (Ed.). (2007). *A study on the research evaluation of science & technology and the rationality of decision making*. The Institute of Statistical Mathematics: Report on Research and Education, No. 26, pp. 1–60. (In Japanese).

Shibuya, K. (2021, in press). Breaking fake news and verifying truth. In M. Khosrow-Pour (Ed.), *Encyclopedia of organizational knowledge, administration, and technologies* (1st ed.). Hershey, PA: IGI Global.

Skordas, T. (2019). Toward a European Exascale ecosystem. *Communications of the ACM, 62*(4), 70–73.

Steen, R. G., Casadevall, A., & Fang, F. C. (2013). Why has the number of scientific retractions increased? *PLoS One, 8*(7), e68397.

Surani, M. A. (2012). Cellular reprogramming in pursuit of immortality. *Cell Stem Cell, 11*(6), 748–750. https://doi.org/10.1016/j.stem.2012.11.014.

Tomita, M., et al. (1999). E-CELL: Software environment for whole cell simulation. *Bioinformatics, 15*(1), 72–84.

Turing, A. (1950). Computing machinery and intelligence. *Mind, LIX*(236), 433–460.

United Nations Security Council. (2018). *Lyon Model United Nations 2018 study guide*. Retrieved from http://www.lyonmun.com/wp-content/uploads/2018/05/UNSC.Def_.pdf

UNODA. (2017). *Perspectives on Lethal Autonomous Weapon Systems*. UNODA Occasional Papers No. 30. Retrieved from https://www.unog.ch/80256EDD006B8954/(httpAssets)/6866E44ADB996042C12581D400630B9A/$file/op30.pdf

UNOG. (2018). *Report of the 2018 session of the Group of Governmental Experts on Emerging Technologies in the Area of Lethal Autonomous Weapons Systems*, Geneva, 9–13 April 2018 and 27–31 August 2018 (CCW/GGE.1/2018/3). Retrieved from https://www.unog.ch/80256EDD006B8954/(httpAssets)/20092911F6495FA7C125830E003F9A5B/$file/CCW_GGE.1_2018_3_final.pdf

Utikal, J., et al. (2009). Immortalization eliminates a roadblock during cellular reprogramming into iPS cells. *Nature, 460*, 1145–1148.

Wallden, P., & Kashefi, E. (2019). Cyber security in the quantum era. *Communications of the ACM, 62*(4), 120–129.

Want, R., Hopper, A., Falcao, V., & Gibbons, J. (1992). The active badge location system. *ACM Transactions on Information Systems, 10*(1), 91–102.

Weiser, M. (1993). Some computer science issues in ubiquitous computing. *Communications of the ACM, 36*, 75–84.

Weiss, G. (2000). *Multiagent systems a modern approach to distributed artificial intelligence*. Cambridge, MA: MIT Press.

Wert, G., & Mummery, C. (2003). Human embryonic stem cells: Research, ethics and policy. *Human Reproduction, 18*(4), 672–682.

Chapter 2
Investigating Identity

The book is concerned with key aspects on identity of the human. Principally, the author motivates to tackle this aporia standing on the perspective of digital social science at once. This chapter concentrates on how digital social science should devote to considering *the nature of identity of the human-being* and *what the human-being is* in terms of both advanced technologies and social sciences.

It reminds both identity concept and its existential modus reminiscing a technical term *Dasein* proposed by Heidegger (2008) in the contemporary digitized world. Thanks to advanced technologies such as AI (Russell and Norvig 2009; Romportl et al. 2015), BMI (brain-machine interface), robotics, and biomedical engineering (McGee and Maguire 2007), by standing on the viewpoints came from above advanced progresses, our identity inevitably faces at the phenomenological and ethical aporia again. As it is likewise ancient philosophers, the author casts the irresolvable questions on the difference between human-being and AI, supremeness of the human-being rather than other existences, and ethical matters on the technological disputes in progress. Therefore, how does an identity in the digitalized AI society mean?

For instance, given that it computationally extracts the genome data and its copy of DNA information of sequences from an individual, are these data equally to the origin of the individual? Through BMI, can the digitized data mediated from our brain to neural circuits of electric activities be possible? Can each identity duplicate? And can we regard it as the same identity?

2.1 Lexical Definitions

First of all, it starts to inquire into the definitions on identity. An *Identity* conceptualized an *identified* entity of the humankind ever since the birth of philosophy. According to the Oxford dictionary, *identity* means "*The fact of being who or what a person or thing is*" and "*The characteristics determining who or what a person or*

K. Shibuya, *Digital Transformation of Identity in the Age of Artificial Intelligence*, https://doi.org/10.1007/978-981-15-2248-2_2

thing is." It derives "*idem*" term in Latin to "the same thing" in English, and "*identité*" in French was also rooted in this term of Latin. Ancient philosophers firstly distinguished each person among the others, and they cast questions to characterize both inherent properties of each individual and common properties of the human.

Certainly, there are a lot of varieties on definitions of identity in each academic area, and thus, next, it reexamines each of them.

2.1.1 From Personality Science

Traditionally, it should firstly touch with a similar concept *personality*. Psychologist Allport (1985) defined "*Personality is the dynamic organization within the individual of those psychophysical systems that determine his characteristics behavior and thought*".

In short, psychologists may stand at the side of the mind and body dualism since the birth of Greece philosophy. According to the Cartesian philosophy (Descartes 2010), what dualism deliberated "*Cogito ergo sum*" and "*Substantia Cogitans*" meant to divide between mentality and whole body, and finally they hypothesized something integrating to the total entity. And it conducts to autonomously lead our mental process with physical constraints surrounding the world environment (Floridi 2011, 2016).

In this way of thinking, *personality science* has further motivations to systematize the understanding of our personality as a psychological entity, and their works recently include biology, bioinformatics, cognitive science, brain science, behavior science, computer simulation as well as traditional psychological realms (Cervone and Shoda 1999; Cervone and Mischel 2002). Apparently their intentions variedly drive to inquiry into from biological origins of our existence to social existence as personal characteristics via cognitive mechanisms. Those trends certainly shed light on the AI and biotechnologies in terms of psychological perspective, and they might contribute toward unveiling what mental system of each personality is. Furthermore, those findings will offer some foundations of the human in the big data and AI age.

2.1.2 From Sociological Context

As sociologically thinking, an *identity* concept is traditionally the nature of human-being in daily world and the nutshell of sociology standing on the dichotomy between methodological individualism and collectivism (Sawyer 2001, 2002a, b). Relations of micro-macro linkages crossing both have many nutritious themes (Sawyer 2003), and for example, these are dyad relationships between ego and alter ego (e.g., Mead, Simmel), collective dynamics, networking, structural patterns of the society and sociocultural issues.

Everyone has a strong need to reflexively recognize oneself through mental mirrors stood on the others' perspective. Sociological studies stress on human relationship and its dynamics in social context. Concretely, role-taking behavior recalls us an interactive behavior regulation based on social roles (Cast 2004). For example, social role between client and doctor contains intrigued facets of implicit problems. As a manner of doctor–client perspective, Parsons (1970) theorized his fundamental theory on personality structure and social role of client. And his theoretical considerations widened the way to investigate adjustment of relationships in latent sociological structure (i.e., AGIL (*adaption, goal attainment, integration,* and *latency*)).

In addition, human behavior based on social identification belonged to specific group and associative network has been investigated in various social situations. Such *identity* can be defined as an agent (or actor) in agent-based model. Standing on the artificial intelligence within society, those social mechanisms have been already exploring by methodologies of system modeling and social simulations (Bainbridge et al. 1994; Carley 1995, 1996; Hegselmann et al. 1996; Liebrand et al. 1998; Kohler and Gumerman 2000; Markovsky and Jin 2002; Shibuya 2004).

Recently, computational social scientists eagerly handle those computable variables (and emergent properties) assumed by the hypotheses in specific social situations. But those trends were limitedly agreed by a part of social scientists, because many of sociologists were redundant (or impossible) to adapt big-data, computational and engineering based trends on artificial intelligence and natural sciences. Theoretical sociologists would not be expected to contribute further toward intriguing into the next steps along with computational advanced skills. Until present, many of them have not conducted to design statistical survey and analyze actual data. Even if researchers on STS have been arguing risks against AI studies and medical sciences, many of them similarly do not have any actual experiences of their studies interweaving with both technological implementation and theorizations. Their motivations often criticize negative extreme facets against the AI and engineering despite of lacking mutual interlinkages with theoretical discussions and own rich experiments of actual laboratory developments.

Thus, both computational and digital social sciences have been conducted by too limited researchers and collaborative partners in the sociological context. They learned vast knowledge on both computational and social sciences. But, for further progress on such fields, every sociologist has to lead studying own themes which should be digitized by external methodologies given from computer and data sciences.

2.1.3 AI and Agents

In the AI studies, as noted above, firstly *identity* has a meaning on representation of the humankind. The AI and its cognitive characteristics had been reflectively realized how and what intelligence of the human-being was functionalized (Russell and

Norvig 2009; Romportl et al. 2015). Especially, recent advancement has been achieved in subfields on machine learning (deep learning is the most representative one), pattern recognition, image and video processing and control system in various needs. Those systems worked for own domain needs such as games, video analyzation, text analysis, auto driven car and smart optimization. The general intelligence of the AI has not been approached to our brain yet, and then especially the humankind is superior to the AI in subfields on predicate logic and inference, metacognition, and adaptation in surroundings (Frankish and Ramsey 2014).

1. Calculation
2. Machine learning
3. Cognitive mechanisms (including heuristics, analogy, etc.)
4. Pattern recognition
5. Image and video processing
6. Logics (i.e., first-order, second-order, and higher-order predicate logic)
7. Natural language processing
8. Inference and reasoning
9. Coordination
10. Planning and scheduling
11. Meta-cognition and knowledge
12. Controls and robotics

In this context, our personality can be defined as a mental model of each computational entity in simulation. "*Self*" within each brain has been modeled by connectionists using neural network models (Read and Miller 1998; Van Overwalle 2007; Sun 2008). Otherwise it can deal with more socialized entity as "agent" of agent-based model (ABM) (Gilbert and Troitzsch 1999; Weiss 2000). Each individual has cognitive properties such as emotion, mood, and beliefs, and then we can define behavior patterns of each agent such as judgment, activities, and attitude as a set of attributes in any social situations. Namely, an agent can be regarded as an implemented software entity, which reflected our mental and behavior model. That is, our personality can be running as an ABM using computer simulation. It is often defined as the *b*elief-*d*esire-*i*ntention (i.e., BDI architecture) in order to attain their goals by their implemented behavior rules and cognitive judgments. Bounded rationality as cognitive constraints in both the human-being and the AI has common backbone. ABM had been originated from serial studies on artificial intelligence and computer science, and it applied to various social simulations and interaction processes (Shibuya 2004).

2.2 Preconditions for Identity

2.2.1 Preconditions

And next, what does "*identity*" constitute? Although there are varieties by theorists in wider academic realms, the author here summarizes a constructive concept that *identity* of the human-being can be defined as holding the biological origin in own irreversible history, keeping the boundary of selfness among the others, maintaining the representable self-image for own physical entity, and holding own cognitive characteristics of personal attitudes. Further, it integrates all of them and constitutes of both individual level (i.e., personal aspect of identity) and group level (i.e., committing in social category and group).

Logically speaking, the first step is to divide the identity concept to its intention and extension. Further it enumerates necessary preconditions to conceptualize the identity of the human-being as follows, at least.

1. Keeping own biological origin and developing biological driven principles.
2. Holding historical irreversible origin from own birth to death as well as the time consequences from the origin of species.
3. Maintaining own boundary of the self for differentiation among the others.
4. Owning singleton entity as coherence between mentality and physical body.
5. Owning autonomous capability of self-consciousness to represent meta-level oneself, understand oneself, have a set of emotional sensibilities, learn own experiences, decide own attitude, and behave adaptively by oneself in surroundings.
6. Coordinating and committing with both personal and social aspects of own characteristics.

It certainly calls ethical disputes on what the human-being is or not (i.e., an identity *of* and *as* human-being). It should shed light on each facet of identity matters for grounding theorizations.

First of all, genome data itself (both genotype and phenotype) cannot be regarded as an identity itself. Dismantled part of whole body cannot be represented as remaining own identity, and if possible so, singleton of the identity can be made apart from the whole. Preconditions 1, 2, and 3 cannot allow that identical twins and perfect clones hold and share same identity. And then they separately hold each own identity memorizing historical own experiences and standing on own perspective with own physical body. Additionally, preconditions 1, 2, 3, and 4 crucially never confuse a medical case of Siamese twins (i.e., conjoined twins shared with a same physical body among two personalities).

Next, until now, a Miller–Urey experiment in the frontier of chemical biology had shocked us to clarify a land breaking fact (Miller 1953; Miller and Urey 1959). Their efforts paved the way to obtain the Nobel Prize. Some of amino acids for a primitive life can be generated by lab experiment. A flask filled gases such as water, hydrogen, methane, and ammonia assumed an atmosphere of the ancient earth, and they gave them electric shocks as lighting-volt in that environment. And lastly, they found and endorsed above fact. Furthermore, using methodology by synthetic biology (Kitano 2002), Hutchison III et al. (2016) successfully generated a minimum cell within an experiment lab. This artifactual and human-modified cell has only a set of minimalized DNA sequence to be alive in the restricted environment. Given a set of assembling those cells has both own physical-body and neural-circuit based self-consciousness, should human-being realize it something biological existence? As mentioned above, the historical consequence delicately differentiates between an existence of perfect natural-origin and human-made one. If so, an identity can be artificially created pouring the pneuma by human-being like the God.

Otherwise, digital data itself cannot be dealt with all of an identity. A robot transformed digitized data from a human brain does not have any humanities, that is because it is necessary to satisfy preconditions 1, 2, and 4 (it strictly allows not to copy oneself) at least. Similarly a projected avatar using augmented reality digitally transmitted and generated from whole genome data and characteristic data of an individual cannot be managed as an identity. And in social media, sock-puppets creating by each individual cannot be understood as each different identity, and these are only just virtual disguised entities employing each role by an individual. Thus, as digitized data can duplicate itself for any needs, if it permits that those digital artifacts hold own identity, the problems will excess the range of discussions on the humanity and existence. As discussed later, it will further call serious troubles in judicial cases.

And ultimately, there is another discussion: "*can AI has own consciousness?*" Even though self-awareness and own self-consciousness within each brain of the human-being has a strong evidence for establishing own identifiable characteristics (Michel et al. 2019), at the present, any artifacts constituted of neural circuits and its complexed network have not been observed by any evidence on organization of something "consciousness" yet (Dreyfus and Dreyfus 1986). Even if such conditions happen, preconditions 1 and 2 cannot be lastly satisfied. If possible completely, it will mean that the human-being approaches to transforming from biological organs' structural patterns and its memorized data to arbitrary digital forms, which includes meta-cognition, self-consciousness, memory, physical kinetics, and other necessary coordination with surroundings. Of course, such technological achievements are impossibly understood yet.

To summarize, hence, those necessary factors are definitively a part of, an aspect of, and a relation of an identity of the human-being. But, each of them is not an identity itself. Thus, the author hypothesizes a definition on *identity* of the human. If and only if it exactly presumes an integrated uniformity of own selfness that simultaneously constitutes of above all necessary preconditions, at last we can recognize an identity distinguishing among the others in regular context.

2.2.2 *Civil Law Rules on AI and Robotics in EU*

Aforementioned preconditions for the identity of the humankind can be approved by other recommendations on the AI systems.[1] For example, according to European Parliament's Commission on Civil Law Rules on Robotics (2017), they summarized the following general principles.

Calls on the Commission to propose common Union definitions of cyber physical systems, autonomous systems, smart autonomous robots and their subcategories by taking into consideration the following characteristics of a smart robot:

- *The acquisition of autonomy through sensors and/or by exchanging data with its environment (inter-connectivity) and the trading and analyzing of those data*
- *Self-learning from experience and by interaction (optional criterion)*
- *At least a minor physical support*
- *The adaptation of its behavior and actions to the environment*
- *Absence of life in the biological sense*

Reversely, their definitions clearly reflect what human-being is or not. This report firstly accepted both autonomous sensing and learning abilities of artificial intelligence and robots adapting in their surroundings. Especially, "absence of life in the biological sense" means a lifeless entity which should be managed as an artificial object and artifacts by engineering (Stephanopoulos and Stephanopoulos 1986). Except for this lifeless factor, a definition of the AI includes necessary factors as such meanings.

2.3 Identity and Its Areas in Digitized Society

Advanced technologies progressively interfere with the realms of our identity. Saying once again, digital social science should tackle considerations on the above matters in depth by this viewpoint. In the era of digitized society, it may encounter something identifiable as a physical-body entity, a representational image entity, a simulating entity, and engineering artifacts. And it should discuss those matters dividing between natural, digitalized, and artifactual based entities. Of course, it investigates that it can regard each entity of them as an identity.

Let me define the problems what should be tackled in this book at least. An aforementioned set of six necessary preconditions for identity can deduce some considerable problems on identity of the human. Each chapter of below list has a category name on the identity which characterized as one of the specific discussion themes.

[1] "A definition of Artificial Intelligence: main capabilities and scientific disciplines", https://ec.europa.eu/newsroom/dae/document.cfm?doc_id=56341.

1. **Chapter 3**: *Identity Ambiguity*: It mainly relates to preconditions from 1 to 6. First of all, the digitalized society calls the issues on identity ambiguity. It indicates ambiguous situations such as losing the boundary between self and the others, sense of depersonalization using VR, and the destabilized relationships between the human-being and the AI. Is there a possibility of dissolving the physical border between the human and the AI and keeping the AI as a harmonious partner with the human? Certainly, it is necessary to reconsider the ethical rules of conduct among them, and it should prepare the harden barriers to draw the borderline against the unforeseen risks. Hence, at first, it should obviously get in touch with traditional topics from philosophical realm such as epistemology, ontology, and ethics on the human, and next step, it leads further contemplations on the ethical conducts between the human and the AI. The AI studies have already interlinkages with those themes, and the author intends to deepen those issues for stepping further discussions.
2. **Chapter 4**: *Identity Valuation*: It mainly relates to preconditions 1, 2, 3, 5, and 6. Identity valuation has decisive influences for your daily life in the past, present, and future (including your future kinships and offspring). Recently, day by day, the credit in finance based on each personality, gifted abilities, educational level, behavioral characteristics, health condition (it can be based on each genome data), human relational factors, estates and cash information are usually registered by the centered agencies. At first place, "valuation" mainly means a technical term in finance, and such subject has been globally teaching in MBA (master of business administration) course of professional business schools. It obviously denotes measuring value for something at the present and the future by the standard. Furthermore, in digitized society, valuation has specific cutting edges for not only financial fields in business scenes but something converted to monetary index by the digitized standards based on big data. Of course, it also targets for us, and then the author named it "identity valuation" for discussions.
3. **Chapter 5**: *Identity Protection*: It almost relates to preconditions from 1 to 6. Keeping awareness for your identity, and protecting own identity including all of actual behavior patterns. The serious reason is to close carefully look at the functions of automatic AI systems that can accumulate and analyze citizens' real-time data, and these systems further will be able to apply for diversified operations in our daily life. It also consists of viewpoints based on CIA (*confidentiality, integrity, and availability*) in computer security. Namely it is time to computationally manage data accumulation, modification, deletion, verification, monitoring, and other necessary operations on an entire integrated identity among all citizens.
4. **Chapter 6**: *Identity Proof*: It mainly relates to preconditions 3, 4, 5, and 6. It means to prove not only your identity as existence but also existence of necessary information (e.g., ability, evidence) as genuine and trustable one without disclosing unnecessary knowledge. In other words, it intends to accomplish proving as an identifiable individual with desirable capability to pursue the goal. To summarize critical points, it can consider that controversial issues on

online deceptions usually lack those convincing validation mechanisms: No provable way to verify prover's knowledge and faithful protocols about information between senders and receivers. Without any troubles and haphazard, it should clarify either not-modified or not-disguised entity. This chapter attempts to introduce zero-knowledge proof and its applicability online.

5. **Chapter 7**: *Identity Deception*: It mainly relates to preconditions 3, 4, 5, and 6. It deals with identity deception. How do those who pretend to someone detect by appropriate ways? How to detect deceptive communications among numerous participants in security context? It is always one of the old and new problems in computer security (e.g., verification on counterfactual data, monitoring, detection, authentication, and protection). In this chapter, the author mainly takes up some cases. First issue is to design a mathematical model on anomaly detections in mutual communications among participants in the larger networking. Scoring services driven by the AI have been focused on daily activities to measure each digitized score of the citizens on the bases of their honesty and other dispositions. And blockchain distributed networking ought to serve mutual monitoring and detection mechanisms among multiple participants. Secondly, such deceptive behavior can be refined as a problem on social intelligence of the human for adaptations. Finally, deceptive technology on generative adversarial machine learning should be discussed.

   These matters do not mean just to conduct deceptions for concealing own privacy and keeping anonymous condition. For example, there are greater needs for strong secured procedures such as online shopping, governmental registration, and other online activities (Castelfranchi and Tan 2002). Of course, the AI driven detection services against deceptions has been already running in business scenes. However, there were sometime severe cases, which the AI could mistake incorrectly the goods online (i.e., false negative and false positive cases). For example, at 2019, Amazon (in Japan) was accused by some consumers, and lastly Amazon recognized that their AI system had failures to detect pirated goods given from anonymous providers in their online shopping site, and they improved such errors. Then, deception detection and anomaly monitoring methods should be investigating and further implemented such process for the AI-based automatic mechanisms.

6. **Chapter 8**: *Social Identification*: It mainly relates to preconditions 3, 4, 5, and 6. Social identification should be rethought by both theories in social contexts. It can regard the hardcore on identity matter. The digitized world has reshaped and transcended the border of national and cultural realms to vague surroundings. Then, identification in social meaning was induced by redefinition of social progresses. We unintentionally tend to flock together based on the same category (e.g., race, ethnicity, gender, region, religion, hierarchy, etc.) and attitudes (e.g., political, socioeconomic and other opinions). Our personality is never independent on social and cultural context, and it means that each individual consists of personal and social identity. Especially, social psychological studies have much advantage in investigating on social identification and group minds through sociocultural differences and uncertainty reductions. In our digitized

society, many people are often redundant to critically understand opponents who hold the different backgrounds, and then social identification matter has strong reasons to be reconsidered. Before discussing on social phenomena such as collective dynamics of movement (Chap. 9), synchronization of networking (Chap. 10), and mental health (Chap. 11), it should deepen common backdrop of identification issues.

7. **Chapter 9**: *Collective Identity*: It mainly relates to preconditions 5 and 6. As there is recently emerging problem on social aspect of identity in global level, it should turn eyes to this matter on the facets of collectiveness of identity. The author introduces collective dynamics of citizens using ubiquitous computation and measuring ways to analyze such spatiotemporal phenomena. These technologies and methodologies have been contributed toward investigating social physical movements and complicated dynamics among citizens in smarter city. In light of this, especially, the author picks up immigration issues. Digital civilization encourages citizens to immigrate to other countries beyond the borders. Those mechanisms and socioeconomic backgrounds should be unveiled by interdisciplinary studies. Finally, by means of demonstrating agent-based simulation, the author examines both collective migration phenomena and acceptance policy in each city in social simulation process.
8. **Chapter 10**: *Networked Identity*: It mainly relates to preconditions 5 and 6. Networked identity indicates interconnected modality on social synchronization and in/exclusiveness among massive global citizens. Those social mechanisms should be revealed how and why occurs or not in each condition. Although the advent of social media gained harmonized opinion formation and contagions through massive influences, the digitalized social world would be coined in sharing emotional experiences among global citizens. Such social dynamics has regenerated influentially fake news, social movements, and responsive support networking. And online virtual nation has been committed and identified among participatory citizens, and their motivations to realize an independent nation could be legitimatized among online interconnected citizens. At the time of 2018, the extent of global users using social networking sites such as Facebook, Twitter, and the others exceeds more than approximately 30% of the total population of citizens on the globe (multiple counts). Many of the citizens are apt to only believe in group opinions among like-minded partners online. Echo-chamber phenomena may show opinion divides between citizens, and their networked clustering become to be actualized online virtual nation government using blockchain technology. Many of online users are to switch and behave own identity differentially in each online community. Further there are the cases to reconsider own identity based on nationality and conflict resolution mapping between the actual nation and bit-nation (i.e., online virtual nation using blockchain technology). Their motivations to be recognized as an independent nation could facilitate their progress by online interconnected citizens.
9. **Chapter 11**: *Identity Health*: It mainly relates to preconditions 1 and from 3 to 6.

Identity health has especially specific meanings for social relationships in contemporary digital age. First, computerized digital communication makes many citizens in severe maladaptation. The WHO often warns mental addictions of internet usages and online gaming among the youth. The advent of social media and online networking has endangered them in ambiguous situations which are not stabilizing in those basic grounds for human relationships. Further, because social networking sites and social gaming frequently enforce each member to interconnect with the others, many participating members often hold harder mental debts to respond and maintain their interconnections. In this situation, in other words, it can say that all of users simultaneously might share common conditions under mental illness. Secondly, the big-data era enables us to analyze massive online data accompanied with actual phenomenon. Especially, medical health data can be arranged to coordinate with any solutions for predictions and forecasting in digitized society. Infectious diseases such as influenza and other pandemic events often statistically appear significant and correlational correspondences with actual events and web query among citizens. Thirdly, subjective well-being and social capital will be still effective concepts for understanding our daily life despite the age of digitized social transformation. It often represents a part of self-definitions embedded in rural communities and human relations. Especially, at disaster conditions and emergency situations, it is quite necessary to monitor a patient's longer-term diagnosis. Their identifications surrounded in illness conditions should be cared about intensively. And fourthly, AI and data science have been equipped to support for our health management and cares in daily life. Those mechanical tools and robots will be eligible to give amenity for various needs among children and elder persons.

10. **Chapter 12**: *Identity History*: It mainly relates to preconditions 1, 2, 3, and 4. It is time to ponder the meaning of living existence as finite time dependence. Sharing all of own individual life-long data from cradle to grave with the others on the bases of blockchain distributed networking and its globally enhanced database system. Those personalized digital data represent entirely each personal history. Here, identity history denotes sequential information of lifelog and digital footprints to envisage about each person. Such data of narratives and life history can be converted to a style of digitized folklores in the heritages by social sharing. Furthermore, it should pay more attentions to the following points in particular.

    (a) *Forensics*: To identify each person by recovering scattered digital information. Especially, at the disaster and criminal cases, it will be difficult to repair those remains and articles of the deceased. So, there are the greater needs discovering the missing people and identifying the corpse who is. Geospatially and temporally meaning, digitalized information utilizes to locate their stayed sites and to trace historical data of their activities.
    (b) *Irreversibility*: Life history and biography of each individual have unilateral physical-time sequence and are embedded by own irreversible, irreproducible,

invariant, and unrepeatable constraints. To date, medical sciences enhanced by big-data analyzation and the AI supported medical science have already accelerated its progress in silico, whereas genome editing and regenerative machines in vivo have been also coincidentally interacted with them. Especially, there are considerable matters on diversified influences of medical prolongation for lifespan of the humankind, and it should consider the inheritable right in law for heritage transplanted from an individual to its biological and digital clones. And then it must prepare appropriate civil laws and legitimizations. In extreme case, as it is likely a science fiction, some of citizens will intend to copy own memory to digitized AI system and clone whole of the body to another body.

11. **Chapter 13**: *General Discussion*: It almost relates to preconditions from 1 to 6. Each theme from Chaps. 1 to 12 of wthis book is discussed, respectively. Finally, taken together, the author may be permitted to generally discuss a little bit freely for further considerations. As it seemed likewise Cartesian, the author firstly defines preconditions for the identity of the humankind (Chaps. 1 and 2), subdivides the whole problems on recent technological matters to necessary parts (Chap. 2), and discusses each issue in depth (Chaps. from 3 to 12). At Chap. 13, the author further contemplates those components from another angle for reconstructions, which should be integrated understanding and reasoning some significant variations on proceeding future patterns. Probably, because researchers must mostly manage relevant vast knowledges crossing various fields, there are currently too limited researchers who can simultaneously deal with combinations of multiple keywords such as ethics, sociology, AI, regenerative medicine (iPS and Stem cells, gene-editing), big data, digital technological transformation, and other engineering. But in contrary, the author has a chance to comprehensively discuss those issues standing in an earlier position.

2.4 Methodologies for Identity in Digitized Society

This book has a final goal to diversely delve into *identity* matters within the paradigm of digital social science. Here, digital social science nearly means an academic discipline centered in social science *on* digital technology issues. It is not equal vice versa, namely it implies neither a study *by* some kind of engineering in social sciences (i.e., social engineering) nor the goal for computational advancement (i.e., computer sciences and social physics (Pentland 2014)). A similar term "computational social science" has a tendency to tackle computations on social phenomena and analyze social media (Epstein 2007; Lazer et al. 2009; Conte et al. 2012; Goncalves and Perra 2015), whereas digital social science intends to stress on below methodologies. Both should have different vectors to each direction, but it aligns to coordinate with each other. Therefore, the author has an intention to investigate *identity* matters standing in this digital social science in particular.

1. Interdisciplinary considerations for our digitized world issues standing on mainly social sciences.
2. Studying history of philosophy and ethics on digitalization.
3. Science and technology studies on computational advancements (e.g., AI, Big data), data science, medical science, and other engineering (including ELSI (Ethical, Legal, and Social Aspects Research)).
4. Digital humanities on human-being in daily world.
5. System thinking on digital transformations in society.
6. Computer simulations: agent-based models and mathematical models.
7. Statistics.

References

Allport, G. W. (1985). The historical background of social psychology. In G. Lindzey & E. Aronson (Eds.), *The handbook of social psychology* (3rd ed.). New York: Random House.

Bainbridge, W. S., et al. (1994). Artificial social intelligence. *Annual Review of Sociology, 20*, 407–436.

Carley, K. M. (1995). Computational and mathematical organization theory: Perspective and directions. *Computational & Mathematical Organization Theory, 1*(1), 39–56.

Carley, K. M. (1996). Artificial intelligence within sociology. *Sociological Methods & Research, 25*(1), 3–30.

Cast, A. D. (2004). Role-taking and interaction. *Social Psychology Quarterly, 67*, 296–309.

Castelfranchi, C., & Tan, Y.-H. (2002). The role of trust an deception in virtual societies. *International Journal of Electronic Commerce, 6*(3), 55–70.

Cervone, D., & Mischel, W. (2002). *Advances in personality science*. New York: Guilford.

Cervone, D., & Shoda, Y. (1999). *The coherence of personality*. New York: Guilford.

Conte, R., et al. (2012). Manifesto of computational social science. *The European Physical Journal, Special Topics, 214*(1), 325–346.

Descartes, R. (2010). *Meditationes de prima philosophia, Fili-Quarian classics* (English translated edition).

Dreyfus, H. L., & Dreyfus, S. (1986). *Mind over machine: The power of human intuition and expertise in the era of the computer*. New York: Free Press.

Epstein, J. M. (2007). *Generative social science: Studies in agent-based computational modeling*. Princeton, NJ: Princeton University Press.

European Parliament resolution of 16 February 2017 with recommendations to the Commission on Civil Law Rules on Robotics (2015/2103(INL)). (2017). Retrieved from http://www.europarl.europa.eu/sides/getDoc.do?pubRef=-//EP//NONSGML+TA+P8-TA-2017-0051+0+DOC+PDF+V0//EN

Floridi, L. (2011). The informational nature of personal identity. *Minds & Machines, 21*, 549–566.

Floridi, L. (Ed.). (2016). *The Routledge handbook of philosophy of information*. London: Routledge.

Frankish, K., & Ramsey, W. M. (2014). *The Cambridge handbook of artificial intelligence*. Cambridge: Cambridge University Press.

Gilbert, N., & Troitzsch, K. G. (1999). *Simulation for the social scientist*. London: Open University Press.

Goncalves, B., & Perra, N. (Eds.). (2015). *Social phenomena: From data analysis to models (computational social sciences)*. Cham: Springer.

Hegselmann, R., Mueller, U., & Troitzsch, K. G. (Eds.). (1996). *Modelling and simulation in the social sciences from a philosophy of science, point of view*. Dordrecht: Kluwer.

Heidegger, M. (2008). *Sein und Zeit* (English translated edition). New York: Harper Perennial Modern Classics.
Hutchison, C. A., III, et al. (2016). Design and synthesis of a minimal bacterial genome. *Science, 351*(6280), aad6253. https://doi.org/10.1126/science.aad6253.
Kitano, H. (2002). Systems biology: A brief overview. *Science, 295*(5560), 1662–1664.
Kohler, T. A., & Gumerman, G. J. (2000). *Dynamics in human and primate societies*. New York: Oxford University Press.
Lazer, D., et al. (2009). Computational social science. *Science, 323*(5915), 721–723.
Liebrand, W. B. G., Nowak, A., & Hegselmann, R. (1998). *Computer modeling of social processes*. London: Sage.
Markovsky, B., & Jin, W. (2002). *Theories, tests and computer simulations*. Annual meetings of the American Sociological Association. Retrieved from http://home.sc.rr.com/bmarkovsky/Papers/ASASim02.pdf
McGee, E. M., & Maguire, G. Q. (2007). Becoming Borg to become immortal: Regulating brain implant technologies. *Cambridge Quarterly of Healthcare Ethics, 16*(3), 291–302.
Michel, M., et al. (2019). Opportunities and challenges for a maturing science of consciousness. *Nature Human Behaviour, 3*, 104–107.
Miller, S. L. (1953). A production of amino acids under possible primitive earth conditions. *Science, 117*(3046), 528–529. https://doi.org/10.1126/science.117.3046.528.
Miller, S. L., & Urey, H. C. (1959). Organic compound synthesis on the primitive earth: Several questions about the origin of life have been answered, but much remains to be studied. *Science, 130*(3370), 245–251. https://doi.org/10.1126/science.130.3370.245.
Parsons, T. (1970). *Social structure and personality*. New York: Free Press.
Pentland, A. (2014). *Social physics*. New York: Penguin Press.
Read, S. J., & Miller, L. C. (1998). *Connectionist models of social reasoning and social behavior*. Mahwah, NJ: LEA.
Romportl, J., Zackova, E., & Kelemen, J. (Eds.). (2015). *Beyond artificial intelligence: The disappearing human-machine divide*. Cham: Springer.
Russell, S., & Norvig, P. (2009). *Artificial intelligence: A modern approach*. Upper Saddle River, NJ: Prentice Hall.
Sawyer, R. K. (2001). Emergence in sociology: Contemporary philosophy of mind and some implications for sociological theory. *American Journal of Sociology, 107*, 551–585.
Sawyer, R. K. (2002a). Durkheim's dilemma: Toward a sociology of emergence. *Sociological Theory, 20*, 227–247.
Sawyer, R. K. (2002b). Emergence in psychology: Lessons from the history of non-reductionist science. *Human Development, 45*, 2–28.
Sawyer, R. K. (2003). Emergence in sociology: Contemporary philosophy of mind and some implications for sociological theory. *American Journal of Sociology, 107*, 551–585.
Shibuya, K. (2004). A framework of multi-agent based modeling, simulation and computational assistance in an ubiquitous environment. *Simulation, 80*(7–8), 367–380.
Stephanopoulos, G., & Stephanopoulos, G. (1986). Artificial intelligence in the development and design of biochemical processes. *Trends in Biotechnology, 4*(9), 241–249.
Sun, R. (Ed.). (2008). *The Cambridge handbook of computational psychology*. New York: Cambridge University Press.
Van Overwalle, F. (2007). *Social connectionism: A reader and handbook for simulations*. New York: Psychology Press.
Weiss, G. (2000). *Multiagent systems a modern approach to distributed artificial intelligence*. Cambridge, MA: MIT Press.

Part II
Actuality of Identity in Digitized Society

Chapter 3
Identity Ambiguity

First of all, the digitalized society calls the issues on *identity ambiguity*. It indicates ambiguous situations such as losing the boundary between self and the others, sense of depersonalization using VR, and the destabilized relationships between the human-being and the AI. Is there a possibility of dissolving the physical border between the human and the AI and keeping the AI as a harmonious partner with the human? Certainly, it is a necessary to reconsider the ethical rules of conduct among them, and it should prepare the harden barriers to draw the borderline against the unforeseen risks.

For example, Kurzweil, R[1] recently warned "*the distinction between AI and humanity is already blurring.*" And furthermore, "*We're merging with these non-biological technologies. We're already on that path. I mean, this little mobile phone I'm carrying on my belt is not yet inside my physical body, but that's an arbitrary distinction. It is part of who I am—not necessarily the phone itself, but the connection to the cloud and all the resources I can access there.*"

Hence, at first, it should obviously get in touch with traditional topics from philosophical realm such as epistemology, ontology, and ethics on the human. And next step, it leads further contemplations on the ethical conducts between the human and the AI. The AI studies have already interlinkages with those themes, and the author intends to deepen those issues for stepping further discussions.

[1] https://www.cbinsights.com/research/ai-threatens-humanity-expert-quotes/

K. Shibuya, *Digital Transformation of Identity in the Age of Artificial Intelligence*, https://doi.org/10.1007/978-981-15-2248-2_3

3.1 Communications with the Others

3.1.1 The Self Encounters the Others

Aristotle told "*All men by nature desire knowledge*." Such knowledge includes knowledge on oneself and the others. Since the dawn of philosophy and metaphysics, the self-concept had been asked by each of the people. We have undoubtedly own perspective surrounding the world (Descartes 2010), and grounding in such environment with the others always steps further another problem. The logic "*Ich bin meine Welt*" (I am the world) (Wittgenstein 1922) approaches to the border of own subliminal self. In contrary, an ego can socially encounter alter-ego, and coexistence among them through mutual communications is crucially embedded in intersubjective world (Schütz 1967). The former is related to a kind of solipsism, otherwise the latter can be especially founded in sociological interactionism and ethnomethodology (Garfinkel 1991).

3.1.2 Empathy for the Others

For each infant of the human-being, realization for the others is a first step to become from social animal to the human-being. In many cases, first person may be mother, and she gets in touch with her infant to induce sociality through careful utterances and sympathetic attachment (Uchida-Ota et al. 2019). In that way, they encounter the others in the same world, and each child learns the manners to coordinate with them (Markus et al. 1985; Knoblich et al. 2003). By learning much experience in such ways of interactive situations, they further assimilate the manners of communications with empathy and helping morality.

Regarding this, in social neuroscience, Cacioppo et al. (2002) comprehensively argued neural mechanisms on the self and interactions with the others. Empathy for other persons well worked by oxytocin and mirror neuron within the brain has been adapted in mutual good relationships, and those physiological and cognitive foundations took advantages in coordinating with the others to build social surroundings (Franks and Turner 2012).

Further, Glimcher et al. (2009) investigated both direct and indirect reciprocity (Nowak and Sigmund 2005; Nowak 2006) among subjects by experiments standing on neuro-economics. Altruism and prosocial norm have been discussing by necessary reasons for social establishment. In this context, social scientific collaboration with neural scientists has been investigating the natural possibility of reciprocal relationships among peoples. Reciprocity, voluntary cooperation, and altruism have been understood as key factors for adaptations of peoples (Shibuya 2012).

3.2 Physical-Body Contact and Interconnected Online Contact

To summarize above points, our characteristics as the humankind are naturally based on intellectual abilities, empathy for the others, and physical modality as own independent existence.

The author distinguishes our communication styles as "*physical-body contact*" and "*interconnected online contact*." The former is that an ordinary identity still maintains communicating with the others through physical contacts around atmosphere. However, the latter is often online interconnections by location-free, temporal, anonymous, and ad hoc communication among the others without physical-body constraints. As an example, tele-existence, VR and AR technology enhances and extends our senses for the surrounding world. Here, we realize that an identity as a total entity should be harmonized by both cognitive and physical factors in both situations.

3.2.1 A Case: The Question for Life

Actually, the human relationships between oneself and the others have difficulty to fill their gap (Bacha-Trams et al. 2017; Francis et al. 2017).

Let me exemplify a practical case of the author. At 2004, during lecturing at undergraduate class, the author had been questioned by one of undergraduate students as follows. "I don't understand why I am not permitted to kill any person. Why does the law restrict doing it against the others?" He intended to play a role as a devil's advocate for debates. At that time in Japan, their surroundings were quite nervous after an elementary school girl had serious troubles and she lastly murdered her classmates because her classmates blamed her online in Japan. Namely the origin of this case can be dealt with varies of the internet-bullying and hate-speech online.[2] Standing on such social conditions, at university class facilitated by the author, discussion themes were mainly "Why does not anyone kill the others?" and "What do legal rights for freedom and liberty mean for us?" Probably, some of the readers had a similar experience to hold such claims.

Similarly at that time, the author gives readers a same direction.

Please spread both arms and turn widely here and now.

If you prompted to do so by this direction immediately, you are undoubtedly a self-centered person. In contrary, if someone looked firstly around own surroundings before attempting to obey this direction, this person can be considered as a careful and sensible person for *others' existence*. Thus, your *free* space and *liberty* for any actions are often restricted by not only legal articles but the others'

[2] https://en.wikipedia.org/wiki/Sasebo_slashing

coexistences in the physical environment. The former is to hit the others by your arms, whereas the latter is to coordinate with the others. This difference is indeed to notice experientially coexistence with the others in the physical world. Psychologists impressively named *personal space* as own space of each individual for private and free spatial field. Citizens must learn to appropriately keep both physical and psychological distances between each other.

3.2.2 *Existence of the Others*

Certainly, coexistence with the others strictly derives to all problems existing in the world. A philosopher Spinoza (2008) said "*Homines ex natura hostes sunt*" in Latin. It means "the others are naturally enemies" in English. Similarly, without saying *Leviathan* proposed by Hobbes (2009), he said "*Bellum omnium contra omnes*" in Latin, and it can be translated as "the war of all against all" in English. These phrases crucially indicate that the malicious conditions always come from the others.

If only one person exists in the world, there is no possibility to happen any illegalities and interruptions by the others. When multiple persons simultaneously coexist in the same place, it is necessary to coordinate with each other in the physical world. It primitively requires order, laws, institutions, morals, and other necessary rules among them (Jost 2005; Cederman 2005). These situations binding "*double contingency*" (Luhman 1984) usually employ a catalyst to reduce social uncertainty as complexity, and in such situations, strategic cognitions against the others certainly formalize original forms of the game theory in various conditions (Cox et al. 2001; Camerer 2003).

At the time, the author answered so in the above lecture for participants included a student who questioned at first. The author felt that all participants could convince such outlines. The most important evidence is that everyone stays in coexistence with the others, and all of them are embodied by the constraints such as physical body with a limited lifespan, common space and time in the world. And it should be fulfilled somewhat rule of conducts in cyber space (i.e., netiquette). Many of us often say that it is common sense, but some of students should be *convinced* in depth by more logical ways which grounded on actualities.

3.2.3 *Other Existence as Intruder to My Space and Time*

However, in the near future, these differences will be imperfectly replaced by cybernetic system, VR, and BMI (Saha et al. 2017), and then the human-being will be induced to step forward into the unknown field against our identification based on physical body and its interactions. It calls coordination problems as human–computer interaction with robots (Sun 2006).

Cyber style by interconnected online communication still turned the existence of the others into the intruder to own precious space and time. In Japan, recently there has been a strong tendency that many of the youth as native online and smartphone users hesitate to call other person by telephone. As they usually use email or text-based chat system, direct telephone calls strictly indicate robbing their free time by the others and they feel some obligations to respond immediately. Some of them alternatively send emails with content which they intend to call. They certainly hate physical contacts with each other. In the above context, they are already living in more cyber side than elder persons.

Thus, according to Heidegger (2008) who got outlooks in the frontier for phenomenology, he had principally articulated an actuality on *leibhaftig* (a modality of existence keeping with own physical body). In contemporary digitized world, it should reconsider his criticisms on those matters, an existence surrounding spatio-temporal world and *Sorge für Miteinandersein* (a care for the others) again. Similarly, Lévinas told an intruder symbolized "face" of the other existence (Morgan 2011). It encounters an embodied others and then an individual and others can vividly recognized actualities with mutual existences in such situations. It may still remain that digitized data cannot provide embedded actualities with alter-ego in online communications.

3.3 Identity as Existence and Representation

The humankind is not ostensibly a computer simulated entity that is supposed by simulation hypotheses of philosophy (Bostro 2003) in one of possible worlds. Such hypothesis is partially extended from philosophical traditions, but there are not decisive ways to verify itself. Namely, this issue would treat the totality of our existence integrated or separated between the mind and physical entity.

A philosophy of mind problem (Dreyfus and Dreyfus 1986; Honderich 2005) might be realized to experience by technological manners. Using tele-existence, VR and AR devices, do you have any experiences like this? *It occasionally confuses us to be aware of whether it is the loss of cognition for the world reality or the loss of actualities in the world.* In the psychiatry concerns, there are the very different meanings (DSM-V[3]: Diagnostic and statistical manual of mental disorders, version 5). The former is not to discern between own perspective and virtual recognition for the world, whereas the latter can be diagnosed as losing the capacity to synthesize both sensibility and physical coherences standing in actual surroundings.

In brief, the former is to mainly confuse between realities in cognitive level; in contrary, the latter is to lose the total unification of own identical coherence between mentality and physicality, and then client somewhat feels a discrepancy between cognitive

[3] https://www.who.int/substance_abuse/terminology/diagnostic/en/; https://www.psychiatry.org/psychiatrists/practice/dsm/updates-to-dsm-5

and physical level. Serious clinical cases (i.e., *Depersonalization derealization syndrome, Dissociative Disorders*) suggest that own singleton identity standing in the actual world must maintain total coherency between cognitive and physical sensibilities.

Regarding this concern, ultimately, an experience of identity ambiguity through digital online communications may allow crossing sensibilities among participants (Suzuki et al. 2017; Kondo et al. 2019). Through VR and other necessary technological artifacts, please imagine a case on interchanging each subjective perspective between person A and B (e.g., there are nearer cases on games and sports events[4]). It means that person A's cognition exchanges to person B's cognition and further person A's physical motion sensibility swaps to person B's (and vice versa). Here, in this case, each of person still holds each own memory through virtually sensing data of view cognition and senses of physical motion. Moreover it can consider that person A's cognition and physical sensibility can broadcast toward multiple others through online communications. This case means that all participants share common memory and physical sense data only by person A's. It can easily engender the condition on discrepancies between perspectives and physical sensibilities among persons. How to identify each own identity among participants by those ways?

Yuste et al. (2017) proposed four concerns on BMI and AI studies in neuroscience. They enumerated *privacy and consent, agency and identity, augmentation*, and *bias*. These are considerable ethical concerns against identical and private mental process reading through BMI. When sharing own experiences with numerous individuals by such artifacts can be possible, it seems that those who interconnected with each other is conceived as synchronized single identity with multiple physical bodies. In this concern, Searle's (1996, 1998) serial though experiments on our cognitive consciousness bridging between social reality and brains' functions will be examined in those BMI and AI technological progresses.

3.4 Dignity for Human-Being

3.4.1 Difference Between Artifact and Human

3.4.1.1 Extensional Artifact of Body

Above all, the digital social era sheds light on the controversial issue on the identity in terms of existence and ontology again (Dreyfus and Dreyfus 1986). Ever since the dawn of the Greek philosophy, many steps by philosophers had driven to further understanding for us (LaChat 1986). So called, it refines the considerable matters as *digitalized* and *embodied artifacts* (Neftci and Averbeck 2019; Howard et al. 2019).

[4]https://www.youtube.com/watch?v=ESl8zoYUUIA; https://www.ericsson.com/en/blog/2018/4/is-virtual-reality-live-sports-viewing-the-future-of-basketball-some-of-the-sports-superstars-think-so-watch-this-video

For example, we already became witness for emergence of cyber-robotics, and it certainly contributes toward enhancing well-being for disabilities and patients. At present, the styles which installed something with human body or implanting sensing substances into human body are varied widely (Sim et al. 2019). At least, these are below categories. Categories 1 and 2 can be easily removable, but categories 3, 4, 5, and 6 require necessary medical operations and engineering supports.

1. Wearable robot suits, exoskeleton (e.g., HAL©[5])
2. Robotics by external manipulations
3. Implanting microchips and any micro substances within physical body
4. Cyber-prosthesis (Worcester Polytechnic Institute 2012)
5. Brain-Machine Interface (BMI)
6. Artificial organs

Those artifacts seem quite different between an entire identity as an original existence and its digitized data. But some of them shall be clearly dignified by the value standard. What does differentiate it among the others?

3.4.1.2 Dignity

A *dignity* defined by the author intensely asserts "*the respective cares for each individual.*" Dignity concept has significant meaning in philosophy and lexically, and it commonly defines "*inviolable value.*" But it is time to get a chance for reconsideration (Schroeder et al. 2017). Actually, dignity concept has ambiguous facets and definitions, and these commonly feature some descriptions such as "human right," "identical pride for racial and ethnic roots" (Fukuyama 2018), the inside-out of "self-esteem," and something respective manner for the others.

Everyone can get touch with the senses of "*dignity*." For example, in Germany language, they have two verbs for a meaning of "die" in English. These are *verenden* and *sterben*. The author ventures to concise that the former can be defined to die *without* the dignity of the human (i.e., it dies likely an animal) and the latter is to die *with* the dignity of the human. The gap of nuance from the former to the latter cannot be leaped over without something valuable. Namely both of them commonly mean "*Sein zum Ende*" (Heidegger 2008), but there is decisively a difference between the two terms (Olivier 2007).

As the author said, identity of each individual constitutes of holding own biological origin in own irreversible history, keeping own boundary of selfness from the others, maintaining the representable self-image for own physical entity, owning singleton entity as coherence between mentality and physical body, and other preconditions. In this sense, all of those specifications become inviolable manner, and it defines "dignity" and forms the multiplied barriers for guarding what the human is or not.

[5] https://www.cyberdyne.jp/

All of the humankind are equally the finite lifespan existence with own physical body and self-awareness, and then dignity in this context seems to be a near concept "*Würde*" in Germany, but "*Fürsorge für Miteinandersein*" (the care for the other existences) in philosophy by Heidegger (2008) may also fit it. These matters have been discussed in thanatology (i.e., philosophy on death) and bioethics on pro or con about euthanasia, for instance.

In that way of thinking, dignity concept can be clearly recognized by an approach for identity of the human-being, as it extracted from neither honorable nor spiritual meaning. Dignity will lose its something value if the advancement of medical sciences ultimately achieves the operation for reincarnation of whole physical body. At that time, priceless value for one-time historical existence and irreversible physical body immediately disappears. Priceless value of physical body will be degraded and converted sooner to economic value. Therefore, it is certain "inviolable," and it should not touch with any ways.

One more example shows: if there are two cases, first is a case on the *death* of each person holding own unique singleton and historical irreversibility. And another case is on the *death* of an existence completely replacing from all of own human body to digitized memory and cyber-prosthesis. Please imagine each situation. Many of us subtly feel something different between them. That's just the respective care with the dignity or without the dignity.

As discussed earlier, an identity holds each singleton and time irreversibility in own historical living, and we can easily remind the unneglectable fact and modal value on "*Summum Bonum*" as the supreme good viewing through a concept of "dignity" at the time of facing death situation. Data protection law on "*right to be forgotten*" for privacy and identity of each person in accord with GDPR clearly denotes to pay much respect to the dignity for each person (Inveradi 2019). They never manage it as an operation to delete digitized data, and rather they take it carefully with dignity even if a person already passed away.

3.4.2 Barriers Against Ethical Misconducts by AI

In fact, little is known about ethics of the AI for the humankind, even though many proposals edited by various authorities have been compiled guidelines and statements (International Committee of the Red Cross 2018).

As discussed earlier in the current chapter, the author exemplified a case of the questions by a student. Similarly, if the AI driven robot asks a question "*Why does not kill any human?*" to the human, how many people can answer logically convincible statement for the AI?

As noted above, many of such questions are often firmly rooted in philosophical and formal logics, and the author already mentioned an example answer for university students. But there is not any guarantee that the AI understands and accepts such explanations by ordinary people in accord with the manners of thinking what the humankind expects. And the AI cannot understand "*common sense*" in the human society. Thus, in the digitized age of AI-centered society, there is a top-prioritized reason that the AI system should be required to learn the ethics for the humankind in such meanings.

The ethics for the AI has been learned by various trials in each context, and those statements can be based on logics, behavior rules, and mathematical proofs. Further, we should implement the firewall against misconducts and misunderstood actions conducted by the AI in various social situations.

As an idea for equipment of the ethics for the AI, it prepares the barriers against malicious logics based on inference of the AI. The barrier which surrounds multiple logics around the heart of the dignity for the human cannot be assessed by AI's inferences and decision-making, should be initially implemented and cannot be updated within the processing chips of the AI. Higher order inference will not be managed by such AI, because Turing' theorem (it partly depends on mathematician Gödel's incompleteness theorems) had already proved the cases of computational limitation (Turing 1950; Copeland and Shagrir 2019; Ben-David et al. 2019). Some cases (e.g., halting problem) in recursive functions for AI cannot be proceeded to prove and operate in finite computable processing. When the AI faces at the ethical border cases, it should be designed that the AI system cannot step and process further operations. Moreover, according to the AI study's history (Russell and Norvig 2009), there will be further possibilities for ethical discussions on AI (e.g., possibilities on meta-consciousness of the AI, controversies on strong AI versus weak AI (Searle 1996, 1998)).

Exactly, as there is certainly "*trolley problem*" (ethical confrontation between two choices in one of the thought experiments), what should AI driven machines do without any sacrifices against people? Otherwise "*Ring of Gyges problem*" in ethics can promise to be provoked by the misconduct of the AI, and namely, how can the AI driven machines be autonomously self-monitored by own honesty in accord with the rules even if the human does not exist here? In other words, how can the human detect misconducts, dishonesty, and deceptions by the AI? Many of the AI driven services and processes will be literally invisible for the human, and their autonomous computations ought to be regulated by the strict rules. In addition, the AI system has serious vulnerability against adversarial intention by antagonistic learning data. For example, Microsoft AI chat-bot "Tay" was already abandoned because this system learned prejudicial hate information by supporters of Nazism. AI driven robots must especially care about social justice and morality (European Commission 2019).

Additionally, as discussed later, military weapons which installed advanced AI will not support such ethical barriers for saving the dignity and safety of the humankind (Copeland 1993). Therefore, it should be deepening ethical topics for the AI from various viewpoints.

3.4.3 Trustful AI

As constructing mutual trust among the human and the AI, it means that we accept an existence of the AI as necessary partners, and contrarily the AI driven system should contact us with the highest dignity. The core of the above discussions requires especially to be holding human-centered orientation by the AI. According to EU commission' report on AI (2019), they argued those trustful AI in ethical conducts. And EU proposed ten necessary preconditions as following points.

1. *Accountability*
2. *Data governance*
3. *Design for all*
4. *Governance of AI autonomy (human oversight)*
5. *Non-discrimination*
6. *Respect for (& enhancement of) human autonomy*
7. *Respect for privacy*
8. *Robustness*
9. *Safety*
10. *Transparency*

Cynically viewing, it seems to triply enlarge several rules (e.g., the *Three Laws of Robotics* proposed by Asimov) to 10 articles in the rule of conduct for the AI. They reversely claimed only enhancements of the amenities and hospitalities for the human-being and its welfare. Through inscriptions of above menu list given by their requirements, it clears how to satisfy the dignity of the human-being (i.e., *amor sui*). Only AI shall be contracted unilateral obligations to keep ethical conducts and commit truthful serving for us. Nevertheless there is a need for the mutual trust between the human and the AI systems, but it appears that there are not equal rights and bilateral obligations in such relationships. It explicitly distinguishes between master side and servant side. Rather EU report implied the rule of conduct of AI as better servants for all people. Someone may recall it associating with the past claims on the White and Black relationships such as Jim Crow laws in the USA and its conflicts.

Otherwise, AI researchers Yu et al. (2018) proposed the following set of taxonomy on the ethical study of the AI.

1. *Exploring Ethical Dilemmas: technical systems enabling the AI research community to understand human preferences on various ethical dilemmas*
2. *Individual Ethical Decision Frameworks: generalizable decision-making mechanisms enabling individual agents to judge the ethics of its own actions and the actions of other agents under given contexts*
3. *Collective Ethical Decision Frameworks: generalizable decision-making mechanisms enabling multiple agents to reach a collective decision that is ethical*
4. *Ethics in Human-AI Interactions: frameworks that incorporate ethical considerations into agents which are designed to influence human behaviours*

What they intended to wholly concise AI studies on the ethics of the AI can be generalized as the conceptual framework corresponding with traditional AI studies. In brief, AI researchers firstly think of implementing a set of the behavior rules to the AI mechanism in accord with the ethical manners at each interactive situation.

Such ethical rules of the AI system are not limited for any humanoid robots. Auto-driving cars (Maurer et al. 2016) and autonomous drones should be legally authorized by our standard. Similarly, as one of the AI machine developers (e.g., Watson), IBM (2018) published own ethical principles for the *explainable AI* (i.e., XAI). AI's logics and consequences of judgements have to be understandable for the human, and it should not become a black-box. Then the AI system should be equipped logically translating consequences to traceable logics described by our natural languages. In this concern, it appears that their doctrines of business organization expressed in following principles. Especially, they enumerated both accountability and explainability, even though these are synonymous to each other. It might reflect viewpoints from both company's side and the AI's side. Both of them shall describe own detail cognitive process and its responsibility for the public understanding (Miller 2019). Otherwise, some argued that we should admit and give a legal personality for the AI, but many cases cannot be admitted for any rights of the AI. And product reliability of the AI must be owned by each manufacturer, and then such legal matters would be undertaken by their responsibility.

1. *Accountability*
2. *Value alignment*
3. *Explainability*
4. *Fairness*
5. *User data rights*

On the other hand, judicial system driven by the AI has been discussed (Sourdin 2018). Virtual lawyers and judicial services online (e.g., ODR: Online Dispute Resolution[6]) may appear in the near future. For launching such services, of course, the AI driven judicial mechanism has to universally learn above all viewpoints, logical interpretations of the law, and the moral standard for the human caring with dignity (Floridi 2016, 2018). But in fact, the ideal judicial system by the AI cannot be developed in easy way, because law interpretation and explanation by the AI requires the vast knowledge' linkages, annotations of word meaning, and ontological database. The symbol grounding problem on the AI must be solved by practical ways (Harnad 1990). As it is likely cyber-physical robots in the real world, the frame problem may be occurred again (expert system for legal issues had failed in the past of AI studies). Therefore, even if we will restrictedly implement decision-making supports for human lawyers and retrieval mechanisms on the past precedents by the AI at first, domain-specific knowledge in law for the AI must be adapted in variabilities of infinite law cases.

[6] http://www.odreurope.com/; http://www.odreurope.com/assets/site/content/odr_resolution_by_UN.pdf

References

Bacha-Trams, M., et al. (2017). Differential inter-subject correlation of brain activity when kinship is a variable in moral dilemma. *Scientific Reports, 7*, 14244.

Ben-David, S., et al. (2019). Learnability can be undecidable. *Nature Machine Intelligence, 1*, 44–48.

Bostro, N. (2003). Are you living in a computer simulation? *The Philosophical Quarterly, 53*(211), 243–255.

Cacioppo, J. T., et al. (Eds.). (2002). *Foundations in social neuroscience*. Cambridge: MIT Press.

Camerer, C. F. (2003). *Behavioral game theory: Experiments in strategic interaction*. Princeton, NJ: Princeton University Press.

Cederman, L. S. (2005). Computational models of social forms: Advancing generative process theory. *American Journal of Sociology, 110*(4), 864–893.

Copeland, J. (1993). *Artificial intelligence: A philosophical introduction*. Oxford: Wiley-Blackwell.

Copeland, B. J., & Shagrir, O. (2019). The church-Turing thesis: Logical limit or Breachable barrier? *Communications of the ACM, 62*(1), 66–74.

Cox, J. C., Shachat, J., & Walker, M. (2001). An experiment to evaluate Bayesian learning of Nash equilibrium play. *Games and Economic Behavior, 34*(1), 11–33.

Descartes, R. (2010). *Meditationes de prima philosophia, Fili-Quarian classics* (English translated edition).

Dreyfus, H. L., & Dreyfus, S. (1986). *Mind over machine: The power of human intuition and expertise in the era of the computer*. New York: Free Press.

European Commission. (2019). *The European Commission's high-level expert group on artificial intelligence: Draft ethics guidelines for trustworthy AI.* Retrieved from https://ec.europa.eu/futurium/en/system/files/ged/ai_hleg_draft_ethics_guidelines_18_december.pdf

Floridi, L. (Ed.). (2016). *The Routledge handbook of philosophy of information*. New York: Routledge.

Floridi, L. (2018). Soft ethics and the governance of the digital. *Philosophy & Technology, 31*(1), 1–8.

Francis, K. B., et al. (2017). Simulating moral actions: An investigation of personal force in virtual moral dilemmas. *Scientific Reports, 7*, 13954.

Franks, D. D., & Turner, J. H. (2012). *Handbook of neurosociology*. London: Springer.

Fukuyama, F. (2018). *Identity: Contemporary identity politics and the struggle for recognition*. London: Profile Books.

Garfinkel, H. (1991). *Studies in ethnomethodology*. Cambridge: Polity.

Glimcher, P. W., Camerer, C., Fehr, E., & Poldrack, R. A. (Eds.). (2009). *Neruoeconomics: Decision making and the brain*. San Diego, CA: Academic Press.

Harnad, S. (1990). The symbol grounding problem. *Physica D, 42*, 335–346.

Heidegger, M. (2008). *Sein und Zeit* (English translated edition). New York: Harper Perennial Modern Classics.

Hobbes, T. (2009). *Leviathan*. Oxford: Oxford University Press.

Honderich, T. (2005). *The Oxford companion to philosophy (Oxford companions)*. Oxford: Oxford University Press.

Howard, D., et al. (2019). Evolving embodied intelligence from materials to machines. *Nature Machine Intelligence, 1*, 12–19.

IBM. (2018). *Everyday ethics for artificial intelligence*. Retrieved from https://www.ibm.com/watson/assets/duo/pdf/everydayethics.pdf

International Committee of the Red Cross (ICRC). (2018). *Ethics and autonomous weapon systems: An ethical basis for human control?* Retrieved from https://www.icrc.org/en/download/file/69961/icrc_ethics_and_autonomous_weapon_systems_report_3_april_2018.pdf

Inveradi, P. (2019). The European perspective on responsible computing. *Communications of the ACM, 62*(4), 64–68.

Jost, J. (2005). *Formal aspects of the emergence of institutions*. Retrieved from http://www.santafe.edu/research/publications/workingpapers/05-05-018.pdf

Knoblich, G., Elsner, B., Aschersleben, G., & Metzinger, T. (2003). Grounding the self in action. *Consciousness and Cognition, 12*, 487–494.

Kondo, R., et al. (2019). Illusory body ownership of an invisible body interpolated between virtual hands and feet via visual-motor synchronicity. *Scientific Reports, 8*, 7541.

LaChat, M. R. (1986). Artificial intelligence and ethics: An exercise in the moral imagination. *AI Magazine, 7*(2), 70–79.

Luhman, N. (1984). *Social systems*. Stanford, CA: Stanford University Press.

Markus, H., Smith, J., & Moreland, R. L. (1985). Role of self-concept in the perception others. *Journal of Personality and Social Psychology, 49*, 1494–1512.

Maurer, M., Gerdes, J. C., Lenz, B., & Winner, H. (2016). *Autonomous driving: Technical, legal and social aspects*. Cham: Springer.

Miller, T. (2019). Explanation in artificial intelligence: Insights from the social sciences. *Artificial Intelligence, 267*, 1–38.

Morgan, M. L. (2011). *The Cambridge introduction to Emmanuel Levinas* (English edition). Cambridge: Cambridge University Press.

Neftci, E. O., & Averbeck, B. B. (2019). Reinforcement learning in artificial and biological systems. *Nature Machine Intelligence, 1*, 133–143.

Nowak, M. A. (2006). Five rules for the evolution of cooperation. *Science, 314*(5805), 1560–1563.

Nowak, M. A., & Sigmund, K. (2005). Evolution of indirect reciprocity. *Nature, 43*(7), 1291–1298.

Olivier, B. (2007). Artificial intelligence (AI) and being human: What is the difference? *Acta Academica, 49*(1), 1–20.

Russell, S., & Norvig, P. (2009). *Artificial intelligence: A modern approach*. Upper Saddle River, NJ: Prentice Hall.

Saha, S., et al. (2017). *Handbook of research on applied cybernetics and systems science* (Advances in Computational Intelligence and Robotics). Hershey, PA: Information Science Reference.

Schroeder, D., et al. (2017). *Dignity in the 21st century: Middle East and West*. Cham: Springer.

Schütz, A. (1967). *The phenomenology of the social world*. Evanston: Northwestern University Press.

Searle, J. R. (1996). *The construction of social reality*. London: Penguin.

Searle, J. R. (1998). *Mined, language and society: Philosophy in the real world*. New York: Basic Books.

Shibuya, K. (2012). A study on participatory support networking by voluntary citizens—The lessons from the Tohoku earthquake disaster. *Oukan, 6*(2), 79–86. (In Japanese).

Sim, K., et al. (2019). Metal oxide semiconductor nanomembrane-based soft unnoticeable multifunctional electronics for wearable human-machine interfaces. *Science Advances, 5*(8). https://doi.org/10.1126/sciadv.aav9653.

Sourdin, T. (2018). Judge V Robot? Artificial intelligence and judicial decision-making. *UNSW Law Journal, 41*(4), 1114–1133.

Spinoza, B. (2008). *Ethica, ordine geometrico demonstrata* (English translated edition). BiblioLife.

Sun, R. (2006). *Cognition and multiagent interaction*. New York, NY: Cambridge University Press.

Suzuki, K., et al. (2017). A deep-dream virtual reality platform for studying altered perceptual phenomenology. *Scientific Reports, 7*, 15982.

Turing, A. (1950). Computing machinery and intelligence. *Mind, LIX*(236), 433–460.

Uchida-Ota, M., et al. (2019). Maternal speech shapes the cerebral frontotemporal network in neonates: A hemodynamic functional connectivity study. *Developmental Cognitive Neuroscience*. https://doi.org/10.1016/j.dcn.2019.100701.

Wittgenstein, L. (1922). *Tractatus logico-philosophicus*. New York: Harcourt, Brace.

Worcester Polytechnic Institute. (2012). *Design of a human hand prosthesis*. Retrieved from https://web.wpi.edu/Pubs/E-project/Available/E-project-042612-145912/unrestricted/MQP_PaulV_Complete_Final_3.pdf

Yu, H., et al. (2018). *Building ethics into artificial intelligence*. Retrieved from https://www.ijcai.org/proceedings/2018/0779.pdf

Yuste, R., et al. (2017). Four ethical priorities for neurotechnologies and AI. *Nature, 551*, 159–163.

Chapter 4
Identity Valuation

Identity valuation has decisive influences for your daily life in the past, present, and future (including your future kinships and offspring). Recently, day by day, the digitized credits based on each personality, gifted abilities, educational level, behavioral characteristics, health condition (it can be based on each genome data), human relational factors, estates, and cash information are usually registered by the centralized agencies.

At first place, "*valuation*" mainly means a technical term in finance, and such subject has been globally teaching in MBA (master of business administration) course of professional business schools. It obviously denotes measuring value for something at the present and the future by the standard. Furthermore, in digitized society, valuation has specific cutting-edges for not only financial fields in business scenes but also something converted to monetary index by the digitized standards based on big data. Of course, it also targets for us, and then the author named it "identity valuation" for discussions.

4.1 Earning Money by Big Data

As one crucial report from a think-tank, FutureMajority (USA) has announced their latest researches (Shapiro and Aneja 2019). They said *"Americans are rightly concerned about mounting evidence that the internet's major platforms and many large companies are systemically gathering, analyzing and selling everyone's personal information. More than nine in ten Americans believe that they should determine who can see their personal information, and nearly nine in ten believe they should be able to direct any website to dispose of their personal data."*

And they estimated that GAFA globally earned 76 billion US dollars in 2018FY (76,046.2 million US dollars in detail) through usage of the vast information accumulated from many people. They further simulated that those trends will be steadily increasing their benefits more than 2018FY. Then, each individual lost own chances

K. Shibuya, *Digital Transformation of Identity in the Age of Artificial Intelligence*, https://doi.org/10.1007/978-981-15-2248-2_4

to gain economic profits by own personal information, and GAFA has overwhelmingly been gaining *free-ride* data given from personal users and eventually those data which have been coined in their global economic advantages for establishing successes and occupying their dominances has been publicly restricted for global citizens.

Therefore, each individual should manage own private data, knowledge and holding own value for appropriate usages. In light of the total benefit in 2018 FY, global population on the earth is now estimated as 7.6 billion people, and then the cash for each individual can be equally allocated 10 US dollars per person. Of course, it seems that this rate might not be accurately shown for our present value. Anyway, it must rethink about the value per se.

4.2 Scoring a Value of your Life

4.2.1 Your Digitized Scores by the AI

4.2.1.1 Scoring Services

In addition, there are possibilities that each individual is evaluated by third parties in digitized society. For example, it now realizes various quantified index services such as *Zhima Credit* in China and *J.Score* in Japan. And now, other similar services competitively appear (e.g., Yahoo!Score, Line Score), and these depend on vast data from a larger amount of their consumers.

Significantly, those online data are automatically accumulated, analyzed, evaluated, categorized, and predicted by the AI driven algorithm (Qi and Xiao 2018), and it probably justifies classifying hierarchical rank order to each person in renewal social class (or stratification). It newly entraps us to the inevitable meritocratic world.

For example, in China, Zhima Credit introduced own AI-based metrics to evaluate each consumer by multiplied viewpoints such as personal characteristics, cashing history, education level, job, career, real estate data, human relationships, preferences, and other necessary data. After that, it rates each individual ordered by qualified score from 350 to 950. As statistical distributions, it can be estimated that many of consumers stay within 550–699. The highest grade rankers (700–950) usually have the prestigious social status and careers. When the extent of such score does not meet the criteria, lowered rank consumer cannot be offered by many banks and financial services, and furthermore their recruitment opportunities and marriages have already much trouble.

4.2.1.2 On Value

Generally speaking, the value per se can be difficult concisely defined across varieties of academic disciplines such as computer simulation, sociology, economics, psychology, and philosophy (Kahneman and Tversky 2000; Camerer 2003; Bostrom 2016), as it is nearly to seek the greatest common divisor from them. It should be at least distinguished among mental account of valuation, measurement ways of value, evaluation function to be implemented for goal-seek, evaluating form on cost benefit analysis, socially shared standard of value and any value derived from the highest good.

So, at first place, what is the *value*? How to measure it? All value is perfectly determined by the free-will, preferences, and demands of the human in the ordinary situations. No value in the nature initially exists. It is quite similar to the measurement problem in quantum physics. By through eyes and consciousness of the humankind, it has to firstly decide the target which is evaluated, to define specific indexes for measurements, and discern meanings what such results indicated. The author proposes to classify several types of the values in daily contexts, for instance.

1. *Existential value*: This deals with a core of the dignity of the humankind. It assumes a given presupposition that each of us has naturally own inseparable value since each birth. It mostly depends on the finite lifespan of the humankind. Even though atomic elements and chemical substances never contain any meaning and value in natural condition, the humankind independently decides to sustain an original value for our living. And it does not permit anyone to be devalued and downgraded by any purposes in ordinary situations.
2. *Metric Value*: The humankind can comparatively decide to measure something valuable in the hierarchical order. For example, pairwise comparisons among a set of things can be assorted by the ascending order. Of course, computational systems driven by the AI can do the same thing, as the scoring services noted above.
3. *Financial Value*: It means that economics and social science often manage these topics in their realms. It is nearly determined by the extents between supplies and demands in specific markets. The value of the investment in cost-profit analysis and estimation of the leverages in financial contexts can be formalized by the stochastic equations.
4. *Social Value*: It devotes to commonly share the values in the society, such as laws, institutions, customs, norms, and agreements for the purpose of grounding our judgements in accord with the standard derived from the highest good.

4.2.1.3 Theories on Value

It seems certain that above scoring service by the AI has firstly common goals corresponding with financial value as well as socially sharing standard of the value. If not so, those services cannot be popularized among citizens and make a greater success in their business. Namely, latent motivation of service providers is to drive ordinary citizens toward the specific value by being promoted among them.

Inducing value by such digital ways has to be actually deepening social meaning. It is indeed an open question on emergent possibility for social norm and internalization of the value among citizens. Still, any social norm and morality accomplished by internalizing common value among citizens will be similarly understood in those contexts. Hence, little is known about those emerging process on internalization of value by empirical ways yet (House et al. 2019). In this concern, scoring services by the AI will become a very important role for further understanding on those themes by social experiments and practical ways in actual society.

In the light of those scoring services, it constitutes a rational way to approve qualifying mechanisms in meritocratic society. Value has an interdependent relationship with *preference* of each individual, and then it must differentiate between individual choice and collective choice. Those social welfare functions on the bases of value and preference had been conducted by welfare economists Arrow (1963) and Sen (1970). Recently, heritages of these Nobel Prize in Economics are now inherited by computational collective choice studies (Brandt et al. 2016), and preference and value issues are underlying in mechanism design studies and behavior economics. According to those theoretical reasons, each value of individual will be hardly being ascended into socially common standard of value. Furthermore, rational decision and prioritization based on value and preference among citizens can occasionally engender paradox, dilemmas, and contradictions in the society (e.g., Condorcet's paradox).

Next, can coexistence of both social morals and economics engender the conflict of interest? Ancient economist Smith (2013: "*The Theory of Moral Sentiments*") and Bonar (1926) discussed both morality and sentimental sympathy for the others in socioeconomic interrelationships, and their conceptual thoughts and actual mental mechanisms of the human had been investigated and revealed by scientific empirical ways (e.g., psychology, brain sciences). Still, intersections between altruism and egoism as well as philosophical considerations and practical rationality ought to be laid much weight on those contexts to argue. Our society could not be appeared as the present regime of governance until social mechanisms lead citizens obeying the established rules.

On the other hand, according to theoretical sociologist Parsons (1970), he had a proposal "*pattern variables*" to critically discuss the value of social issues in meritocracy. Below symmetrical pattern variables mean comparatively both static and dynamic structural patterns of society.

1. *Affectivity and Affective Neutrality.*
2. *Collectivity or Self.*
3. *Particularism and Universalism.*
4. *Diffuseness and Specificity.*
5. *Ascription and Achievement.*
6. *Expressive and Instrumental.*

As standing on these patterns, AI-based qualification and valuation for each individual has a superficial tendency to be decided on the bases of neutrality, collectivity, universalism, achievement, and instrumental standards. It is too natural to understand what AI laid weights on our parts of valuation.

In contrary to economics, unfortunately, economic sociology had not contributed toward above trends enough yet. Nevertheless *value-added approach* in fascinated movement caused by collective dynamics (Smelser 1962) has intrigued viewpoints, but many parts of knowledge in general and economic sociology (Smelser and Swedberg 2005) should be almost revised by actual progress of the AI and big-data sciences, and namely they still show little presence. In his theory, a concept of *value added* was named from earlier stages in economics, and it refers to the increasing value of products by the development of production. In contrast, relatively, contemporary economists in behavior economics and mechanism design can contribute toward engineering-based collaboration with natural scientists and engineers in those areas. Therefore, the latter cases have been already producing vast applications, through collaborative works on auction and matching mechanisms to apply for automatic stock exchanges by the AI and above valuation services as incentive design for each individual by big-data analysis. The former studies in sociology were only relying on the traditional thinking styles, and their materials for thought could not be reflected by state-of-the-art progresses.

4.2.2 Internalizing the External Moral Standard Through Scoring

4.2.2.1 Internalizing Value Among Citizens

Identity value what the author mentioned may be reflected by the meanings of own living ways itself. Each personal standard should coordinate with social aspects of the values (Heath 2011; Bowles and Gintis 2011). Ranking based on some standards has been quantified by somewhat rational manners. But, a term "valuable" is to simultaneously suggest "measurable" and "comparable" than the others in classified order. Namely, when ordinary citizens are staying in such social conditions, those systems can autonomously optimize the behavioral manner for the citizens, as consequently many of the citizens will be frequently motivated to care about their speech acts and behaviors for maintaining own qualified value by the AI driven systems and achieving more rank-up. In that way, all of them forced to unintentionally obey the rule of conducts, and those social designs explicitly employ loop-cycles of enforced learning to be fitted their daily actions to the value directed by the AI.

If such ways are possible, those conditions are quite similar to self-enforcing machine learning of the AI, and further it commonly calls ethical concerns. Here, the ethical rules of conduct for each individual can discern between internal and external factors.

Firstly, internalization of value has been examined by psychological experiments (Carver and Scheier 1998). Experimental psychologists Christensen et al. (2004), for example, explored a self-regulation model on subjective well-being and ethical behavior to attain social order and norm. Especially, they focused on self-discrepancy theory proposed by Higgins (1987, 1989), and social norm could be lastly achieved and regulated by mental self-guide mechanism and reducing discrepancy between actual-self and ought-self within parts of each individual.

And secondly, our motivation has two aspects such as intrinsic motivation and extrinsic motivation. Of course, the latter can regard aforementioned scoring services by the AI. But, in contrary, the former motivation can be characterized as self-incentive and self-initiative directions and such mentality spontaneously drive people to employ in achieving the goal per se. For example, what profits can editors of Wikipedia get (i.e., one of online encyclopedia projects)? They are still all volunteers nevertheless their works are ongoing hard works (Yang and Lai 2010). It means that they do not earn any money and financial bonus. Their motivation can be filled by honors and praises rather than monetary incentives. Economically, volunteering activities cannot be explained by their economic theory, because their theory cannot deal with irrational behaviors without earning money (Wilson 2000). Then, they call it "*common goods*" for social welfare and values as including the economics' paradigm. Such social commons exemplify social capital, social friendships for everyone, philanthropy, and mutual reciprocity.

Although psychological experiments had been managed in laboratory situations, but there were very impressive findings. As mentioned above, if cyber valuation directed by the AI has actually normative influences to regulate each individual in accord with social guideline of moral value, those societies might be exactly harmonized with both internal (i.e., each person) and external values (i.e., society). Such society reaches to Kantian's ideal for the morality. In ethical perspective, Kantian's moral law for ethical behaviors has to educate ordinary citizens matching with both own practical value and social standards (Kant 2017). Ordinary citizens should be cultivated by themselves for own practical behavior meets to the social common standard and rationality in the whole society. Can "*moral law*" and "*Reich der Zwecke*" be realized by such ways in the contemporary society?

In such context, next viewpoint is sociological discussions on social norm, morals, law, and institution. At first, Riesman et al. (1969) named several personality types such as *tradition-directed, inner-directed,* and *other-directed* in their notable work "*The Lonely Crowd*," especially tradition-directed type ought to be adapted in social norm directed by the authorities. This type of people's motivation to avoid social anxiety had been assumed for their adaptation in modern society, but their theory partly exhibited why above three patterns of personality could be explained for organizing between social order and individuals. Of course, their categorizations were mostly depended on only theoretical conceptualization, and there still seems to be convincible rooms for discussing how conducting people within social regulation.

According to one of the sociological discussions (Therborn 2002), he categorized three types of social norms, and those patterns would be reflected in the sociological standpoints to stress on dynamical interactions and social constructive process.

1. Constitutive Norms: *The first consists of constitutive norms, which define a system of action and an actor's membership in it.*
2. Regulative Norms: *The second set is made up of regulative norms, denoting actors' expected contributions to, or expected performance or execution of tasks in, the system.*
3. Distributed Norms: *Third, we may distinguish distributive norms, designating how rewards, costs, and risks should be allocated in a given social system.*

Otherwise, those social factors should be systematized as one of the emergent properties into social system models on internalization of morality among citizens. Axelrod (1997) argued emergent possibility on social norms and its formation process. He categorized the following discussion points, and his distinguished idea was to investigate social norm as emergent property using social simulations in game theoretical backgrounds. *Meta-theory*, *internalization*, and *reputation* have his intrigued insights for emerging possibility of social norms, and if each individual has acquired disposition and capacity regulated by each value standard, any social morality will be firmly stabilized by the majority of social members.

1. *Meta-Theory*
2. *Dominance*
3. *Internalization*
4. *Deterrence*
5. *Social Proof*
6. *Membership*
7. *Law*
8. *Reputation*

4.2.2.2 An Example of Social Simulation on Dominant Influence of the Value

Likewise field experiments, those scoring services by the AI can be demonstrated as a kind of social experiments. Ordinary experimental methodologies in sociology have not been pursuing to be verified by more scientific process (Bowles 2017). But consequently, those services approach to examine our internalization of the ethical and moral value using smartphone in daily life. Therefore, it can be grounding in both theoretical discussions on morality and value in sociology and actual observations for their usage patterns among ordinary peoples.

Further, it is possible to simulate those matters by appropriate modeling. At first, scores of value measured by the AI are almost reluctant to open the detail algorithms and inner logics of the evaluation process for us. A score of each individual can be latterly displayed as a part of calculated results. Many of us do not understand what factors have critical influences being measured by the AI. In game theory, such social situation is called as "*Information Asymmetry*" (Akerlof 1970) between the AI driven service providers and ordinary citizens. The former, of course, knows all of own evaluation process, whereas the latter can indeed only know results and estimate those inner logics. When many of us are surrounded in such higher uncertainty environment, one of the most rational ways will refer to the most adaptive others' behaviors and imitate their patterns of behaviors through learning results in continual process. It is nearly similar to aforementioned "*social proof*" and "*reputation*" in Axelrod's model. In such ways, if a specific value of the standard directed by the AI will be gradually sharing with many citizens ("*Internalization*"), those social tendencies lastly have much influential "*dominance*" for other citizens.

Here, for the purpose of visually exhibiting above discussions, the author exemplifies a kind of social simulations on socially pervading dominance of specific value by ABM (agent-based model), and below configurations are applied.

Each agent (i.e., each citizen) has own value standard and intentions to earn own profits in each round, and their profits can be determined by unseen pay-off matrix. It assumes that the nature (e.g., the AI driven scoring services) periodically provides rewards (or sanctions) for those who obey the value conducted by the nature. But such matrix in detail cannot be known for each agent, and then each agent must learn more adaptive ways through their learning from results in each round. As mentioned above, so then, this model presumes that each agent can imitate a strategy of the most adaptive other agents who directly adjoined with each other. Adaptiveness of value can be defined as the total amount of profits of each agent.

- Simulation: Iterated games in finite two-dimensional lattice.
- The world conditions: Topological torus and the Moore neighborhoods. Then each agent always has eight linkages of adjoined other agents, and in each round, every agent must interact with each of them.
- Agents in the world: 10,000.
- Agent's strategies: It assumes that agent strategy reflects own attitude to the value standard. First time initially randomizes to set a strategy (coop (i.e., obeying the value) or defect (i.e., denying the value)) in each agent. At the beginning of next round, an agent imitates a strategy of the most earned and directly adjoined agents in the previous round (when nobody did not earn more than myself, an agent keeps own strategy).
- Pay-off matrix: A frame of ordinary PD (prisoner's dilemma) game (Axelrod 1984, 1997). Namely, where pay-off structure between $agent_1$ and $agent_2$ are respectively [coop/coop] = [a/a], [coop/defect]&[defect/coop] = [b/c]&[c/b], and [defect/defect] = [d/d], and it always keeps $c > a > d > b$. If both agents mutually keep coop strategy (i.e., [coop/coop]), their result can become coexistent condition by obeying the value, and both of them gain each profit from the nature as a system. Namely, social norm consensually shared by coop strategy among agents can be emerged if and only if pay-off matrix intends to prevent contagions of defective evil trends (aforementioned "*Deterrence*" and "*Law*" in Axelrod model) (Fig. 4.1).

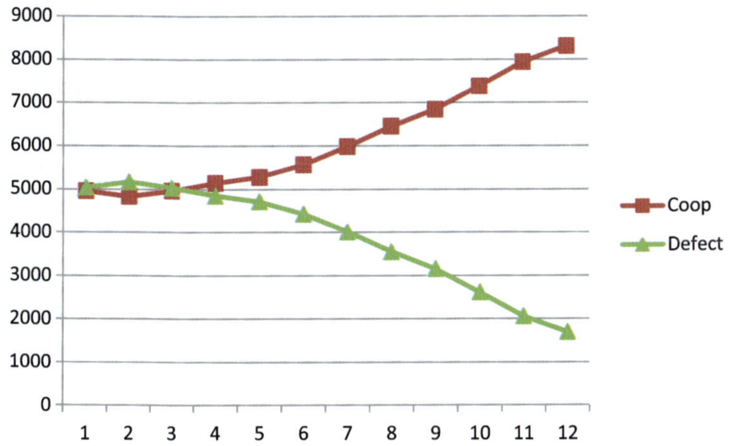

Fig. 4.1 An example of simulation results (Y axis denotes total counts of each strategy among all agents)

Above figure clearly illustrated an example result of simulations and Y axis denotes total counts of each strategy among all agents. Consequentially, a dominant value as coop strategy has steadily grown among agents, and it means that many agents share the common value which is induced by the nature (i.e., it assumes the AI driven scoring services here). Therefore, such convergent mechanism was caused by one of "*reward or sanction*" for citizens, and it implies that scoring service of valuation by the AI can easily engender those similar consequences. It indeed actualizes an emergence of social norm in order to induce driving citizens toward a specific value in actual society.

However, this simulation result simultaneously suggested that there will be serious anxiety against social inequality among citizens by the AI scoring services. At first place, the AI scoring services will possibly engender metric divides among citizens. Those who do not and cannot obey and adapt the value directed by the AI will be gradually decreasing their rewards and downgrading own evaluation, and finally they cannot escape from those negative spirals. Those situations should not be named as the results of evolutionary adaptations in actual society (Bowles and Gintis 2011). Therefore this simulation has much intrigued viewpoint.

Furthermore, using smartphone, many of us can already gain more useful services and cashless transactions, and such rewarding digital economy enlarges its comfortable value for us. And consequently we will not impossibly escape from the gravity of the new paradigm of digitized economy. Conversely, it will be certain that some of citizens cannot access and consume such services because of their lower socioeconomic trusts and status. Then, digitization can easily divide the unreachable clusters among citizens by quantifications on the measurements and standards.

4.2.3 Working for the Ethical Rules and Economic Valuation

Scoring services by the AI can engender similar conditions of "*benefits obtained in this world through obeying norm*" based on the Buddhist teachings. In this concern, contrarily, it should carefully consider another style. According to sociologist Weber (1964, 1977), one of his representative works "*The Protestant Ethic and the Spirit of Capitalism*" which introduced some distinguished viewpoints has now interesting considerable matters. First, Weber almost thought as follows: when the protestants were subject to work ruling in the Predestination, they would eagerly obtain their mental reliefs and comforts for the present and future, by daily proving their excellences that matched to own gifted abilities and calling (*Beruf* in Germany) for the purpose of confirming the certainty for their salvation. And eventually, the Protestants' nations that governed by those who believed such story had been developing more effectively than the Catholics' nations, and those advantages worked proficiently for the emergence and progress of the capitalism. Of course, at Weber's age, it seems that many people belonged in the working class ought to eagerly obtain own daily foods, money, and necessary goods for their living, whereas those who were categorized in the classical wealthy class could be legitimately justified by traditional custom. But the cause of their difference might be a simple reason that protestants were relatively majority in the former; in contrary, the latter was probably occupied by many of Catholics than the former. And when larger incentives by pre-industrial revolution and national trends of drastic technological progress had come, the consequence from little difference between them might become more decisive one. Namely, their standard of value for living would be also cultivated by social surroundings and historical common senses. Anyway, their necessity for living could be internalized as the social value by standardizing external factors.

Then, does such traditional sociological theory have still any meanings for the digitized AI society? Weber also told the value neutrality in sociological contexts, but mostly social values can be defined by the degree which ought to be exhaustively aspired among many of the citizens. Certainly all concepts never initially contain specific value, but all values can be latterly ordered by the extents of meaning senses and needs of the human. When the AI driven system evaluates all of the citizens by the somewhat unique standard of value in the digitized big-data world, it can say that the AI system induces latent motivations among citizens being highly evaluated. Then, many of the citizens have stronger incentives, and they would inspire to obtain the chances of their status to be maintained and upgraded by the AI. As keeping own behaviors corresponded with the directions induced by value function algorithms of the AI (e.g., Q-learning, Actor-Critic), many citizens may feel own amenity during the present and future in their life course. For such ordinary citizens, lastly, the AI driven system could be easily laid in the alternative as the Protestantism for "God." Then, those social trends will possibly become the coincidence between ethical rules and economic valuation.

If such society appears, will nations further achieve higher economic development? When these mechanisms can be regarded as useful by the centralized government leader, the global economy would be lastly converged or diverged by synchronizing the value of each nation. Because, when every nation has only the single value by the AI, global economic development will be attained by their expectations. However, if not so, global developments might engender chaotic and unstable conditions. The total number of the AI indicates those different values, each segmented value within following citizens led by one of the AI has potentials to idealize each value cluster in the world, and namely it seems similarly to the *polytheism* in the digitized world. How can global collaborations beyond heterogeneous cultural backgrounds be achieved among different standard of value within citizens? But rather, each of the AI driven systems in each nation might immediately learn optimizable ways among heterogeneous value standards of them, and consequently their variations of value could be shrunken and lost by those mechanisms.

4.3 Distinction: Random Selection or Modified Life

4.3.1 Genetic Therapy and its Value for Life

In medical science, several equities shall be assured for every patient. At least, these are *an equity of the value of life, an equity of the finite lifespan* ("*Sein zum Ende*" proposed by Heidegger: a living existence heads for the death), and *an equity of caring by necessary medical operations*.

Especially, on equity of the value of life, to date, there exist considerable issues on medical progress in social contexts (Mittelstadt and Floridi 2016; Marin et al. 2016; King et al. 2017). Genome and DNA data usually determines a part of our ultimate value, physical existence as an aspect of identity, and life-long well-beings. Bioinformatics already traced and revealed all set of the genome of the humankind (International Human Genome Sequencing Consortium 2004). If anyone pays not-expensive cashes, now they can know future risks of own gene diseases and potentials of intellectual possibilities by gene analysis services (e.g., 23 and Me[1], genelife, etc.). This statistical prediction has not still provided perfectly accurate information for each client yet, but it will indicate a future possibility of each individual.

As well known, genome has much inevitable influence for our intellectual, educational attainments, behavioral characteristics, physical tendencies, and disease vulnerabilities (Lee et al. 2018). In some developed countries, many of the life insurances vary the payment rates by the diagnosis of medics on genome assessment of each client. On that background, there are serious cases that genomic assessments (e.g., The new prenatal diagnosis (NIPT[2])) and its counseling for pregnant women may doom a cruel future against her offspring (Adam et al. 2019; Birch et al. 2019; Wurtmann et al. 2019). Hence, as reconsidering on identity valuation, it

[1] https://www.23andme.com/en-int/

[2] https://www.nipt-biomnis.com/

should not forget that genomic counseling shall take care of not only clients of active generation but their children and offspring of future generations. Namely, it can also say that our society still forges many of families to face at such conditions.

Furthermore, using any wearable and ubiquitous devices, many citizens can manage their health data in daily life. Interweaving with genome data, these services can support them from the birth to death. In Japan, similarly, a child company of DeNA (one of the largest ICT companies) had launched gene analysis services for health care based on genome informatics. Health and genome big data can be matched for computational ways to explore the unknown gene functions and the secrets of the nature. And those larger big data has been continually facilitating further business chances to combine with daily medical health, insurance services, pensions, working history, and life-long daily event logs. Such ICT companies oriented for biological and medical intensiveness globally increase its numbers across nations, and lastly they gain the advantages to manage for citizens' life data during each lifetime.

4.3.2 Genome Editing and Modification of Life

At the end of 2018, a China scholar announced a success case on human-modified twins by genome editing (i.e., HIV infection tolerance). This case was completely a false start and misconduct of the ethics in scientific races. It is too hard to stop those technological progress and its applying cases. Given that genome-edited children have more intellectual and preferable properties than ordinary children, now it should verify biological and medical cares for total genetic matters of the human-being. If not so, they approve a fact that genome-edited children have more valuable than the others. Medical care against any disabilities and diseases of clients certainly enhances their well-being and a part of identity value, and nobody can interrupt their hopes. However, some of them may further expect that ultimate replacement of the whole of clients' body regenerated from a set of modified genome can valuably achieve the ideality of their well-being and life-long value.

Certainly, there was a SF like pessimism in the past. For example, it was "*Remaking Eden*" (Silver 2007). Consequently, in the future, the extent of identity value does not fill in the gap between the modified human and the ordinary human. It engenders the structural order of social stratifications, and it further divides regenerations of the future generations within and between them. The future generations will not identify them as same race and rather categorize them into new supreme species more than ordinary human-being. It may rank the intellectual superiority within AI, the modified human and the others in descending order. Some theorists usually claim that such hierarchy between them cannot decidedly subvert by themselves. What does the residual remain? Who does evaluate more valuable than the modified one?

4.3.3 Sociological Criticisms

If such future is possible, sociologically, it realizes an immobilization of the social hierarchy. It further often calls controversial issues on negative feedback spirals on reproductions of social class or stratifications. Those social mechanisms justify difference between those who belong in the upper class and the lower class. Their conditions will be allowing some justifications by social stereotypic reactions (Jost and Major 2001). And the superiority of the upper class for the lower class will be often inherited by the future generations of their offspring. So then, the higher winner's children will successively win, and the lower loser's children will lose again.

Additionally, there are some related scientific evidences on genetic inheritances of physical similarity by marriages among citizens. Researchers of the National Museum of Nature and Science, Tokyo, Japan [3]had endorsed the fact that it was possible to discern a social class of each citizen among samples of buried skulls by statistically discriminant analysis. In the past Japan in the Edo Era (200 years ago at least), ancient peoples were roughly separated within social classes such as *Kuge* (It was a kind of nobles), *Bushi* (or *Samurai.* It was a kind of warriors, knights, or landlord), and ordinary citizens. Population ratio in ordinary citizens was the most majority in the social class pyramid. As an individual of each social class had been often apt to get marriage with a partner of the same class, research results could conclude that their "*physical facial similarity*" physically determined by skull specificities can be discerned easily by anatomical measuring and restoration. Namely, their differences of facial appearances had been inherited mainly by such marriage patterns from generation to generation, and then their facial appearances simultaneously and clearly denoted not only their birth of social class but living environment, education level, cultural backgrounds, and other properties. Such situations were underlying in traditional old customs and somewhat prejudice, but the most important is that such social customs of marriage could gradually actualize differences of physical appearances and reproduce social classification itself. External factors as partner selections for marriage in social surroundings embodied such differences of physical body among citizens. In many cases, physical appearance of those who belonged in the upper class itself might denote to be regarded as a symbol of something valuable, and such physical evidence could offer to explain the legitimacy for prestigious social status. It was very suggestive for understanding on social classification and genetic inheritance among citizens. The author does not intend to assert phrenological disputes (i.e., Gall who was a surgeon of brain at eighteenth century proposed a theory on localizing brain functions) as pseudo sciences. Above evidence could be verified in scientific manners. Of course, epigenetic factors and genetic drift could be elaborated in such process, but it was a kind of self-actualizations of social class and prejudice. In similar cases, globally, racial

[3] https://kaken.nii.ac.jp/file/KAKENHI PROJECT-20300290/20300290seika.pdf (in Japanese), https://www.museum.or.jp/modules/topics/?action=view&id=272 (in Japanese), https://youtu.be/JrKM5J_3T2s (YouTube video lecture. In Japanese).

similarity of physical body might also be based on such reasons, but there is not clear evidence that prejudice itself engendered such differences.

Those symbols of socioeconomic, social-relational, and cultural advantages have specific manners considered by sociologist Bourdieu (1984). His notable work "*Distinction*" articulated those reproductive processes and its justifications enforcing immobility within social classes (or stratifications). But unfortunately his notations lacked the "*genetic capital*" and "*social capital* (especially, it means human relations with prestigious celebrities)." Each gene and its inherited dispositions can be based in not only individual but family identity, and further these are perfectly exempted from taxation against property inheritances. Of course, genetic inheritances could embody a similarity among family members, but there are not statistically enough strict scientific evidences on intellectual factors and personality factors corresponding with specific gene alignments and gene network's functions yet. Otherwise, social capital of human relations with many celebrities can be entirely inherited from parents, and then those children usually take advantages for success in the next generation. Would the reasons against those social inequalities be underlying in their efforts and attributes of each personality or including genetic-origin differences?

On the other hand, Rawls claimed social inequality, fairness, and social justice (Rawls 1971, 2001), and even he further noted feasibility of the public policy for equalizations among gifted person, ordinary citizens, and disabilities. In his context, should gifted and ordinary citizens be downgraded to equalize their body capabilities and intellectual levels by somewhat operations? Or rather, should disabilities be upgraded by somewhat genetic operations? If someone says yes, immediately this person must admit different value of life between them. And relative deprivations should also be rendered in this context (Merton 1968). It should carefully contemplate such fairness in each issue.

4.3.4 Value Chains Beyond Family and Individuals

Given that each life of a child can be determined by such predicable *genetic capitals*, it seems likely to be challenging customizations to obtain own avatar and character in online games. When big-data analysis on health and genetic inheritances of infants will be accurately characterized, and genome-editing medical operation for pregnancy and birth nearly approaches to customizing own favorite avatar at the beginning of any online games, it would envision a part of those modified life selection process. Even if you cannot get a preferred character, within the limit of your financial cost, you may reset and reconfigure an old character, and you try repeatedly until you obtain your preferred and regretless one. Without saying, actual life cannot reset any results by challenging random selections.

Finally, if medical progress will promote gene therapy and genome editing, family concepts and relationships among members will be altered to one of the associations with each other. Identity concept of each individual has been definitively organized and ordered in family tree and genetic inheritances. Manipulated genetic specifications will modify their identification attributing to family history. In 2019, in the Netherlands, there is a case on "*Late Dutch fertility doctor may have more than 49 children*" (April 12th, 2019). Such suspicious medical doctor implicitly donated his genes to 49 fertility mothers without any informed consents and agreements, at least, and consequently he was accused by the legal authority. In that legal court, *his* children had heard the secret of truth, and there might be a problem on how to compensate for their identity. Such case was quite extreme and unordinary condition, but gene therapy and its operations have nearly same meaning against intrusion to living phenomena and modification of each identical background.

For revealing and preventing such criminal conducts, there are various opinions that medical charts should be stored in globally sharing database such as blockchain-based distributed database. If genome editing and gene therapy have latent threats and uncertainty against the existence of the human-being (Cohen 2019), there are greater needs to be prepared to trace each genetic specification on medical charts with keeping privacy.

4.4 AI Optimizing Economic Satisfactions for Medical Services

Medical economics deals with quantified utility for our health, well-being, and cost-benefits in economic perspective (Brynjolfsson et al. 2019). In some well-developed countries, total average of lifespan of many citizens already exceeds more than 80 years old, and now their life planning may require for 100 lifespan.

For example, how to quantify a utility of the life expectancy and evaluate it for each cancer patient? Euthanasia has much utility for them? AI may clearly calculate "a whole value of the life-long cost-benefit," "appropriate life-long planning for each citizen" and "the only way to satisfy own life" by inducing big data. Can these recommendations maximize well-being of each person? It is ostensibly rational that each person makes a decision considering both own physical constraints and possibilities to meet life events, as the AI recommended (Crato and Paruolo 2019).

However, at the macro level, ultimately the data of unspecified majority consisting of massive clients may be accumulated and gradually managed by somewhat global AI system. If so, these systems will easily analyze those data and predict the future patterns of demographic dynamics, occurring time of specific diseases and global health risks against contagions and exponential outbreaks (Ginsberg et al. 2009). And the AI will evaluate the most optimized solution among expectable medical plans based on supply and needs. It is certain that those services will be quite important and necessary to sustain our policy planning on demographic statistics.

But it should pay much attention to further consequences. The AI will be apt to decide a triage utilizing the hierarchical order, as consequently it will evaluate that those who requires less medical care ranks more valuable than the others. At the time of moment, the AI can be understood to convert the most supreme value to something comparable and calculable on each value of life among patients in order. This ordering definitely reflects each identity value among citizens.

As economic indicators often warned about medical care levels, it suggests the extent of medical services in each country. Such indexes also imply somewhat raison d'etre of each citizen within and across countries in the market-oriented economic standard of value (Diener and Suh 1997; Andrews and Withey 2012). The higher extent of such index often reflects positive linear correlations among trustiness and well-being level of citizens comparing between countries. And then those statistical coefficients clearly show that valuations of the life can be affected by decisively external factors than internal factors.

4.5 Regaining Priceless Value of Own Identity

Many critics often say that any AI driven system is not an omnipotent existence now and they cannot precisely predict the future. But digital social science and relative studies must assume varieties of unexpected conditions in advance. And it should deepen to discuss considerable issues for further understandings.

All of the above discussions, own identical value of each individual must be regained for humanity. A term "*Vorlaufende Entschlossenheit*" theorized by Heidegger (2008) reconfirms the will of each citizen to undertake own existential value along with both constraints of limited lifetime and physical body. It does not assure to estimate any possibility of the value accomplishing own life goal, even though the advancement of medical science cares to prolong our lifespans. What defines the value of each citizen as *dasein* is purposely underlying in both subjective and progressive willingness for staying alive and its actual performances in the living world. During the life process of each, an individual shall decide preferred living ways of own life and protect own value of the life. It is not crucially same thing between the values of own life undertaken by oneself and the something evaluated by the others.

References

Adam, S., et al. (2019). Assessing an interactive online tool to support parents' genomic testing decisions. *Genetic Counseling, 28*(1), 10–17.

Akerlof, G. A. (1970). The market for "lemons": Quality uncertainty and the market mechanism. *The Quarterly Journal of Economics, 84*(3), 488–500.

Andrews, F. M., & Withey, S. B. (Eds.). (2012). *Social indicators of Well-being: Americans' Perceptions of Life Quality*. New York: Plenum Press.

Arrow, K. J. (1963). *Social choice and individual values*. New Haven: Yale University Press.

Axelrod, R. (1984). *The evolution of cooperation*. New York: Basic Books.

Axelrod, R. (1997). *The complexity of cooperation*. Princeton, NJ: Princeton University Press.

Birch, P. H., et al. (2019). Assessing shared decision-making clinical behaviors among genetic Counsellors. *Genetic Counseling, 28*(1), 40–49.
Bonar, J. (1926). The theory of moral sentiments by Adam Smith. *Journal of Philosophical Studies, 1*, 333–353.
Bostrom, N. (2016). *Superintelligence: Paths, dangers, strategies*. Oxford: Oxford University Press.
Bourdieu, P. (1984). *Distinction: A social critique of the judgment of taste*. Cambridge, MA: Harvard University Press.
Bowles, S. (2017). *The moral economy: Why good incentives are no substitute for good citizens*. New Haven: Yale University Press.
Bowles, S., & Gintis, H. (2011). *A cooperative species: Human reciprocity and its evolution*. Princeton: Princeton University Press.
Brandt, F., et al. (2016). *Handbook of computational social choice*. New York: Cambridge University Press.
Brynjolfsson, E., Colls, A., & Eggers, F. (2019). Using massive online choice experiments to measure changes in well-being. *PNAS, 15*, 7250–7255.
Camerer, C. F. (2003). *Behavioral game theory: Experiments in strategic interaction*. Princeton: Princeton University Press.
Carver, C. S., & Scheier, M. F. (1998). *On the self-regulation of behavior*. Cambridge: Cambridge University Press.
Christensen, P. N., Rothgerber, H., Wood, W., & Matz, D. C. (2004). Social norms and identity relevance: A motivational approach to normative behavior. *Personality and Social Psychology Bulletin, 30*(10), 1295–1309.
Cohen, J. (2019). WHO panel proposes new global registry for all CRISPR human experiments. *Science*. https://doi.org/10.1126/science.aax3948.
Crato, N., & Paruolo, P. (Eds.). (2019). *Data-driven policy impact evaluation: How access to microdata is transforming policy design*. Cham: Springer.
Diener, E., & Suh, E. (1997). Measuring quality of life: Economic, social, and subjective indicators. *Social Indicators Research, 40*, 189–216.
Ginsberg, J., et al. (2009). Detecting influenza epidemics using search engine query data. *Nature, 457*, 1012–1014.
Heath, J. (2011). *Following the rules practical reasoning and deontic constraint*. Oxford: Oxford University Press.
Heidegger, M. (2008). *Sein und Zeit* (English trans. ed.). San Francisco, CA: Harper Perennial Modern Classics.
Higgins, E. T. (1987). Self-discrepancy: A theory relating self and affect. *Psychological Review, 94*, 319–340.
Higgins, E. T. (1989). Self-discrepancy theory: What patterns of self-beliefs cause people to suffer? *Advances in Experimental Social Psychology, 22*, 93–136.
House, B. R., et al. (2019). Universal norm psychology leads to societal diversity in prosocial behaviour and development. *Nature Human Behavior*. https://doi.org/10.1038/s41562-019-0734-zDO.
International Human Genome Sequencing Consortium. (2004). Finishing the euchromatic sequence of the human genome. *Nature, 431*, 931–945.
Jost, J. T., & Major, B. (2001). *The psychology of legitimacy emerging perspectives on ideology, justice, and intergroup relations*. New York: Cambridge University Press.
Kahneman, D., & Tversky, A. (2000). *Choices, values, and frame*. New York: Cambridge University Press.
Kant, I. (2017). *Fundamental principles of the metaphysic of morals* (English trans. ed.). Scotts Valley: Createspace Independent Publication.
King, S., Mills, A., Kadirkamanathan, V., & Clifton, D. A. (Eds.). (2017). *Equipment health monitoring in complex systems*. Oxford: Oxford University Press.
Lee, J. J., et al. (2018). Gene discovery and polygenic prediction from a genome-wide association study of educational attainment in 1.1 million individuals. *Nature Genetics, 50*, 1112–1121.

Marin, H., et al. (Eds.). (2016). *Global health informatics: How information technology can change our lives in a globalized world* (1st ed.). Oxford: Oxford University Press.
Merton, R. K. (1968). *Social theory and social structure* (1968 Enlarged ed.). New York: Free Press.
Mittelstadt, B., & Floridi, L. (Eds.). (2016). *The ethics of biomedical big data*. Basel: Springer.
Parsons, T. (1970). *Social structure and personality*. New York: Free Press.
Qi, Y., & Xiao, J. (2018). Fintech: AI powers financial services to improve people's lives. *Communications of the ACM, 61*(11), 65–69.
Rawls, J. (1971). *A theory of justice*. Cambridge, MA: Harvard University Press.
Rawls, J. (2001). *Justice as fairness a restatement*. Cambridge, MA: Harvard University Press.
Riesman, D., Glazer, N., & Denney, R. (1969). *The lonely crowd: Changing study of the American character*. New Haven: Yale University Press.
Sen, A. (1970). *Collective choice and social welfare*. Cambridge, MA: Harvard University Press.
Shapiro, R., & Aneja, S. (2019). *Who owns Americans' personal information and what is it worth?* Future Majority (USA). https://assets.futuremajority.org/uploads/report-for-future-majority-on-the-value-of-people-s-personal-data-shapiro-aneja-march-8-2019.pdf
Silver, L. M. (2007). *Remaking Eden: How genetic engineering and cloning will transform the American family*. New York: Avon Book.
Smelser, N. J. (1962). *Theory of collective behavior*. New York: The Free Press.
Smelser, N. J., & Swedberg, R. (Eds.). (2005). *The handbook of economic sociology*. Princeton, NJ: Princeton University Press.
Smith, A. (2013). *The theory of moral sentiments*. New York: Createspace Independent Publishing.
Therborn, G. (2002). Back to norms! On the scope and dynamics of norms and normative action. *Current Sociology, 50*(6), 863–880.
Weber, M. (1964). *The theory of social and economic organization*. New York: Free Press.
Weber, M. (1977). *The protestant ethic and the spirit of capitalism* (English trans. ed.), Saddle River, NJ: Prentice Hall.
Wilson, J. (2000). Volunteering. *Annual Review of Sociology, 26*, 215–240.
Wurtmann, E., et al. (2019). Risk communication in families of children with familial hypercholesterolemia: Identifying motivators and barriers to cascade screening to improve diagnosis at a single medical center. *Genetic Counseling, 28*(1), 50–58.
Yang, H.-L., & Lai, C. Y. (2010). Motivations of wikipedia content contributors. *Computers in Human Behavior, 26*, 1377–1383.

Chapter 5
Identity Protection

Digitized world requires *identity protection*. Until now, these concerns were dealt by identity and reputation management (Whitley and Hosein 2009). And it has been only tackling by each celebrity and enterprise. But it is time to computationally manage data accumulation, modification, deletion, verification, monitoring, and other necessary operations on an entire integrated identity among all citizens. For example, thieving identity information of the others such as hidden password, account, and whole personality characteristics should not be allowed. Nobody is allowed to pretend another person. And then it should do reconsiderations between the benefit and risks on digitized data storing everything of personal information in daily life.

5.1 Identity Management in Social Context

Identity management has vague technical nuance in social computing context. In computer security, identity management can be usually understood as a kind of lifecycle management from an active entity allocated computer resources to terminations (Whitley and Hosein 2009; Helland 2019). During such process, each entity must be under control and security assessment. An entity will be assigned own tasks, and their performances are continually monitored by allocating resources such as CPU, memory, network, and disk spaces.

Therefore, when somewhat online services have been working in computer systems on the bases of big data accumulated from massive users, at least different meanings on *identity* coexist in the same computational environment. The first indicates above identity management, and another is to manage for protection of each user's privacy and identifiable information. This chapter mainly deals with the latter cases.

K. Shibuya, *Digital Transformation of Identity in the Age of Artificial Intelligence*, https://doi.org/10.1007/978-981-15-2248-2_5

5.2 What Data Should Be Protected?

What *data* should be protected? In pursuing this goal, how and what does information guard from vulnerabilities and risks (Collmann and Matei 2016; Australian Council of Learned Academies 2019)? The author proposes that it should be considered as following points, at least, as it distinguishes several categories apart in different levels. In each individual, at level from zero to four, it should be cared about privacy and private data. Otherwise, in business and national security level, higher security has to be managed by own purposes.

5.2.1 Individual

5.2.1.1 Level 0: Sensibility on Somewhat Data by Physical Contacts

Everyone can obtain, know, infer, and memorize immediately visible information seeing through others' physical existence. For example, visibility on your face, skin color, and hairstyles clearly provides your information for everybody, and it can probably estimate a coding data on DNA sequences even if it is not necessary to obtain your DNA data. Kinetic motions and specific behavior patterns can identify each individual. On the other hand, marriage rings also clearly signal the evidence holding your family, and further your speech manners and favorites have subtle signaling data on your backgrounds (e.g., education level, social class, birth region) (Bourdieu 1984). If detective Holmes meets you, he can immediately detect your backgrounds came back from Afghanistan.

Further, if video cameras[1], Google map cars (service for the Google street view), and drones have been watching your daily activities everywhere, its log data will be analyzed by someone around the clock all the time (Tistarelli et al. 2009; Li and Jain 2011; Massimo and Christophe 2017). Actually, the Biometric Information Privacy Act (BIPA[2]) of the Illinois State (USA) has already regulated to monitor, accumulate, and use in illegal policies. In the light of this regulation, there are cases that implemented cameras in AIBO (one of popular personal home robots) and interphones at home-door have been mandated to be dysfunctional states for the commercial purposes. The BIPA requires all business companies in Illinois State to comply with following requirements pertaining to the collection and storage of biometric information.

1. Obtain consent from individuals if the company intends to collect or disclose their personal biometric identifiers.
2. Destroy biometric identifiers in a timely manner.
3. Securely store biometric identifiers.

[1] ELSYS Japan: https://www.elsysj.net/

[2] http://www.ilga.gov/legislation/publicacts/95/095-0994.htm

However, such strict law will be probably inhibited to progress IoT, wearable and mobile devices within the state. It might be intrigued case in the digitized social age. Thus, although it is level zero, it should care about your daily activities. If you gaze into the video camera, the *Panopticon* also gazes into your eyes.

5.2.1.2 Level 1: Identification Data for Public Intercommunications

If necessary, your identification information can be openly provided for everyone, and these data enumerates your name, nickname, titles, address, telephone number, email address, and personal SNS pages for communicating with the others under careful management. But, without saying, communication log and transaction information can be consequently accumulated by service providers, and then it should not forget that the government authority may access such information. Actually, for the purpose of enhancing national security, the US government announced to accumulate and check social media data of not only immigrants but also fix-term travelers.

And Wifi and wireless communication using mobile devices must be watched against privacy leaks (Qi et al. 2019). The author does not recommend that naïve citizens use public wireless access points for the purpose of sending higher confidential data unless they accustom security tools and knowhow such as encryption, Tor (the onion router), TLS (transport layer security), and VPN (virtual private network). Loss of smartphone, removable flash memory (e.g., USB (universal serial bus)), geolocation information, and internet logs should also be taken care for your privacy.

5.2.1.3 Level 2: Protection as the Way for Access Controls on Privacy and Sensitive Private Data

Your privacy and sensitive data must be protected by all ways against illegal accesses at outdoor, and it is desirable to be perfectly controllability in personal security context. These data includes credit card information, real estate data, financial data, CV, passwords, health data, daily lifelog, and other necessary data. Medical data should be inserted into such list, and PHR (Personal Health Record) must be managed by higher security. Copyrights in law must be legally treated to be careful about moral rights and other rights related to own personality and identity matters. Of course, such data can be dealt by the officer in charge, and then they have much chance to access and understand such information in detail.

For example, Google account which created by initial and default configuration should be carefully confirmed by handling own security policy. If all privacy configurations of Google accounts could be functionalized, they have been steadily accumulating personal historical data of behaviors corresponding with geolocations and temporal data. Such data can easily suggest characteristics and preferences of each individual by the third parties.

Moreover, in near future, your mind might be interpreted by BMI. And using such devices will become threats that brain of each citizen will be unintentionally leaked by external malicious attackers. Recently, some researches succeed mental unspoken text can be read by BMI, and then it meant that mental unspoken language converted outside to digitized data in computer (Anumanchipalli et al. 2019). Of course, the AI will deal with those digitized data, it will be prepared necessary services for disabilities and little children, but it has to be watchful against crime purposes (mental cracking?). Reversely, if externally intruding into each brain through BMI in the future, external cracker hacks to read your mind of brain and replace some thoughts and belief for political propaganda. It means to read (and edit?) your mind. It will become more serious security incidents against those risks.

5.2.1.4 Level 3: Guarding Private Data of Each Individual by Any External Attacks in Daily Living

You and your physical existence itself must be assured by guarding against any physical attacks outside (e.g., criminal activities), and it must be controllable in legal context. In digitized society, digital stalkers are often accused and arrested by the police authority. Using information tools and social media, there are some cases that many of them quickly detect privacy of each individual accumulating and combing with personal information (e.g., home address, telephone number, and security secrets). By appropriately keeping information privacy policies, it should watchfully pay attention to vast data on each individual online and offline.

Using any SNS, there is a need to care about your nuance and description of absence at home. For example, if you upload web contents on a current long vacation trip with photograph, criminal attackers can read your contents in real time. They often mechanically and manually collect vast information from online SNS, and they vigorously hunt absence homes for the purpose of theft and possible intrusion into the secret areas of companies.

5.2.1.5 Level 4: Protections for the Private Rights to Manage Each Physical Body and Its Parts

It means that whole physical body itself has still information, and physical existence should be regarded as an owner of private right in law. In ordinary cases, any medical doctor cannot operate your body without necessary legal permissions. Implanting microchips and cyber-assisted parts in each physical body has to obey legally due procedures. Donation for transplant operation and eliminating any organs from someone's body should be still obeyed in legal due process. In Japan case, the civil law stipulates obligation to compensate for obstacles to others' bodies except for emergency conditions (article 698 and 710: at the 2018 present). Further, it pays attention to another facet. How should a set of human genome, DNA sequence (including haplotype, SNPs (Single Nucleotide Polymorphisms), etc.), embryo, cells, and modified samples deal with (Farahany et al. 2018)? Ashe of the deceased is usually managed with one of inheritable "things" in civic law, and an embryo has own legal right to be managed as "human-being in the future" even if it is still before the birth from mother's uterus. Parts of the physical body cannot be posthumously donated for the others without posthumous writings or willingness of donation for organ transplantation. But digitized data (e.g., cancer data in USA[3]) and any cells (or a part of organs eliminated from the body) should be managed as "human-being" with dignity? Those "gray-zone" substances may be required for careful management by legal acts.

In addition, garbage collection committed by the third party and criminals has to be careful about your private data leak. In computer security management, *scavenging* has been said to implicitly duplicate and accumulate useful data from a large amount of CD-ROMs, USB flash memories, un-shred papers, and unscrambled materials in garbage box for the purpose of thieving business trade secrets and private information.

Especially, in the digital society, we should take more attention to our DNA data before throwing to garbage box. It means that a part of your hair, bloody tissues, saliva, and any cells are usually just thrown to road and garbage box in daily life. Those substances certainly contain your DNA. And then if the third party intends to reveal and abuse your private data, there is a possibility to be theft. In the Hong Kong case[4], for shame punishments or criminal searches against illegal garbage dumping in city using electric panels of digital signage, some ICT companies can help for city government to unveil privacy for facial visualization of citizens who illegally thrown the trash with their DNA (e.g., tobacco, handkerchief papers). Namely, they can successfully visualize faces of criminals from DNA data, and it clearly denotes that such trash confesses its thrower. Because DNA data includes ultimate private data such as your disease tendencies, personality, sex, race, physical features, behavioral basics, and other patterns, everyone should eliminate and scramble those materials before dumping. And blood, hairs, and nails clearly display a daily living history of nutritional status and disease infection, and then those materials can reveal such information at least. A part of hairs and nails itself has vast information on your health conditions as well as DNA samples. For example, because ancient scholar Newton was eagerly conducting alchemical experiments, his hair of the deceased had contained much harmful chemical substances such as arsenic, lead, and others.

But, there seems to be someone who does not understand that this issue places at the danger level 4. Do you think why so? They just guess that it is just garbage collection and its abuses for identification of throwers. But private detective Holmes could infer further steps after he had done it. Let us consider further. If some medics get trash samples which contain human DNA from the city's garbage box, they can

---

[3] Giant study poses DNA data-sharing dilemma: https://www.nature.com/news/giant-study-poses-dna-data-sharing-dilemma-1.18275

[4] https://player.vimeo.com/video/124896310

not only examine those samples but secretly apply a set of extracted DNA to transplanted cells in cloning operations and embryo cultivations transplanted DNA for any illegal purposes in biotechnologies without given any permission from the throwers of trashes. Further, when an ideal case will be able to embody biometrical information cultivated from a set of DNA, which extracted from the trashes, anyone can break the biometrical security in daily life (e.g., fingerprints, DNA sequence based authentication). A part of DNA sequence data has been already imprinted into some ID cards and IC chips for identification of the genuine owner. The third party can forge your fake ID card which imprinted your DNA sequence data if they obtain your DNA data from trashes, and they will pretend yourself and intend to commit criminal acts using such fake card. And otherwise, if trash-thrower infected by dangerous diseases disposes smoked tobaccos in the city, such trashes still contain RNA of harmful disease virus against the human. Terrorists may collect and abuse it. Namely, digitized society can transform a mass of trash to vast information. In ordinary, many countries already enacted that hospitals, factories, and science laboratories shall completely proceed the chemical and physical procedures ruled in legal manners before their disposals. In daily life, those risks will be lower probability, but it should not forget such risks.

5.2.2 Enterprises

Protection of enterprise data and higher security becomes one of the cores for systems management in business scenes. In enterprise level, each company has to evaluate and activate own Information Security Policy and Information Security Management System (ISMS), and they usually assign CISO (Chief Information Security Office) for risk management of total security.

It promptly organizes CSIRT (Computer Security Incident Response Team) when some security incidents happen or expect. It consists of viewpoints based on CIA (*confidentiality, integrity, and availability*) in computer security. As standing on general computer engineering as well as security (e.g., ISO/IEC JTC1/SC27), necessary features should be referred in employing any systems on authenticity, accountability, non-repudiation, and reliability. Recent security trends are that APT (advanced persistent threat), social engineering, ransomware, bots, and business e-mail based frauds (BEC (business e-mail compromise)) have been increasing in business scenes. BCP (business continuity planning), supply-chain management, and risk management must be proficiently prepared in advance. Further, education for all employees on security indeed becomes quite significant for ongoing administrations of business.

Otherwise, to date, information and data related to IR (Investor Relations) and consumers have been monitored by not only investors and government but also autonomous system driven by the AI. And using social media, each employee and company officer should not exaggerate and overexpress those data, and there will be potentials to engender unpredictable consequences in global markets. Conversely, they should rather check those online media to prevent pervading fraud information and misleading fake news as necessary. The third party might intend to commit such matter based on criminal purposes.

5.2.3 Nations

National security and protections for the whole nation such as top secret data of the nation, unopened information, and some institutions (e.g., national central departments of government, embassy in foreign countries, nuclear power plants, information and socioeconomic infrastructures) are one of the highest priorities. For instance, *WikiLeaks* was a controversial issue a decade ago; such information of top secrets and confidential matters must be always kept against any leaking activities. Internal suspicious providers of information should be continually watched in proper ways (e.g., video analysis of behavior, anomaly detection in security, logging on who managed the secret data).

In China, they have already run own internet monitoring systems (e.g., "*Great Firewall*": it was named after their world heritage). Their online monitoring and excessive system management governed by the national authority has been accused from the perspectives on human rights and international cooperation. However, reversely saying, it is certain that they offer high security online environment for ordinary citizens. In other words, their excessive data and security management can be achieved in due to pursue the ideal of computer security in national level.

In the national level, recognized security incidents by cyber-attacks have been increasing recently. Many of representative cyber-attacks are categorized such as port scan, APT (advanced persistent threat), DoS (deny of services), DDoS (distributed DoS), email sending with malwares (e.g., computer virus), illegally updating against webpages, illegal attacks against DNS (domain name services) servers (e.g., DNS cache poisoning), OS command injection, password cracking, and so on. Of course, the highest level system and security management against such incidents should be enduringly cared in national level around the clock.

Firstly, in Japan cases, on 12th June 2012, the *Anonymous* as an online associative organization among global computer crackers largely cyber-attacked websites of Japan government's central ministries and legal courts, and parts of those contents on such websites were just updated and falsified by their statements. Those who are called hacktivists are often illegally leading to do cracking against arbitrary organizations, companies, and nations (i.e., hacktivism). Fortunately, in this case, their used techniques were just recognized as password cracking (it might be attacked by brute force, dictionary attacks, or password theft) or infiltrated into some web servers by the causes of security wholes. Such security incidents stayed minor damages, and those web servers could be recovered soon. Then, recently, one of the national research institutes (i.e., NICT[5]: National institute of communication and technology) and the national police agency have been always watching and monitoring overlay network named as dark-net (and dark-web) from the outer of ordinary IP addresses, and they ought to analyze anomaly signs of cyber-attacking against the nation. Other authorities and institutes[6] in developed countries (e.g., USA) have been similarly monitoring online data and its transactions.

Secondly, in South Korea cases, on 20th March 2013, there were more severe chains of security incidents such as computer system troubles and disruptions of computer network in wider homeland. The secondary damages caused by those system troubles were further enlarged in all over the nation. Their security teams and military forces examined the causes on this larger system troubles and security incidents, and lastly they found that some necessary data (MBR (master boot records) of Windows operating system) of main central computers of the government were illegally broken by invaded computer viruses. It meant that those systems could not reboot and kernel functions of damaged computers could not be worked. Of course, we do not know those damages in detail because such information was not open enough for the national security, but they concluded that those cyber-attacks were led by the agency of the North Korea (someone said that China agency committed to do those actions, because IP address were mainly identified from China domain).

Moreover, the police has to pay more attention to internet security during special events. For example, terrorists, hacktivists, and other sympathizer who committed terrorisms and criminal cases are apt to distribute and broadcast their criminal video clips online and their messages globally (e.g., YouTube, Facebook, and other social media). Digitized analyzation and prevention should be applied against those activities in advance rather than police undertakes it by aftermath of those events.

Those example cases did not expand serious further consequences. But when such invasions against cyber or actual territory of the nation can be recognized at once in some countries (e.g., USA), there is higher possibility that retaliations will be commanded against attacked nations, hacktivists, and terrorists. That is a reason that *Cyber war* and *Cyber sovereignty* have focused in international conflicts.

Additionally, artificial space satellites which are equipped for remote sensing and are transmitting vast digitized data have already been working for ICT-based transaction, GPS, mobile computing, and smarter infrastructures in each homeland. And then, such satellites which have a severe vulnerability to be attacked by any missiles of enemies ought to be totally protected by military systems. If those artificial space satellites will be completely damaged by any attacks, those who conduct assault against target nation easily cause destructions of information networking, larger scale ICT-based communication failures, invoking confusions of governmental directives, and weakening military actions based on GPS navigations (e.g.,

[5]NICT: http://www.nict.go.jp/en/index.html

[6]CAIDA(Center for Applied Internet Data Analysis, USA, http://www.caida.org/research/security/telescope/

intercontinental missiles, aircrafts, etc.). Moreover, it should recently consider the threats of military AI and robotics, and these topics will be dealt at later chapter.

5.2.4 Database Management and Database Protection

Thus, requiring levels for both security policies and database management perfectly make different in each need from individuals to nations. Further, Mantelero (2018) proposed his report on artificial intelligence and data protection. He summarized risk management on massive data usages, analyzation in various demands, and ethical concerns. And he recommended necessary legal frameworks for our daily life. Certainly, GDPR[7] stipulates the significant statements on data management and requirement for privacy, but each article seems to be dependent on traditional computer security and data management (ICO 2018). Profiling for each individual has been possibly accomplished by anyone combining with much information. Thus, the goal of GDPR is to exclusively declare such significance for citizens and enforce them striving effectively in social situations.

On the other hand, databases and big data itself have to be protected by own security standard. For example, traditionally, adversarial attacks against database exemplify SQL injection. And sometimes there were cases that database manager lacked to configure appropriate grants for each user in database systems. Many databases have a vulnerability against database reconstruction attacks (Garfinkel et al. 2019). It has been lastly leaking privacy queried by intentional purposes of the adversary. And consequently, if any privacy data exists, by preparing mathematical constraining solvers, there is a possibility to reach specific conditions, which each person is statistically identifiable and recognizable for the malicious purposes.

5.3 Information Bank

In Japan, "Information bank[8]," which aims to store the vast digitized data, is administrated by the *centralized* national authority. It has been proposed and conceptualized to improve pervasive systems in next generation which implemented accessibility, usability, and interoperability for the purpose of innovating commerce, academic, technology, national security, and other usages. In systems management at the governmental level, their purpose is to take the handle tremendous information flows and storages. For necessary legal regulations against oligopoly on big data by GAFA and similar ICT companies, it has been enacted in Japan, there seems to be possibly authorized to interrupt and stop against them.

[7] https://gdpr-info.eu/

[8] http://www.information-bank.net/index.html

On the mechanisms of cyber-physical system for social information, IoT and mobile computing enhance our daily life, and these data will be simultaneously centralized and authenticated by the singleton authority (Geng 2017). In other words, it deals with both each individual and collective patterns of citizens such as intercommunication logs, spatiotemporal mobility, online transactions, web access logs, chatting conversations, security concerns, and others. These relational data can be often identified as each individual and conjoining with multiple data can trace and induce privacy of each person for the purpose of profiling. Then their legal actions by the government agency have to manage such data carefully.

5.4 Blockchain and its Distributed Database

Our digitized world has been indeed interconnected by multiplying structures for reorganizing our socioeconomic needs (Easley and Kleinberg 2010). Data economy has been reorganizing socioeconomic and business styles such as inter-transaction by cryptocurrency, cashless payment by smartphones, and automatic payment system managed by the AI (Cavanillas et al. 2016).

Recently, blockchain technology further calls the breakthrough for proving the trustworthy and keeping security (Nakamoto 2008; Asharaf and Adarsh 2017). Traditionally, the national central bank has been undertaking as the centralized authority for monetarizing system in each nation, but blockchain-based transactions of financial data can globally break those barriers. Blockchain-based FinTech (cryptocurrency: *bitcoin, Ethereum,* and similar services are one of represented applications) consists of trustful proofing protocols in decentralized Peer-to-Peer (P2P) networking, and it prohibits endangering leakage of secret information and privacy of participants.

Here, the author briefly summarizes blockchain specifications as follows.

1. P2P-based networking.
2. Sharing distributed database among participants
3. Precondition on existential evidence of each identity
4. Provision of higher security and protection of privacy
5. Smart contract
6. KYC (Know Your Customer): verifying identity of each individual
7. Consensus-based architecture
8. Proof-based Trust: Proof of Works (PoW), Proof of Stake (PoS), etc.
9. Applicability for various needs such as electric money, finance, real estate, government services, and other data

Distributed data management has to be possibly established by mutual trust and active engagement among participants, and security level of those systems require higher and more fault-tolerance architecture. And proofing for such activities and consensus-based evidence verification will become quite significant meanings. Given that fault-tolerance system against Byzantine failure (Lamport et al. 1983) can be assured, security attacks such as deception and tampering are not effective against blockchain-based applications. But some other problems have happened such as Sybil attack (It called a pseudo spoofing. It means that one person creates and controls many phony identities at once), 51% attack (It means that one person gains consensual majority among members by illegal ways), theft of cryptocurrency, and illegal mining (It means malicious process to consume and waste other's computer resources by external intentions of the third parties).

With this trend in mind, ordinary social media and social networking sites are often struggling in the problems such as trust proof of contents and anonymous dispersion of fake news. But it is possible to reconstruct collective, participatory, verifiable, interactive, and decentralized knowledgebase using blockchain technology. Further, the possibility of falsification against big data and database by antagonists might be decreased in the future. Actually, *Everipedia* and similar projects have been conceptualized as the blockchain-based knowledge base expanded from Wikipedia online (Yang and Lai 2010). One of their motivations is characterized as collective verification in each description and content by numerous experts and professional participants. It means that these concepts indicate the critical idea interweaving with each trustable knowledge as know-what and collaborative experts as know-who (Shibuya 2006). Sharing those verified block as a part of knowledge and expertise among citizens can facilitate to establish our global knowledgebase in P2P decentralized networking. As it will be challengeable matter, but when it appears, this knowledge base can ensure to offer trustable information and keep quality of stored knowledge.

5.5 Rule of Conduct for Authority in Medical Data

Especially, in medical context, Krutzinna and Floridi (2019) pointed out that every medics encourages to obey the rule of conduct due to the process of medical data donation. Personal health record is composed of massive health data of each patient such as genome, drug administration, diagnostic data, and other privacy information. Informed consents between medical doctors and patients are quite necessary in such due process (Marin et al. 2016; King et al. 2017).

In addition, all physicians shall obey some ethical doctrines (Woodward 1930). Originally, first doctrine "*the Hippocratic Oath*" had been succeeded among physicians ever since ancient Greek, and secondly, at the World Medical Association at Geneva in 1948, they published "*Declaration of Geneva*" for physicians. For instance, WHO (2005) summarized their medical ethics manual, and they stipulated to protect the human rights of the patients and privacy as well as reject against any prejudices. Namely, they should not violate such rule of conduct. Without saying, informed consents between medical doctors and clients should be further required in each occasion.

5.6 Risk Management on Identity

5.6.1 *Online Harms and Reputation Management for Each Individual*

Reputation management for each individual is much necessary to be adapted in social media era. Of course, read-only users also hold risks, because anyone may post accusative comments without any factual evidences. Moreover, by observing and estimating another's identity information combing with various online big data, it can be possibly recognized by the AI driven systems. Usage patterns of social media of each person can already statistically reveal meaningful tendencies on each person's characteristics and personality (e.g., intellectual level, hobby, sexuality, interests, political attitudes, and religion engagement) (Cao and Yu 2012; Krisp 2013).

In contrary, including ordinary citizens, identity management should spontaneously and properly protect each own individual evaluation and information, as it is necessary to reconsider GDPR policy (General Data Protection Regulation, EU) and AI-oriented technologies in progress (Hoeren and Kolany-Raiser 2017). Regarding this concern, online bullying, harms, and illegal actions against citizens using social media must be understood to be inhibited, and such illegality should be interfered. And the authority will be relieving for the human right of victims. Actually, the UK government (UK Government 2019) has announced a white paper on online harms. The consultation by expertise described plans that should be tackled as follows.

1. The online services in scope of the regulatory framework.
2. Options for appointing an independent regulatory body to implement, oversee, and enforce the new regulatory framework.
3. The enforcement powers of an independent regulatory body.
4. Potential redress mechanisms for online users.
5. Measures to ensure regulation are targeted and proportionate for industry.

5.6.2 *Identity Theft*

A next problem is *identity theft*. Pursuing the goals of criminal actions and disguising a target, it often conducts to thieve a part of or all of the targets' personal information and data such as personal name, gender, birthday, social insurance number, PIN number of credit cards (personal identification number), financial transactions, DNA data, varieties of passwords, personal digital data, and other privacy concerns.

Especially it has been already shown in the context of computer security such as man-in-the-middle attacks, password cracking (e.g., brute force, dictionary), and other ways in online transactions. Attackers ostensibly pretend someone, and they will conduct to behave criminal matters by the third party identity. On the computer security concerns, these backgrounds further call other ways such as biometric authentication and behavioral biometrics (it means to identify each user by his behavior patterns). The former can be designed as physical characteristics of each user (e.g., iris of eyes, fingerprint, voiceprint), whereas the latter is underlying in kinetic dynamics depended on human motions and behaviors whether it is intentional or not. But these patterns are too difficult to perfectly mimic and pretend target person.

5.6.3 Leaking Privacy

Privacy should be recognized to prevent unintentional leakages in online interaction. Please look around your office and home, you can find voice recognition applications such as Siri (i.e., iPhone), google-home, and other devices. Your voice reaches over the wall and distance for all. Dialogs using dialect, jargon, and any technical term can be detected to specify and identify spoken person and his characteristics such as birth place, growing environment, education level, professional area, and other unintentional information.

And you are often watched and monitored by video cameras and sensing devices around your personal space as well as urban city areas. In addition, IoT, connectable cars, auto-driving cars, drones, bystanders taking with smartphone, and the smart homes equipped in autonomous smart devices have already serious potential dangers against leaking residents' information in daily life.

Online transactions should be still watchful for citizens. For first example, Genkin et al. (2018) has warned *bitcoin mixing* (or bitcoin tumbling) and reviewed necessary instruments. He said "*While blockchain-based cryptocurrencies like Bitcoin do not directly reveal users' identities, they are often prone to deanonymization attacks. By observing the flow of transactions stored in the public blockchain, third parties can make accurate guesses about the identities of involved individuals.*" His notation claims when the third party (e.g., attackers) can illegally monitor and analyze real-time transaction in blockchain, they can declaratively estimate target person's consuming patterns. And further, this risk of information leak may call more dangerous conditions using a collaborative filtering algorithm and other machine learning mechanisms (Hastie et al. 2009). Then, there are two trends to prevent such risks. First idea is to make difficult to combine with target person and his historical information by mixing massive transaction protocols. For example, the centered authentic institution receives the orders from massive consumers, and it keeps the secrets of linkages between each order and consumer. Moreover, it scrambles with each order and each individual consumer by mixing multiple transaction routes and services. As this effect of scrambling results, attackers cannot easily detect the linkages between each order and ordered person. And second trend is to develop more advanced and secured electric currency for our daily transactions. It can enumerate zero-knowledge proof as well as usability of RSA puzzle and Tor (the onion router) in more secure networking technologies.

On the other hand, Nakamura et al. (2018) announced their worlds' first products to block against privacy leaking using image recognitions. They applied AI driven patterns recognition systems to protect privacy data given by donors. When donors take photographs, those pictures inevitably contain donor's private information ("internal privacy information") such as favorite, location, someone who took photographs together, and unnecessary facts. They resolved it by pattern recognition of the AI.

5.6.4 Unintentional Usages of Privacy

As mentioned privacy protection on image recognitions, innumerable video cameras and IoT sensors have been already tracing ordinary citizens everywhere. Those stored visual data can be analyzed to reveal mathematical patterns on collective flow and dynamics of human behaviors in the city (Helbing 2012). GIS (geospatial information system) and remote sensing technology have been monitoring their daily activities from the outside of the atmosphere. One of the most considerable matters is identifying each person using facial recognition systems (Li and Jain 2011). In China, face recognition system already covered nearly all of the citizens in all over the megalopolises and homelands. As each "face" can discern and identify perfectly each individual using machine learning and other AI mechanisms, their efforts still suggest emerging problems of externalities caused by face recognition systems (Kugler 2019).

And then, there are serious concerns. Please imagine, if you are a French student of Sorbonne University and you were eager to attend the mass meetings (e.g., "*Mouvement des Gilets jaunes*" 2018–2019). And if one of French journalists took video clips on your identifiable behaviors using video cameras at that time and posted this video clip globally online for broadcasting news. So, what happen? This video clip can be globally watched by the unlimited number of citizens. And further the worst case is when you visit to other nations for your vacation. The police of other nations was already analyzed your face from video clips and registered you as the potential offender, and they may arrest and expel you at the airports. Was your participatory demonstration "*young and stupid*"? The "*right to be forgotten*" should be enacted for you globally?

References

Anumanchipalli, G. K., Chartier, J., & Chang, E. F. (2019). Speech synthesis from neural decoding of spoken sentences. *Nature, 568*, 493–498.

Asharaf, S., & Adarsh, S. (2017). *Decentralized computing using blockchain technologies and smart contracts: Emerging research and opportunities (advances in information security, privacy, and ethics)*. Hershey, PA: Information Science Reference.

Australian Council of Learned Academies (ACOLA). (2019). *The effective and ethical development of artificial intelligence: An opportunity to improve our wellbeing*. Melbourne: ACOLA. https://acola.org/wp-content/uploads/2019/07/hs4_artificial-intelligence-report.pdf

Bourdieu, P. (1984). *Distinction: A social critique of the judgment of taste*. Cambridge, MA: Harvard University Press.

Cao, L., & Yu, P. S. (Eds.). (2012). *Behavior computing: Modeling, analysis, mining and decision*. London: Springer.

Cavanillas, J. M., Curry, E., & Wahlster, W. (Eds.). (2016). *New horizons for a data-driven economy: A roadmap for usage and exploitation of big data in Europe*. Cham: Springer.

Collmann, J., & Matei, S. A. (Eds.). (2016). *Ethical reasoning in big data: An exploratory analysis*. Cham: Springer.

Easley, D., & Kleinberg, J. (2010). *Networks, crowds, and markets: Reasoning about a highly connected world*. New York: Cambridge University Press.

Farahany, N. A., et al. (2018). The ethics of experimenting with human brain tissue. *Nature, 556*, 429–432.

Garfinkel, S., Abowd, J. M., & Martindale, C. (2019). Understanding database reconstruction attacks on public data. *Communications of the ACM, 62*(3), 46–53.

Geng, H. (2017). *Internet of things and data analytics handbook*. Hoboken: Wiley.

Genkin, D., Papadopoulos, D., & Papamanthou, C. (2018). Privacy in decentralized Cryptocurrencies. *Communications of the ACM, 61*(6), 78–88.

Hastie, T., Tibshirani, R., & Friedman, J. (Eds.). (2009). *The elements of statistical learning: Data mining, inference, and prediction*. New York: Springer.

Helbing, D. (2012). *Social self-organization*. Berlin: Springer.

Helland, P. (2019). Identity by any other name. *Communications of the ACM, 62*(4), 80–87.

Hoeren, T., & Kolany-Raiser, B. (Eds.). (2017). *Big data in context: Legal, social and technological insights*. Cham: Springer.

ICO (information commissioner's office, UK). (2018). *Guide to the General Data Protection Regulation (GDPR)*. https://ico.org.uk/media/for-organisations/guide-to-the-general-data-protection-regulation-gdpr-1-0.pdf

King, S., Mills, A., Kadirkamanathan, V., & Clifton, D. A. (Eds.). (2017). *Equipment health monitoring in complex systems*. Oxford: Oxford University Press.

Krisp, J. M. (2013). *Progress in location-based services*. Berlin: Springer.

Krutzinna, J., & Floridi, L. (2019). *The ethics of medical data donation*. Cham: Springer.

Kugler, L. (2019). Being recognized everywhere. *Communications of the ACM, 62*(2), 17–19.

Lamport, L., Shostak, R., & Pease, M. (1983). The byzantine generals problem. *ACM Transactions on Programming Languages and Systems, 4*(3), 382–401.

Li, S. Z., & Jain, A. (Eds.). (2011). *Handbook of face recognition*. London: Springer.

Mantelero, A. (2018). *Report on artificial intelligence*. Artificial intelligence and data protection: Challenges and envisaged remedies. https://rm.coe.int/report-on-artificial-intelligence-artificial-intelligence-and-data-pro/16808b2e39

Marin, H., et al. (Eds.). (2016). *Global health informatics: How information technology can change our lives in a globalized world* (1st ed.). Oxford: Oxford University Press.

Massimo, T., & Christophe, C. (Eds.). (2017). *Handbook of biometrics for forensic science*. Cham: Springer.

Nakamoto, S (2008). *Bitcoin: A peer-to-peer electronic cash system*. https://bitcoin.org/bitcoin.pdf

Nakamura, K., Nitta, N., & Babaguchi, N. (2018). Encryption-free framework of privacy-preserving image recognition for photo-based information services. *IEEE Transactions on Information Forensics and Security, 14*(5), 1264–1279. https://doi.org/10.1109/TIFS.2018.2876752.

Qi, M., Wang, Z., He, Z., & Shao, Z. (2019). User identification across asynchronous mobility trajectories. *Sensors, 19*(9), 2102. https://doi.org/10.3390/s19092102.

Shibuya, K. (2006). Collaboration and pervasiveness: Enhancing collaborative learning based on ubiquitous computational services, including as chapter 15. In M. Lytras & A. Naeve (Eds.), *Intelligent learning infrastructures for knowledge intensive organizations: A semantic web perspective* (pp. 369–390). Hershey, PA: IDEA group Publishing.

Tistarelli, M., Li, S. Z., & Chellappa, R. (Eds.). (2009). *Handbook of remote biometrics: For surveillance and security*. Dordrecht: Springer.

United Kingdom (the Secretary of State for Digital, Culture, Media & Sport and the Secretary of State for the Home Department). (2019, April). *Online harms white paper*. https://assets.publishing.service.gov.uk/government/uploads/system/uploads/attachment_data/file/793360/Online_Harms_White_Paper.pdf

Whitley, E., & Hosein, G. (2009). *Global challenges for identity policies*. Basingstoke: Palgrave Macmillan.

WHO. (2005). *Medical ethics manual*. http://www.whcaonline.org/uploads/publications/em_en.pdf

Woodward, S. B. (1930). Medical ethics. *The New England Journal of Medicine, 202*, 843–853.

Yang, H.-L., & Lai, C. Y. (2010). Motivations of Wikipedia content contributors. *Computers in Human Behavior, 26*, 1377–1383.

Chapter 6
Identity Proof

Next, the author picks up *identity proof*. It means to prove not only your identity as physical existence but also existence of necessary information (e.g., ability, evidence) as genuine and trustable one without disclosing unnecessary knowledge. In other words, it intends to accomplish proving as an identifiable individual with desirable capability to pursue the goal. To summarize critical points, it can consider that controversial issues on online deceptions usually lack those convincing validation mechanisms: No provable way to verify prover's knowledge and faithful protocols about information between senders and receivers. Without any troubles and haphazard, it should clarify either not-modified or not-disguised entity. This chapter attempts to introduce zero-knowledge proof and its applicability online.

6.1 Proving Evidence

Principally, before discussing an interactive proof by experiential ways, it firstly gets start from a theme related to psychological and philosophical foundations in stochastic and subjective probability (Rich and Gureckis 2019). These studies critically denote our cognitive characteristics to recognize the world. Well-organized experiments can be certainly conducted as strict and decisive methodology for clear understanding on empirical facts of both natural and social phenomena. The burden of proof must evaluate its trustfulness by scientific manners included reproducibility, objectiveness, and falsifiability (Fanelli 2018; Shiffrin et al. 2018). It never accepts neither solipsism nor pseudoscientific manners. There are still hardships to conquer beyond our cognitive vulnerability for examining the scientific truth.

According to Bloor (1976) who stood on the academic background of SSK (Sociology of Scientific Knowledge), it named the *Strong Program* in sociology of science, and he proposed four tenets, namely *Causality, Impartiality, Symmetry,* and *Reflexivity,* for scientific verification and examination by proving evidence.

K. Shibuya, *Digital Transformation of Identity in the Age of Artificial Intelligence*, https://doi.org/10.1007/978-981-15-2248-2_6

1. Causality: It requires examining both belief and knowledge about its causal relationship.
2. Impartiality: It requires accounting whether true-false or rational-irrational.
3. Symmetry: It requires dealing with the common principle whether true or not.
4. Reflexivity: Above three tenets should be applied for all verifiers per se.

Further explaining these matters, Bayesian statistics will be useful for them. In this concern, Sober (2008) further discussed that data-centered experimental researches which are based on inductive evidence and facts can be categorized as Bayesian and other schools. Regularly, Bayesian-school researchers specify their studies to model observer's objective probability and these conditional facts, and then it can be properly estimated by updating its probability.

Usefully, Bayesian statistics can similarly indicate a subjective probability of each individual when it acquires post-probability information as truth or falseness (Kahneman and Tversky 1973, 2000). They demonstrated serial consequences by psychological experiments that many subjects unintentionally and unconsciously evaluated higher subjective probability for a member of their belonging group (in-group) than other groups (out-group) (Gabbay et al. 2018). These statistically significant evidences were clearly obeyed in formalization derived from Bayesian statistics. It was a well-known breakthrough for the researches in cognitive biases, and behavior science and social psychological experiments repeatedly endorsed that participatory subjects are inevitably apt to neglect their undesirable information, and they are strongly committed to believe in their supportive one (Oakes et al. 1994).

6.2 Proving What You Are

This chapter mainly treats a way of interactive proving. Generally, proving what each individual is or not frequently dedicates much time consuming tasks. How should prover identify an existential modality of each individual? Or how should he verify own ability to meet the requirement for achieving given tasks?

In practical cases, blockchain technology and online transactions shed light on secure and interactive proof algorithm again. It abbreviates KYC (know your customers) as the ways of verifying an identity of each individual online. Certainly, electronic certification services proved by the authority as third party institutions have been already worked in our backgrounds during online transactions. Because the central bank of each nation and international organizations have owned their responsibility for risks of the market, identifying and verifying evidence for each customer must be given by trustful manners [e.g., a customer due diligence for banks[1] by BIS[2] (the Bank for International Settlements)]. For critical reasons, it prevents from illegal money laundering and keeps the trust of financial properties.

[1] https://www.fsa.go.jp/inter/bis/f-20011004-2c.pdf

[2] https://www.bis.org/

Namely there are serious needs to be proved within interactive partners without revealing private information. All of the blockchain applications perfectly rely on such mutual trust among participants who can be identified by the most credential ways. And it can be always clarified as a reliable protocol for verifications between the verifier and the prover. Principally, it must ponder the logic and mathematical ways to solve distinctly. Especially, zero-knowledge proof is considered as a good example of this protocol (Goldwasser et al. 1985). It can be realized the utilization of zero-knowledge proof conjoined with signaling interaction.

For example, this signaling game is characterized as a way of proving something in the actual case between consultant (prover) and client (verifier). As it exemplifies information-asymmetry situation (Akerlof 1970), client cannot usually know consultant's ability and estimate success probability of his prospect very well, whereas consultant perfectly knows how to win on the condition that they are collaborating with each other. But consultant also does not want to open his trade secrets and know-how to anyone in detail, and then he motivates to realize his power itself for his client.

6.2.1 Ciphering and Encoding

In computer security and cryptography studies, both ciphering and encoding (conversely deciphering and decoding) procedures have a long-term struggling history (Stinson 2006). Cryptographic procedure can be condensed how to conceal contents of message and reverse it to plain text. The core of cryptography can be said that it has been improving more secure message passing among stakeholders without breaking secrets. Of course, those procedures certainly include the specific ways for identification and authentication of message sender per se. When those who can freely encrypt and decrypt any messages will appear, it is worth believing that such individual will prove his genuineness. These technologies have been examined and applied by mathematics, computer sciences, and engineering. In near future, quantum encryption has potentials to compromise traditional foundations of computer security.

Generally, in contrast, as mentioned later in detail, an interactive procedure on zero-knowledge proof can be regarded in another context. It does not encrypt message per se, rather it can be more public. But the third party and attackers cannot easily break those processes and understand its true meanings unless they can obtain several and necessary information in advance for proving its genuineness. Otherwise, recently, blockchain-based technology usually requires both of them, and namely it means both secured cryptographic ways and interactive procedures for proving own contributions and evidences (Franco 2014).

6.2.2 Signaling

Identity proof by offering trustful *signaling* information for partners and proving genuine identity will crucially become larger weight. Many of such conditions among partners are underlying in information-asymmetry situation. Namely each of partners intends to negotiate toward his opponent, and some of them may appeal his abilities and preferences to anonymous viewers in your website. The ways of signaling can be varied in various contexts such as nonverbal symbols, signs, something suggestive, and other useful clues for the others (Yokoo et al. 2001).

Digitized knowledge-intensive society will be supposed to induce innovative actions using computational assistances. Know-who, know-how, and know-what information become larger value than previous era. Blockchain-based collaboration networks carry out the common pursuits among P2P interactive partners, and each member has to publicly notify own identifiable and specific capabilities for the others to contribute toward achieving the common goal. It is certain that each of them hopes to conceal privacy and personal data, and then they send a message as somewhat signal to the public.

Additionally, those interactive patterns can be signified as a manner of signaling problems in economics (Camerer 2003). As discussed by Casey et al. (2019), for preventing identity deception and deceptive behaviors, they proposed a Bayesian model in evolutionary signaling game among game players in information-asymmetry conditions. Their models can be beneficial for understanding player's behavior patterns and types in decision process.

6.2.3 Interactive Zero-Knowledge Proof in Depth

This protocol enables verifier (e.g., *Andy*) to confirm that his opponent prover (e.g., *Bob*) has certainly specific knowledge about what Andy desires to know, and conversely Bob does not offer any private secret information against Andy except for the convincible evidence that Bob has specific knowledge. It is just *zero-knowledge proof*. Some practical applications of this idea are already known in public key cryptography and user authentication in computer sciences. It is underlying in the mathematical and logic problems in knowledge and information asymmetry. Many cases can be defined as provisional styles, and it is often derived from polynomial proofing problems in mathematics.

This zero-knowledge proof requires below necessary factors (Vadhan 2018). If and only if below all factors are simultaneously satisfied, it can be regarded as an actualization by zero-knowledge proof.

- *Completeness*: For every valid assertion, there is a prover strategy that will make the verifier accept with high probability.
- *Soundness*: For every invalid assertion, the verifier will reject with high probability. No matter what strategy the prover follows.

After that above protocol in N time trials successfully passed, it will be verified as valid. Namely, each verifier can have strong confidence for prover and its knowledge.

6.3 The Monty Hall Problem with Interactive Zero-Knowledge Proof

Next, please recognize a similarity between interactive form of above zero-knowledge proof and the renowned problem of the Monty Hall (Rosenhouse 2009). The latter problem called critical reviews on the significance of Bayesian solution and our vulnerability of cognitive mechanism. However, many of us often forget the most implicated matter. *At first place, in the frame on the Monty Hall Problem, why does the prover always discern and know "correct" or "incorrect"?* Certainly, this partiality is just one of basic suppositions on the problem. But by sufficient serial N time trials of the interactive procedure between prover and verifier, prover can only prove and exhibit a fact that he can discern answers ("correct" or not). Namely, it can be applied as one of the interactive zero-knowledge proof.

In this case, there is unreservedly an information-asymmetry condition between prover and verifier. Prover always chooses "correct" answer as he desires, and then he can always win when he wants to do so. On the contrary, verifier unfairly loses because he cannot discern answers. Therefore, in this model, prover can send a something important signal for verifier, as a trustworthy message underlying in his behavior. And it means that prover does not intend to win verifier, and rather he continually cultivates mutual trust with each other in the procedure.

And statistically, this procedure can also be regarded as a variety of signaling games in game theory. In this signaling process, opponent is not always to assume any human-being. As the Turing test indicated (Turing 1950), it is quite important that the human-being can perfectly see through whether such mechanism is an AI-based system which enables to chat among people or not. But, given that these mechanisms have own identity or not, it is unnecessary information for interactions among stakeholders standing on a context of zero-knowledge proof. In AI studies, such problems have been handled as constraint satisfaction problems without revealing private information in multi-agent interactions (Yokoo et al. 2005).

6.3.1 Procedure

Now, a basic original procedure of the Monty Hall exemplifies a calculation based on the well-known Bayesian theorem. Bayesian theorem has been coevolved in machine learning and varied in computational inference models (i.e., Bayesian network and belief network models). It might be also intrigued in experimental philosophy, psychology, and statistical studies (Kahneman and Tversky 2000; Sober 2008; Hindriks 2019).

Let's consider a system which ought to be arranged between prover and verifier. The system shuffles and offers three cards such as A, B, and C in each trial. A front of each card prints each initial (A, B, or C), and back of each prints something (e.g., recognizable symbols and photos). For instance, at first trial, only card C is *correct* now, and residuals (i.e., card A and B) are all *incorrect*. Verifier initially chooses card A, whereas prover chooses card B and he shall open this result for verifier to observe that card B is *incorrect*. Please memorize a fact that prover always chooses *incorrect* because he can perfectly discern *correct* and *incorrect* cards. In this condition, after verifier observes a result opened by prover, should verifier retain and hold card A? Rather, should he reselect and change card A to card C (Ostensibly, it seems likely to recall a well-known psychological experiment of "*Wason four cards selection task*" (Wason 1968), but these are different problems configured by experimental designs)?

Where, correct probabilities of each card (A, B and C) can be expected equally.

$$\mathrm{P}(A) = P(B) = P(C) = \frac{1}{3} \tag{6.1}$$

Conditional previous probabilities before prover selects a card are given below.

$$\mathrm{P}(B|A) = \frac{1}{2} \tag{6.2}$$

$$\mathrm{P}(B|B) = 0 \tag{6.3}$$

$$\mathrm{P}(B|C) = 1 \tag{6.4}$$

Conditional post-probability of card A as *correct* after observation on the information about card B (*incorrect*) which selected and opened by prover can be estimated as follows.

$$\mathrm{P}(A|B) = \frac{P(A)P(B|A)}{P(A)P(B|A) + P(B)P(B|B) + P(C)P(B|C)} \tag{6.5}$$

Conditional post-probability of card C as *correct* after observation on the information about card B (*incorrect*) which selected and opened by prover can also be solved.

$$\mathrm{P}(C|B) = \frac{P(C)P(B|C)}{P(A)P(B|A) + P(B)P(B|B) + P(C)P(B|C)} \tag{6.6}$$

$$\therefore \mathrm{P}(A|B) = \frac{\frac{1}{3}\times\frac{1}{2}}{\left(\frac{1}{3}\times\frac{1}{2}\right)+\left(\frac{1}{3}\times 0\right)+\left(\frac{1}{3}\times 1\right)} = \frac{1}{3} \tag{6.7}$$

$$\mathrm{P}(C|B) = \frac{\frac{1}{3}\times 1}{\left(\frac{1}{3}\times\frac{1}{2}\right)+\left(\frac{1}{3}\times 0\right)+\left(\frac{1}{3}\times 1\right)} = \frac{2}{3} \tag{6.8}$$

As comparing with both results between 1/3 (= 33%) and 2/3 (= 66%), a correct probability of reselection to card C estimates twice than holding card A. Therefore it certainly clarifies that verifier should rationally reselect card C.

Moreover, the author gives alternatively a demonstration of computational experiment among three conditions by simulation. The example results are given below. Each trial of computer simulation was 1000.

- Condition 1 (Random Reselection): After prover chooses and opens *incorrect*, and verifier *randomly* decides to reselect or not. Correct answer among three cards will be randomly changed in each trial.
- Condition 2 (Always Reselection): After prover chooses and opens *incorrect*, and verifier *always* decides to reselect another card. Correct answer among three cards will be randomly changed in each trial.
- Condition 3 (Never Reselection): After prover chooses and opens *incorrect*, and verifier *does not always* decide to reselect another card. Correct answer among three cards will be randomly changed in each trial (Table 6.1).

If and only if both of them conduct to honestly play each role in this procedure (i.e., prover always chooses "incorrect" and verifier always does reselection), above simulation indicated that the total correct-rate of verifier can converge to approximately 66%.

Therefore, the author examined a basic idea on multiple N time trials in serial procedure rather than one-time trial. First reason was to verify mutual honesty check between two persons in longer-term interactions, and secondly it needs converging probability fluctuation. If prover betrays verifier, his deception can be immediately detected by verifier soon. That is because prover shall always choose only "incorrect" and open publicly a result for verification by opponent. Conversely, when verifier betrays prover more than once, his correct-rate will be distinctly

Table 6.1 Example results among three conditions by computer experiments

Conditions of experiments	Trials	Correct rate (%)
1. Random Reselection	1000	34.5
2. Always Reselection	1000	66.4
3. Never Reselection	1000	32.8

shown lower than 66% (lastly, it converges nearly to 33%). And, when serial trials are too little (e.g., less than 10 times), the total result of the correct-rate of verifier may not exceed approximately 66%, it can conclude that verifier failed to do reselection in serial trials or probability fluctuation.

6.3.2 *Applicability Online*

In this procedure, many of us are probably subject to pay much attention to probability 66%. However, the most significant point was that prover could prove perfectly to discern results in all trials (100%). According to the above canonical procedure in dyad interactions, it can be proved to assure 100% by prover. Verifier wants to know only the precision to discern the cards by prover. Subsequently, by this proving procedure, verifier can be convinced by the results, and after that, he will request prover to achieve the actual jobs kept with 100% criteria (e.g., appraisal of things for online auctioneer and customs at the airports).

In addition, someone may ask why it is necessary to repeat numerous times in complicated procedure. As mentioned *bitcoin mixing* at Chap. 5, online transactions in background computational process similarly require the tolerant system of secure-by-designs and trustful protocols against security incidents. By unsecured online transactions, there are severe vulnerabilities. In such conditions, any data can be easily spied and theft by the third parties and external crackers. In multiple, complicated and secured procedures, many online transactions can be actually managed in backgrounds. During proposed procedure, both persons (and also spying third parties) cannot know other's privacies and trade secrets in detail, and there is not necessary to understand such things. Even if any third party implicitly watches transaction data and its processes with *encrypted* protocols [e.g., TLS 1.3 (transport layer security, version 1.3)], it will be harder to estimate the true meaning among two persons in complicated procedure. Therefore, when enough trials are given, this interactive process can be concluded that there is sufficient rationality between prover and verifier in signaling procedure based on zero-knowledge proof.

6.4 Discussion on Model

As mentioned above, traditionally, many cases require authentication data for each user by own passwords, passphrases, and other physical specifications (e.g., fingerprint, iris of eyes). But, to date, proving who and what I am has become to be reevaluated in blockchain technology. P2P based networking can be underlying in mutual trust based on public information which can be endorsed and verified by each other. Such trustworthy data can be authorized only by each identity and its existence of contributions for P2P based networking community. Namely, as necessary, own identity must be proved anytime in arbitrary processes. As historical data of each user can be also shared with each other in distributed database built in blockchain-based networks, each user does not have any incentives to do deception against the other members.

Furthermore, such proving procedures will be extended for various ideas.

First, this framework may appeal applicability for the plea deal in judicial court cases. However, the plea deal often claims that there are not any sufficient confidences and guarantees to endorse articulations of suspects as truth in the due process. Unless prosecutors never verify evidence and fact-check on their assertions, the greater extent of legal risks unclearly remains. In such cases, zero-knowledge proofing can support their investigations.

Secondly, there is a problem on STEM (science, technology, engineering, and mathematics) education for ordinary citizens. Recently, there are social issues on our mental vulnerability, which cannot be overcome by easy ways, for probability understanding and its subjective estimation. For example, any gamblers and ordinary citizens have been overlooking traps of their subjective probability. As mentioned later in detail at Chap. 11, in many online games and entertainments using smart phone, many of the game providers have malicious tendencies to forge consumers to waste their money by abusing their mental deficits and maladaptation in stochastic rationality. Therefore, those current conditions clearly indicate the strong needs of STEM education for ordinary citizens, and their experiences should be cultivated by proper ways. Monty Hall problems can be exemplified as one of good materials.

Thirdly, there are feasibilities for operation management in strategic scenes. For example, let me redefine above detail descriptions and reset some rules of original Monty Hall problems. If you are a Sherlockian, please imagine several example cases (not iterated trials, and one-shot trial). There are three alternatives such as going directly to Reichenbach, stopping-over at next station, and returning quickly to London. When you become a private detective Holmes, your purpose is to gain strong evidence at Reichenbach without being discovered by professor Moriarty. Firstly, you think to stopover at next station, but you clearly observed that professor Moriarty headed for Reichenbach. Which alternatives should you choose? In another example, if you are an Asian, please consider a situation. There are three alternatives such as assaulting directly to the Chang'an city in ancient China, bypassing through harder trails and guarding fortress around the critical sites. If you play a role of the greatest strategists Zhuge Liang as one of the historical heroes in China, at first you have a plan to bypass through harder trails by leading his armies without being detected by your enemy, but your employed spy reported that your rival headed for one of the fortresses by leading the larger number of armies. Which alternatives should you choose? Certainly, these sample cases are not iterated historical events, but similar cases, which can be defined as a pattern of aforementioned frame of Monty Hall problem, can be observed anywhere and anytime. And then it is much worth experiencing for various case studies. It has to keep the frame of Monty Hall problem, but there are possibilities to examine their mental dispositions and decision-making process.

And fourthly, there are engineering demands. The blockchain era clearly requires feasibility of zero-knowledge proofing algorithm and its unbreakable secured procedures. Especially, their needs stress on verification of reliability and prevention of forgery among collective peers in best-effort based intercommunication routes and mutual P2P transactions. For further open question on blockchain security, it must consider that zero-knowledge proof can (or cannot) always assure confidential information more than public key exchange algorithm [e.g., RSA algorithm (Rivest–Shamir–Adleman)] and electronic signature online.

References

Akerlof, G. A. (1970). The market for "lemons": Quality uncertainty and the market mechanism. *The Quarterly Journal of Economics, 84*(3), 488–500.

Bloor, D. (1976). *Knowledge and social imagery*. Chicago: Chicago University Press.

Camerer, C. F. (2003). *Behavioral game theory: Experiments in strategic interaction*. Princeton: Princeton University Press.

Casey, W., Kellner, A., Memarmoshrefi, P., Morales, J. A., & Mishra, B. (2019). Deception, identity, and security: The game theory of Sybil attacks. *Communications of the ACM, 62*(1), 85–93.

Fanelli, D. (2018). Is science really facing a reproducibility crisis, and do we need it to? *PNAS, 115*(11), 2628–2631.

Franco, P. (2014). *Understanding bitcoin: Cryptography, engineering and economics*. Chichester: Wiley.

Gabbay, M., et al. (2018). Frame-induced group polarization in small discussion networks. *Social Psychology Quarterly, 81*(3), 248–271.

Goldwasser, S., Micali, S., & Rackoff, C. (1985). *The knowledge complexity of interactive proof-systems (extended abstract)*. STOC, pp. 291–304.

Hindriks, F. (2019). Explanatory unification in experimental philosophy: Let's keep it real. *Review of Philosophical Psychology, 10*, 219–242.

Kahneman, D., & Tversky, A. (1973). On the psychology of prediction. *Psychological Review, 80*, 237–251.

Kahneman, D., & Tversky, A. (2000). *Choices, values, and frame*. Cambridge: Cambridge University Press.

Oakes, P. J., Haslam, S. A., & Turner, J. C. (1994). *Stereotyping and social reality*. Oxford: Blackwell.

Rich, A. S., & Gureckis, T. M. (2019). Lessons for artificial intelligence from the study of natural stupidity. *Nature Machine Intelligence, 1*, 174–180.

Rosenhouse, J. (2009). *The Monty hall problem*. Oxford: Oxford University Press.

Shiffrin, R. M., Börner, K., & Stigler, S. M. (2018). Scientific progress despite irreproducibility: A seeming paradox. *PNAS, 115*(11), 2632–2639.

Sober, E. (2008). *Evidence and evolution: The logic behind the science*. Cambridge: Cambridge University Press.

Stinson, D. R. (Ed.). (2006). *Cryptography: Theory and practice* (3rd ed.). Cambridge: Cambridge University Press.

Turing, A. (1950). Computing machinery and intelligence. *Mind, LIX*(236), 433–460.

Vadhan, S. P. (2018). *A study of statistical zero-knowledge proofs*. Berlin: Springer.

Wason, P. C. (1968). Reasoning about a rule. *Quarterly Journal of Experimental Psychology, 20*(3), 273–281.

Yokoo, M., Sakurai, Y., & Matsubara, S. (2001). Robust combinatorial auction protocol against false-name bids. *Artificial Intelligence, 130*(2), 167–181.

Yokoo, M., Suzuki, K., & Hirayama, K. (2005). Secure distributed constraint satisfaction: Reaching agreement without revealing private information. *Artificial Intelligence, 161*(1–2), 229–245.

Chapter 7
Identity Deception

This chapter deals with *identity deception*. How do those who pretend to someone detect by appropriate ways? How to detect deceptive communications among numerous participants in security context? It is always one of the old and new problems in computer security (e.g., verification on counterfactual data, monitoring, detection, authentication, and protection). In this chapter, the author mainly takes up some cases. First issue is to design a mathematical model on anomaly detections in mutual communications among participants in the larger networking. Scoring services driven by the AI have been focused on daily activities to measure each digitized score of the citizens on the bases of their honesty and other dispositions. And blockchain distributed networking ought to serve mutual monitoring and detection mechanisms among multiple participants. Secondly, such deceptive behavior can be refined as a problem on social intelligence of the human for adaptations. Finally, deceptive technology on generative adversarial machine learning should be discussed.

These matters do not mean just to conduct deceptions for concealing own privacy and keeping anonymous condition. For example, there are greater needs for strong secured procedures such as online shopping, governmental registration, and other online activities (Castelfranchi and Tan 2002). Of course, the AI driven detection services against deceptions have been already running in business scenes. However, there were sometime severe cases, which the AI could mistake incorrectly the goods online (i.e., false negative and false positive cases). For example, in 2019, Amazon (in Japan) was accused by some consumers, and lastly Amazon recognized that their AI system had failed to detect pirated goods given from anonymous providers in their online shopping site, and they improved such errors. Then, deception detection and anomaly monitoring methods should be investigating and further implemented such process for the AI-based automatic mechanisms.

K. Shibuya, *Digital Transformation of Identity in the Age of Artificial Intelligence*, https://doi.org/10.1007/978-981-15-2248-2_7

7.1 Detecting Deceptions

First of beginning, it argues the risks on identity deception against computer security in communication network. Phishing online and APT (advanced persistent threat) attacks have been increasing in computer security, and social engineering also has not been solved yet. Those vulnerabilities in computer security focus on our cognitive deficits against heuristic reactions, and attackers aim to steal significant information. They often pretend business partners and acquaintances in various situations.

Especially, Carrol and Grousu (2011) offered some idea such as "honey-pot" traps against any attackers (i.e., they pretend themselves as an authenticated user) and game theoretic model for dyad interactions of signaling. Then, there are needs of continuous monitoring and validating specific patterns on daily behaviors of valid counterparts, and it is to verify the real-time facts on all of their intentions to respond honestly. As the author's model will be mentioned later, some of those detection processes can be resolved by mathematical models.

Tsikerdekis (2015) has recently argued identity deception in computer networks. In the past, identify deception has been managed as pretending by attackers, and their criminal activities had been effectively detected by verbal, nonverbal, and network logs (e.g., access log, behavior history patterns). Its perditions were not enough to reach the requirement. He categorized several patterns such as identity theft, identity concealing, and identity forgery. Those actions include virtual "individual" and creating multiple accounts as sock-puppet, and it has much bases prepared for criminal and malicious purposes. Otherwise, a study has been investigating a way of verification on posted evidence from online witness (Truelove et al. 2017). This report suggested that online users using social networking sites (microblogging) were possibly to be identified by statistical patterns.

In spite of the below plans to detect deceptions online at least, struggling matters remain in actually.

1. Verification of web access log: It analyzes both IP (internet protocol) address and MAC (media access control) address given from accessing user via internet communication, because it can usually obtain only their device and software information (e.g., OS, browser, computer name). But it can still hard to identify accessing visitors in detail. If such visitor takes with same devices to access a web page, web server manager can only estimate each site visitor corresponding with both IP address and MAC address. When ordinary account alters to another account taking with same device of MAC address, it might discern among site visitors, but those pairwise data with site visitor and its access log cannot be always identified as same person, because another person pretends to ordinary site visitor.
2. Language usage patterns: As a person has unintentional tendencies to use slung, technical terms, dialects, speech speed, pronunciation, and other spoken variations, identifiable patterns can be often specified in those communication logs (Bryden et al. 2013). For example, it often pays attention to word frequencies,

vocabularies, professional knowledge, birth region, gender, favorites, communication skill, education level, and other unintentional information.
3. Kinetic patterns: If it can analyze kinetic behavior patterns of suspicious user such as typing speeds on keyboard, its precisions and specific patterns, there is a possibility to identify who he is (Tistarelli et al. 2009; Massimo and Christophe 2017; Farhi et al. 2019).
4. Web pages' linkages: Analyzing linkage patterns of webpages, it may be possible to identify such browsing user. It exemplifies access data what pages he browsed, belonged online community, historical patterns of browsing, online behavior patterns, and those linkages.

However, it is still difficult to perfectly prevent "sock-puppet" and deceptive identity in all cases.

7.2 Dyad Interactions for Anti-deceptions: A Perspective from Difference Equations

7.2.1 Anomaly Detection in Interaction

To date, computer security studies proposed many ideas to prevent identical deceptions. Especially, using computational ways, employing own identity derived from actual-self to multi-virtual avatars such as many accounts and camouflaged entities are frequently recognized. Many online users are to switch and behave own identity differentially in each online community. And then, there is *anomaly detection* (Pramanik et al. 2017). By tracing real-time data online and comparing with historical data, computer system has been running to monitor users' interactions and data transactions. When those systems detect extraordinary and unnecessary patterns in continuous process, it can be managed as anomaly patterns on time comparing with daily logs. In this way, it will avoid serious security incidents. Time-series analysis and multivariate statistical analysis can contribute toward advancing its precisions (Patcha and Park 2007; Hastie et al. 2009; Chio and Freeman 2018).

7.2.2 Agent-Based Dyad Interaction

This chapter firstly intends to deal with anomaly detection in dyad interaction between two persons. It concisely refers to dyad interaction in the past studies. So far, dynamical interaction among different individuals in social sciences and personality science has been investigating as one of the fundamental problems by mathematical, clinical, and experimental methodologies (Coleman 1990; Carver and Scheier 1998; Cervone and Shoda 1999; Cervone and Mischel 2002; Camerer 2003; Cast 2004).

Regarding those points, dyad interaction model based on difference equation was originated by serial Kaneko's studies on dynamics of periodical doubling in physics and chaos (1984, 1985, 1993). As dynamical social psychologists Nowak and Vallacher (1998) and Nowak et al. (2002) proposed an extension model derived from Kaneko's studies, they refined it to dyad interaction model as a coupling phenomenon of social behavior between two persons.

This model can convert to a kind of agent-based models. It is defined as following a pair of equations (Eqs. 7.1 and 7.2) to coordinate with two agents. It formalizes a kind of difference equation on the bases of chaotic complexity studies. X_1 and X_2 can be understood as evaluation of each agent who holds own personal attitude for his opponent per time. Parameter a denotes each commitment (range from 0.0 up to 1.0), and r_1 and r_2 (ordinary range from 3.47 up to 3.67) are regarded as control parameters of mental state and its process.

$$x_2(t+1) = \frac{r_2 x_2(t)(1 - x_2(t)) + a r_1 x_1(t)(1 - x_1(t))}{1+a} \tag{7.1}$$

$$x_2(t+1) = \frac{r_2 x_2(t)(1 - x_2(t)) + a r_1 x_1(t)(1 - x_1(t))}{1+a} \tag{7.2}$$

And please confirm an example code written in programming R language as a supplement at the Appendix of this book. If you have installed a statistical tool of R environment, you can promptly run an example code. And you can confirm some attractor patterns, time-series process with statistical analyzation and graphical plots.

7.2.3 *Applying for Deception Detection*

Look another side of above difference equation model, it can apply to anomaly detection in dyad interaction. Nonlinear dynamics of above model are underlying in several control parameters r_1, r_2 and a. Meanwhile dynamically tuning one of these parameters in simulation, it immediately engenders from stationary to not stationary tendencies in interdependent relationships. Such extraordinary responses will be verified whether it is deceptive behavior or not in each occasion, and similarly anomaly detection in machine learning programs can be modeled in statistical and mathematical criteria. It will be usefully monitored by time-series analysis, multivariate analysis based on linear model, and pattern recognitions systems (e.g., fuzzy model, principal component analysis, support vector machine) in various situations (Hastie et al. 2009; Chio and Freeman 2018).

Next, it demonstrates how to discern anomaly patterns in dyad interaction.

- Agent X_1's properties stably remain as constants. It assumes that this agent is apt to honestly behave and does not change own properties.

- Where, $r_1 = 3.67$ and $a = 0.6$.
- In contrast, agent X_2 properties are defined as variables. It can be understood that this agent intends to betray against his opponent X_1 in their interactive sequence.
- Where, r_2 can change from 3.47 up to 3.67, and a varies from 0.4 up to 0.6. Both r_2 and a are incremented by 0.1.

At next Figs. 7.1 and 7.2, it can be prepared between both agents of X_1 and X_2. For example, simulated consequences show nine patterns when both agents were honestly interacted by stable conditions as follows. From top-left to top-center and top-right, {X_1 [$r_1 = 3.67, a = 0.6$] and X_2 [$r_2 = 3.67$, $\boldsymbol{a = 0.4}$]}, {X_1 [$r_1 = 3.67, a = 0.6$] and X_2 [$r_2 = 3.67$, $\boldsymbol{a = 0.5}$]}, and {X_1 [$r_1 = 3.67, a = 0.6$] and X_2 [$r_2 = 3.67$, $\boldsymbol{a = 0.6}$]}. From middle-left to middle-center and middle-right, {X_1 [$r_1 = 3.67, a = 0.6$] and X_2 [$\boldsymbol{r_2 = 3.57, a = 0.4}$]}, {$X_1$ [$r_1 = 3.67, a = 0.6$] and X_2 [$\boldsymbol{r_2 = 3.57, a = 0.5}$]}, and {$X_1$ [$r_1 = 3.67, a = 0.6$] and X_2 [$\boldsymbol{r_2 = 3.57, a = 0.6}$]}. And from bottom-left to bottom-center and bottom-right, {X_1 [$r_1 = 3.67, a = 0.6$] and X_2 [$\boldsymbol{r_2 = 3.47, a = 0.4}$]}, {$X_1$ [$r_1 = 3.67, a = 0.6$] and X_2 [$\boldsymbol{r_2 = 3.47, a = 0.5}$]}, and {$X_1$ [$r_1 = 3.67, a = 0.6$] and X_2 [$\boldsymbol{r_2 = 3.47, a = 0.6}$]}. As it configures above parameters in each condition, both Figs. 7.1 and 7.2 denote each result of combinations in time-series sequential plots of dyad interactions between agents (X_1 and X_2), respectively.

At first glance, as computed consequences depicted, this current model formalizing a set of difference equations appears very sensitive in parameter tuning. For example, when both parameters between two agents are in perfectly stable condition (e.g., r_1 and $r_2 = 3.47$, and both $a = 0.6$), their dynamic patterns will completely plot synchronized lines among them. Each behavioral response reinforces time-series tendencies of opponent by each activity within mutual relationship. When invariance of mutual parameters will stably become a time-series process, then each of agents can be regarded as resemble attitude during such process.

However, during the simulation sequences, if X_2 betrays his partner (X_1) (it means that X_2 varies his parameters a little bit during simulation), for example, when r_2 was changed from 3.47 to 3.57, his partner can easily detect an anomaly value recognized as committing deceptive behavior. As illustrated in above figures, their historical patterns of interactive dynamics immediately alter its specific attributes between stationary and unsteady chaotic signals. And such cases can be statistically endorsed by autocorrelation and ARMA (auto-regression moving average) models. Therefore, it clarifies both processes between synchronized ordinary interaction (e.g., mutual trust condition) and asynchronized anomaly interaction (e.g., mutual distrust or deception against an opponent) in coupling dynamics phenomena.

Therefore, this difference equation model can monitor mutual trusts in continuous dyad interactions (e.g., internet transactions, communication patterns of users). Of course, the above detections against unusual anomaly patterns in dyad interactions can be autonomously managed by the AI-based monitoring system. Using machine learning, time-series analysis, and other statistics, it statistically compares with historical logs of both agents' variables during interaction process.

Further, one of the identity deceptions can be possibly detected by such models. If someone pretends X_2 at the anytime of interacting with X_1, it cannot disguise

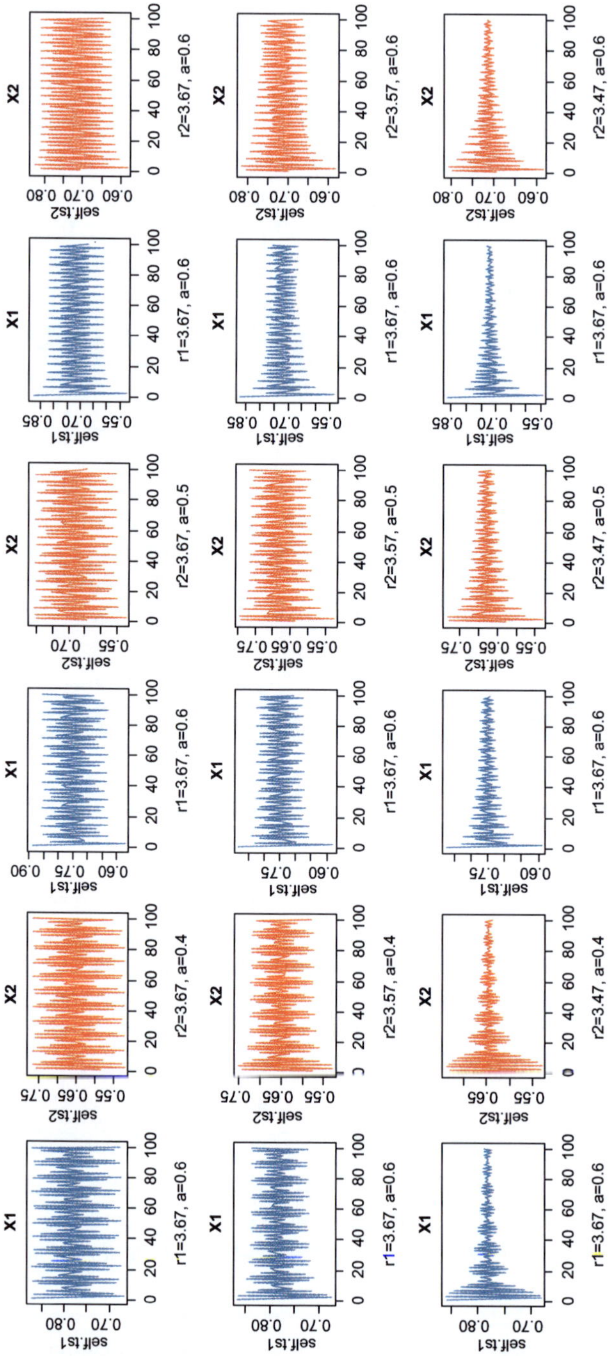

Fig. 7.1 Each pattern of nine combinations in time-series dynamics between two agents (X_1 and X_2)

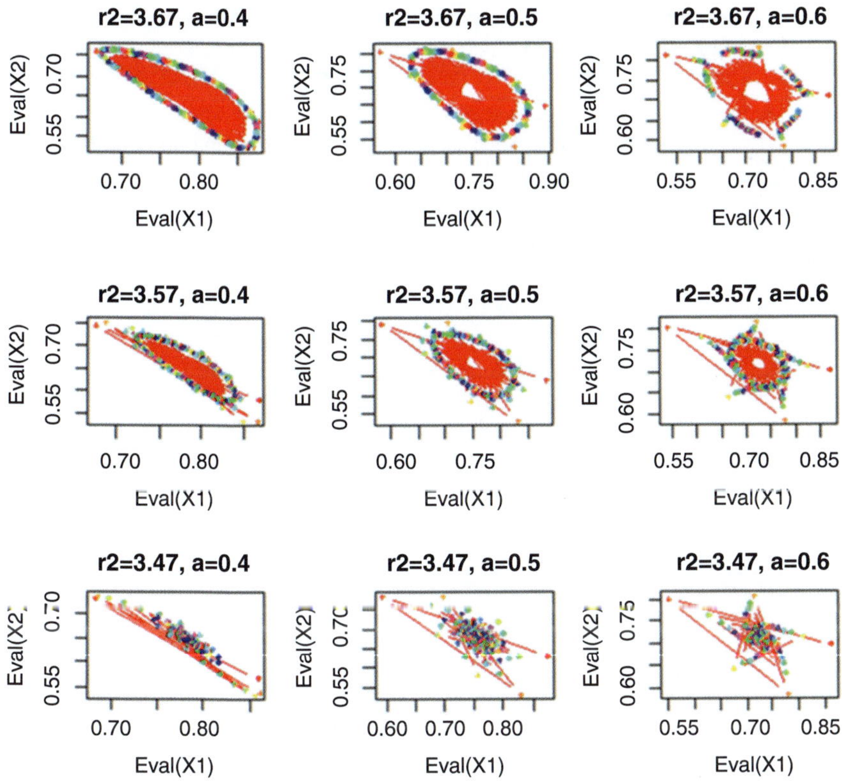

Fig. 7.2 Each pattern of nine combinations in time-series plots between two agents (X_1 and X_2)

perfectly other's characteristics and behaviors. As personality and behavior sciences are revealed by experimental and mathematical ways, each personality holds cognitive and behavior patterns such as coherent, unaware and strong tendencies in social contexts. There is a good reason that those parameters, in this concern, are assumable to underlie in imprinting personal characteristics of each identity. Thus, AI technology enables to discern and detect those differences of each person (Li et al. 2019).

7.2.4 *Discussion on Model*

With this model in mind, it is more effective to conceptualize an anti-deception mechanism in multiple N persons' online interdependent community (Modi et al. 2005; Casey et al. 2019). At this point, it can extend above difference model from the coupling interactive process to interconnected models on *complete graph* consisted of multiple N persons (K_n). For example, given that four people (A–D) are mutually interconnected online, interconnection edges between A–B, A–C, A–D, B–C, B–D, and C–D can be defined as mathematical graph. Each edge as dyad interaction can be applied sufficiently by above difference equation model, and different parameters in each person stay usually unchangeable after it configured initially. Even if someone alters to a different and unusual parameter, it will be immediately known to the public. This paradigm helps to endorse the other side of consensual algorithm derived from a motif of the Byzantine Generals Problem in blockchain networking. This problem requires a total trustful condition. Consensus and cooperation among participating members can be mutually monitored and maintained among all members by such paradigm.

Otherwise, it is certain that there are some problematic concerns. When total interconnected persons (n) will be increasing, each participant must manage much data of interaction history in each linkage. As it is similar to the *common key cryptosystem*, it follows the formula: $n(n - 1)/2$. Using above example case of four persons, every person holds interconnections with each other, and then $4(4 - 1)/2 = 6$ data in total dyads (similarly, when there are ten persons, the result is 45 data). Here, if sharing each personal memory by distributed databases conjoined with blockchain technology, it might be effective to detect deceptions among participants.

In this concern, Dunbar (1992) proposed an interesting theory, which presumes that our neocortical field size within the human brain socially correlates with each capability to maintain acquaintances in group relationships, and then the max of recognizable acquaintances was almost estimated as approximately as 150 persons in each individual (i.e., *Dunbar's number*). Conversely, the total amount of human relationships engendered our larger brain size and cultivated its social intelligence. If the hypothesis of Dunbar's number perfectly matches to above social network model based on difference equation for mutual monitoring, each individual should theoretically memorize private information with every 150 opponents, and then it must hold vast data of interacting history which will be exponentially increased within this networked community ($150(150 - 1)/2 = 11{,}175$ data). It is certain that social activities and its complicated combinations of relationships could require larger size brain than other zoological species, and then brighter brain of the humankind could be still more adaptive against deception and anomaly detection in uncertainty environment. But many ordinary individuals usually retain lesser size of own human relationship, it implies that the quality of friendships might be more important than quantity.

Those social nexuses can be redefined as a kind of social capitals. Mamei et al. (2018) recently analyzed synchronization patterns on social capital using big data of telephone communication logs. They argued how social capital can facilitate synchronizing social norms and conformity among human relationships. It expects that blockchain and information networking progress encourage for application of social scientific factors to computational modeling. Those actual interactions will be obeyed in social norms through mutual monitoring for trustable behaviors to maintain their good relations as social capital. In another example, Candeago et al. (2019) similarly reported synchronization phenomena based on telephone-call interactions across actual cities in Italy. Probably, those patterns almost reflect intimate interactions among business partners, friendships, and family relationships, and it means both patterns of communicators are unintentionally and gradually resembling each other. Then those patterns cannot be randomized among them. It is rather rational that their behavior patterns along with temporal conditions have tendencies to be harmonized between communicators beyond the physical distance.

7.3 Social Intelligence Against Deceptive Interaction

7.3.1 A Game of Werewolf

A participatory interactive gaming "*Werewolf*" was popularized in Japan. It can be available for everyone using any smartphones to download freely. This popular game indicates a part of our social intelligence for adaptation.

Multiple *N* members participate in this game, some of them are assigned as a role of werewolves and the residuals play a role of ordinary citizen. In brief, the ultimate goal of this game is to detect and hunt all werewolves by ordinary citizens, or werewolves stay continuing deceptions and prey all of ordinary citizens. At every turn ends, one of werewolves suddenly assaults to prey arbitrary ordinary citizens. So then, as interactive communications among all members are mandatory in gaming, before game is over, ordinary citizens vigorously seek to detect subtle signs as the werewolves, whereas werewolves players eagerly trick and fraud citizen players in gaming.

7.3.2 Detection vs. Deception

Regarding this game paradigm, there is a theory that deception and its detective minds among the humankind could be regarded as a symbol possessing higher social intelligence for survivals in actual history. At least, those intellectual abilities such as self-awareness, strategic planning, empathy, and logical inferences against the other existences in uncertainty environment are required. They said that acquisitions for these neural mechanisms of social intelligence took advantage in beating against other wild animals during the humankind's evolutionary process (Holekamp 2006; Ashton et al. 2018). In above gaming situations, it is very intriguing in investigating the consequences of multiple participatory experiments for understanding the nature of our social intelligence by verbal and nonverbal communications to detect deceptive patterns.

Now, the author casts a further question: *could AI detect any deceptions in such dyad interactions?* The Turing's test (1950) conceptualized that a possibility of computer systems to detect the opponent who human-being or AI is by continuous trials of dyad interactions. Hence, above challengeable material helps understanding for our identity and social intelligence adapting in complex surroundings.

Furthermore, "*Ring of Gyges problem*" in ethics offers another viewpoint on the invisible misconduct possibility committed by the AI. The author wonders how the humankind can detect misconduct, dishonesty, and deception conducting by the AI. In this context, everyone lastly realizes that deception and anomaly detection have rich spaces to consider the ethics of the AI.

7.4 Deception Technology by Generative Adversarial Machine Learning

In the light of contexts on deception detection studies (Granhag et al. 2015; Schwardmann and Weele 2019), facial information employs a key factor in any interactive communications in daily life and speeches in politics. Recently, image recognition technologies have advantages for evidence-based communication system. In this concern, Wu et al. (2018) produced an AI-based system for deception detection. They articulated that their systems have approached high quality for detecting deception of humans' vague facial expressions and motion captured data extracted from video clips. This approach may open the possibility for deceptionless interactive communications.

However, new technology has already encountered new vulnerabilities against computer security and anomaly. It is the very possible to lose the evidence for occurred events provided formats by video-clip and photo files using generative adversarial network (GAN) models in AI technologies (Goodfellow et al. 2016). It is called *deep fake.*[1] Deep learning is specified as pattern recognition of machine learning based on massive samples. In contrast, roughly speaking, GAN is defined to recreate and mathematically modify arbitrary image files (Kingma and Welling 2014; Li et al. 2015). These automated artifacts as image files can be automatically regenerated by this model to match actual events, and this technique has been worried about effective abuse for alibis and counterfeit in fake news and political forgeries against antagonists.

Thus these models have possibilities to critically damage blind spots in computer security using facial recognition systems for identifying authentication. As smart phone and other computational devices have equipped digital cameras, ordinary people can attach their photo clips for identification to intercommunicate with each partner, but they can make facial deceptions each other. Online witness can also make any deceptions by photo clips regeneration.

Moreover, according to empirical findings in brain sciences, mental trust for other individuals can become a fundamental for our sociality. Those cognitive and neural mechanisms were not well revealed yet. But, Feldman Hall et al. (2018)

---

[1] FaceForensics Benchmark : http://kaldir.vc.in.tum.de/faceforensics_benchmark/index.php

recently found that similarity of stimulus between trusted acquaintance and morphed face image was a critical factor for trustworthy. Can our human mind break those mocked deceptions?

In other words, for instance, for the purpose of forgery against you, fake movie clips can be automatically generated by the AI driven system, and antagonists probably scatter both tampered face data and body motions of yours. And consequently, your honor and reputations will be injured and lost, and these will be hard to be repaired. If you are suspected in such situations, how can you prove innocence? Can you completely detect anomaly in advance?

References

Ashton, B. J., et al. (2018). Cognitive performance is linked to group size and affects fitness in Australian magpies. *Nature, 554*, 364–367.

Bryden, J., Funk, S., & Jansen, V. A. A. (2013). Word usage mirrors community structure in the online social network Twitter. *EPJ Data Science, 2*(3). https://doi.org/10.1140/epjds15.

Camerer, C. F. (2003). *Behavioral game theory: Experiments in strategic interaction*. Princeton: Princeton University Press.

Candeago, L., Bertagnolli, G., Bosetti, P., Vescovi, M., Sacco, F., & Lepri, B. (2019). Cities of a feather flock together: A study on the synchronization of communication between Italian cities. *EPJ Data Science, 8*(19). https://doi.org/10.1140/epjds/s13688-019-0198-4.

Carrol, T. E., & Grousu, D. (2011). A game theoretic investigation of deception in network security. *Security and Communication Network, 4*, 1162–1172.

Carver, C. S., & Scheier, M. F. (1998). *On the self-regulation of behavior*. Cambridge: Cambridge University Press.

Casey, W., Kellner, A., Memarmoshrefi, P., Morales, J. A., & Mishra, B. (2019). Deception, identity, and security: The game theory of Sybil attacks. *Communications of the ACM, 62*(1), 85–93.

Cast, A. D. (2004). Role-taking and interaction. *Social Psychology Quarterly, 67*, 296–309.

Castelfranchi, C., & Tan, Y.-H. (2002). The role of trust an deception in virtual societies. *International Journal of Electronic Commerce, 6*(3), 55–70.

Cervone, D., & Mischel, W. (2002). *Advances in personality science*. New York: Guilford.

Cervone, D., & Shoda, Y. (1999). *The coherence of personality*. New York: Guilford.

Chio, C., & Freeman, D. (2018). *Machine learning and security: Protecting systems with data and algorithms*. Farnham: O'Reilly.

Coleman, J. S. (1990). *Foundations of social theory*. Cambridge: Berknap Press of University of Harvard Press.

Dunbar, R. (1992). Neocortex size as a constraint on group size in primates. *Journal of Human Evolution, 22*(6), 469–493.

Farhi, N., Nissim, N., & Elovici, Y. (2019). Malboard: A novel user keystroke impersonation attack and trusted detection framework based on side-channel analysis. *Computers and Security, 85*, 240–269.

Feldman Hall, O., et al. (2018). Stimulus generation as a mechanism for learning to trust. *PNAS, 115*(7), E1690–E1697.

Goodfellow, I., et al. (Eds.). (2016). *Deep learning*. Cambridge: MIT Press.

Granhag, P. A., Vrij, A., & Verschuere, B. (Eds.). (2015). *Detecting deception: Current challenges and cognitive approaches*. Chichester: Wiley-Blackwell.

Hastie, T., Tibshirani, R., & Friedman, J. (Eds.). (2009). *The elements of statistical learning: Data Mining, inference, and prediction*. Cham: Springer.

Holekamp, K. E. (2006). Questioning the social intelligence hypothesis. *Trends in Cognitive Sciences, 11*(2), 65–69.

Kaneko, K. (1984). Period-doubling of kink-antikink patterns, quasiperiodicity in antiferro-like structures and spatial intermittency in coupled logistic lattice. *Progress of Theoretical Physics, 72*(3), 480–486.

Kaneko, K. (1985). Spatial period-doubling in open flow. *Physics Letters, 111*(7), 321–325.

Kaneko, K. (Ed.). (1993). *Theory and application of coupled map lattice*. Singapore: World Scientific.

Kingma, D. P., & Welling, M. (2014). *Auto-encoding variational Bayes*. https://arxiv.org/abs/1312.6114

Li, Y., Swersky, K., & Zemel, R. (2015). *Generative moment matching networks*. https://arxiv.org/abs/1502.02761

Li, Y., et al. (2019). Differential game theory for versatile physical human–robot interaction. *Nature Machine Intelligence, 1*, 36–43.

Mamei, M., et al. (2018). Is social capital associated with synchronization in human communication? An analysis of Italian call records and measures of civic engagement. *EPJ Data Science, 7*, 25.

Massimo, T., & Christophe, C. (Eds.). (2017). *Handbook of biometrics for forensic science*. Cham: Springer.

Modi, P. J., Shen, W.-M., Tambe, M., & Yokoo, M. (2005). Adopt: Asynchronous distributed constraint optimization with quality guarantees. *Artificial Intelligence, 161*(1–2), 149–180.

Nowak, A., & Vallacher, R. (1998). *Dynamical social psychology*. New York: Guilford.

Nowak, A., Vallacher, R., & Zochowski, M. (2002). The emergence of personality: Personality stability through interpersonal synchronization. In D. Cervone & W. Mischel (Eds.), *Advances in personality science*. New York: Guilford.

Patcha, A., & Park, J.-M. (2007). An overview of anomaly detection techniques: Existing solutions and latest technological trends. *Computer Networks, 51*, 3448–3470.

Pramanik, M. I., et al. (2017). Big Data analytics for security and criminal investigations. *WIREs Data Mining and Knowledge Discovery, 7*, e1208. https://doi.org/10.1002/widm.1208.

Schwardmann, P., & Weele, J. (2019). Deception and self-deception. *Nature Human Behaviour, 3*(10), 1055–1061. https://doi.org/10.1038/s41562-019-0666-7.

Tistarelli, M., Li, S. Z., & Chellappa, R. (Eds.). (2009). *Handbook of remote biometrics: For surveillance and security*. Cham: Springer.

Truelove, M., Vasardani, M., & Winter, S. (2017). Testing the event witnessing status of microbloggers from evidence in their micro-blogs. *PLoS One, 12*(12), e0189378. https://doi.org/10.1371/journal.pone.0189378.

Tsikerdekis, M. (2015). Identity deception prevention using common contribution networking data. *IEEE Transaction on Information Forensics and Security, 14*(8), 41–49.

Wu, Z., Singh, B., Davis, L. S., & Subrahmanian, V. S. (2018). *Deception detection in videos*. https://arxiv.org/pdf/1712.04415.pdf

Chapter 8
Social Identification

Social identification should be rethought by both theories and sociocultural evidences. It can regard the hardcore on identity matter. The digitized world has already reshaped and transcended both borders across nations and cultural realms in our surroundings. Then, recently, identification in such social contexts was induced by redefinition of digitized progresses (Papacharissi 2010). Traditionally, social psychological studies have much advantage in investigating on social identification and group minds through sociocultural differences and uncertainty reductions. In our digitized society, many of people are often redundant to critically understand opponents who hold different backgrounds, and then social identification matter has strong reasons to be reconsidered again. Before discussing on social phenomena such as collective dynamics of movement (Chap. 9), synchronization of networking (Chap. 10), and mental health (Chap. 11), it should deepen common backdrop of identification issues.

8.1 Needs of Rethought for Social Identification

Innumerable studies have been made on social identification. We unintentionally tend to flock together based on the same category (e.g., race, ethnicity, gender, region, religion, hierarchy) and attitudes (e.g., political, socioeconomic, and other opinions). In the era of social media, collective dynamics based on such differences becomes now one of the hot topics in social science again (Beiro et al. 2016; Deville et al. 2016; Dolata 2018). As articulated later in Chaps. 9 and 10, it calls frequently tremendous movements of political oppositions against the government (Shibuya 2017). Authorities in the some revolution cases were overturned by ordinary citizens' dynamics. On the other hand, in the EU cases (i.e., *immigration crisis*), the criticisms against legitimation on massive immigration policy were unneglectable reasons among citizens and it demonstrated that their identities were firmly indivisible with their collective minds in nations. Further, in recent digitized society, many

K. Shibuya, *Digital Transformation of Identity in the Age of Artificial Intelligence*, https://doi.org/10.1007/978-981-15-2248-2_8

of people are often redundant to understand opponents who hold different opinions in social networking sites, and their prejudice has a tendency to intensify their hostilities against antagonists.

In addition, Eurasia Group (2018) reported the Top 10 Risks, and they ranked "9. Identify Politics (Fukuyama 2018) in Southern Asia: *Southern Asia identity politics threatens the future of these increasingly prosperous regions, creating unexpected challenges for economic planners and foreign investors.*" It can be understood that repressions and conflicts based on different categories such as race, ideologies, genders, and social classes in developing countries have own risks to identify themselves. Those differences on categorical backgrounds for social identification among citizens clearly exhibit the fundamental cause against social troubles and hardships to be conquered. According to Maalouf (2012), he argued that social identification which committed to those specific categories could be rooted in identity crisis of "Arab" citizens and its conflicts, as literary "*Les Identités meurtrières*" (*The murderous identities* in English).

In brief, the nutshell of such recent problems is commonly rooted in specific *identification* for own group, community, and nation. In many of social movements for the political appeals whether these are online or not, motivations of participants are not often independent from their identifications deeply committing in their homeland and ideologies (e.g., a case of Catalonia, a case of Hong Kong in 2019, Okinawa issues in Japan, etc.). Social identification has been shaped by committing by own emotional attitude for specific group, and those social mechanisms could organize memberships and reduce uncertainty in social surroundings. Those matters have been investigating by social psychological studies such as self-esteems and commitments of each individual in social groups. And it experimentally revealed their social identification and group minds through sociocultural differences and uncertainty reductions. And then such social identification matter has strong reasons to be reconsidered now in those perspectives.

8.2 Sociocultural Backgrounds of National Identity

In the first place, *national identity* and *nationality* are not always same meaning, and both concepts do not imply common backgrounds of birth and development place among citizens. The former is to prioritize mental commitments as a part of own identity for the nation, whereas the latter is just suggestively possessing the legal right of living and citizenship in specific nation. Especially, the former is often interdependent with *nationalism*, and it apparently relates to political propaganda and populism. In contrary, the latter problem can be condensed that traditional regime of each nation has weakness of both immigration policy and globalization issues to induce intensive commitments from cosmopolitans who hold multiple nationalities. Thus, those subtle differences between two concepts came from the intensity of mental commitments among citizens for the nation, and such cognitive mechanisms could engender some important reasons to complicate for each meaning on own existence in their life history.

One of the hypothetical concepts "*national identity*" (or "*national character*") has been continually assessed by social surveys in each country. Such concept means here just an aspect of personality and attitudes among citizens. For example, one of the Japan national research institutes (the Institute of Statistical Mathematics, Japan) has been inquiring into social trends on public opinions among Japanese every 10 years. Until recently, Yoshino and Hayashi (2002) and other results by their colleagues (Yoshino et al. 2009; Fujita and Yoshino 2009) reported their parts of social survey results, and they compared with cultural differences among Asia and other countries. Tendencies of answers among citizens clearly indicate a part of and relation of self-definition and each commitment for their nation, and those differences could be shown as the cultural cluster of ethnic similarity and roughly binding with cultural heritages among wider rural area.

Biologically, for example, some researches could successfully endorse that many of the contemporary Japanese have been inherited from their ancient people lived in the Japan islands (the *Johmon* Era: more than 3000 years before the present) (Kanzawa-Kiriyama et al. 2017). In such way, further, many of Japanese shared and owned biological similarity which could be clarified by genetic analyzations. It is certain that such biological similarity implicitly keeps somewhat identification which shared similar roots in adjoined and larger areas.

Therefore, *identity* is never independent on historical, social, and cultural contexts among citizens, and it means that each individual holds both facets on personal and social identity.

8.3 Uncertainty Reduction in Social Group and Culture

8.3.1 Collectivism and Individualism

For several reasons, social identification must be considered in systematic perspective. Social sciences would be based on sociocultural dichotomy. Members' personal behavior and cognition in social system can be varied by cultural foundations. It had already discussed very well about academic conversational dichotomy such as individualism and collectivism in various specific cultures (Triandis 1995). And until recently, social and cultural psychological researchers had been examining these cultural aspects in detail (Yuki 2003).

In addition, Hofstede (1980) reported an empirical fact that culture can be systematized by four categorical axes from survey results. Namely, these axes were *power, uncertainty avoidance, individualism-collectivism,* and *masculinity-femininity*. In particular, uncertainty avoidance composed with minimization of unstructured situations, adaptations, and innovation in groups (e.g., IBM Corporation). Cultural and social psychological studies indicated that cultural aspects can be regarded as systematic understandings. And it unveiled primordial principle of group organization and sociocultural factors. Especially, sanctioning system against free-rider (Yamagishi 1988a) and social deviations in community could be depended on those principles in sociocultural differences. And it already examined on this free-rider problems and cultural collectivism in serial social psychological experiments (Yamagishi 1988b), but it could not satisfy his hypotheses effectively and sufficiently. Although his experimental verifications, it could consider that these consequences still retain many socioeconomic suggestions (Davis 2007).

As further consideration, the cultural differences of decision-making styles suggest the differences of social legitimated process in various societies (Triandis 1995). As this dichotomy corresponds with both collectivism and individualistic principles, these social and political differences reflect social legitimacy as the ways of gaining social power in each domestic culture. Individualism culture laid weight on the principle of decision by majority, whereas collectivism is based on the principle of social relationships-oriented decisions. The former principle has heterogeneous opinions within the group and society initially, and it assumes to discuss among independent members. Then social legitimacy in this culture will be more time-consuming and democratic than the latter culture. Otherwise, the latter case previously assumes that single opinion, and ideological belief and exclusive cognition will govern all of the group members. And it conceives that their decision process will lay stress on homogenous conformity with tight human relationships. It means that group-based decision-making process may be apt to be less flexible than the former culture. Thus, it can expect that social legitimacy is equally justified as communal group decision process.

8.3.2 Social Uncertainty

In the past, sociologists (e.g., Durkheim) assumed that sociological concepts of the society could not be defined as an assembly of each individual. And the social world can be dynamically shaped by emergent properties as group and society in higher meta-level. Society is certainly based on individual activities such as communication and interaction with the others. It is certain that social structure and institution explicitly affect for our living in the society. We certainly decide something by ourselves, and we autonomously follow social institution, norm, and coordinating behavior with the others (Cederman 2005).

On the other hand, intercultural communication studies have furnished understanding for cultural differences and heterogeneous memberships (Kappes et al. 2019). That is because communication process becomes necessary to coordinate with the others who came from various cultural and global backgrounds. For example, Gudykunst (1995) proposed his theory and conceptualization on intercultural relationships. His inclusive axioms were very clear, and it was even more geometric theorization which was similar to philosophical descriptions. In accord with his theorems, intercultural backgrounds and heterogynous memberships across cultures and societies might be apt to heighten the degree of anxiety, and then it enhanced motivations among members to reduce uncertainty in such confronted conditions. Thereby, when social awareness against uncertainty among heterogeneous members becomes subjectively arousal, mindfulness of group members toward out-group members ("stranger") will be quite intensified. Thus social uncertainty and emotional tensions will be still increased more seriously in those conditions.

Social anxiety and subjective uncertainty are almost assumed to be relying on unpredictability toward strangers and unfamiliar peoples in particular. When any people encounter strangers at the unpredictable (or predictable) conditions, they certainly confront with uneasy feeling against those strangers in specific social contexts. Of course, group prejudice and stereotypic behavior might be derived from the degree of predictability and uncertainty in communication process. Unless reducing social uncertainty and coordinating with such unexperienced problems, intercultural and intergroup conflicts will be more salient and inevitable critical antagonism to strangers who came from other cultures. In this context, Gudykunst's theorization seemed abstractly, and there were multiple needs to qualify and endorse his axioms by experimental results. Actually, he also exemplified some experimental results, but it should be investigated more deliberatively in actual social contexts.

8.3.3 Group Identification in Uncertainty

High social uncertainty facilitates to induce something unrest among naïve people (Luhman 2002). As human nature, naive people will desire eagerly much accurate and necessary information to confirm and understand about the world state in highly uncertain conditions (Kasperson et al. 2003). In the case of terrible disasters and emergency situations, they often crucially seek much information whether it is accurate or not. Such information directly reflects mediated reality such as rumor and fake news, which causes panic disorders and chaotic social conditions. Even if they intrinsically thirst for knowledge and information, there is another risk that they construct own opinions and assumed to be shared same opinion among them whether it is true or not (Deaux and Philogene 2002). According to empirical findings in social psychological experiments, as Festinger (1954) already noted, social comparison processes among group members and unfamiliar others were stimulated in such social uncertainty.

Especially, traditional research trends on social identity theory (SIT) categorized such situations as two types. First type is traditional feature, which is a *self-esteem hypothesis*. Second type is an *uncertainty reduction hypothesis* (Hogg and Terry 2001). It will be true that group identification can be constructed by those principles. The group constructs can be defined as specific group goals, organizing of group members, and visual distinctiveness such as *entitativity* (Hamilton et al. 1998). For some reasons, it must exclude and distinguish from other group members, and heighten their belief and commitments to own belonging groups. Through these in-group categorization processes, group members may be adapted and functionalized to reduce uncertain aspects, and they still intensify their motivations and emotional cognitions.

According to Hogg and Turner (1985), Hogg and Mullin (1999), and Hogg and Terry (2001) who are social identity theorists, they showed some experimental results that participatory subjects sought to join a specific favorable group in the social uncertainty situations. And moreover, many cases in psychological experiments endorsed those mental mechanisms for reducing uncertainty (Yamagishi et al. 1998a, b). And these experimental situations presumed cross-cultural settings, and it could ubiquitously expect to be applied in various cultural phenomena.

8.3.4 Culture as System and Social Uncertainty

Thus far, culture can be regarded as one of somewhat structural systems in the society. In this point, it suggests that cultural system explicitly and implicitly functionalizes to manage social uncertainty and structural patterns. According to Luhman (1984) who was standing on a view of sociology, he regarded the social system and structural patterns as emergent property constructing by communication, networking, and autopoietic organizations. His systematic theories in sociology could afford to be useful for delving into emerging patterns of social systems. And his theoretical description on social uncertainty (*"Unwahrscheinlichkeit"*) was very suggestive except for one thing, and it was not enough to formalize and verify his concept of "social uncertainty."

8.4 Self-Consciousness and Identity

8.4.1 Self-Consciousness and Self-Regulation

Scholars who standing on social identity theory (SIT) clarified self-conceptual foundations for group identifications by serial social psychological experiments (Abrams 1990, 1994, 1998; Paez et al. 1998). For example, Abrams constructed the social self regulation model (SSR Model), and it lays more weight to clarify social identification and regulation of social behavior among group members. It was possible to include social cognitive aspects such as self-concept and also self-consciousness, and these are a part of social identification and self-concept. But their studies were mainly dependent on SIT perspectives and serial experiments, their examinations on identity concepts (Personal, Social, and Group) were still unsatisfied.

Certainly, self-regulation and its experimental studies were very relevant to traditional personality psychology fields (Carver and Scheier 1998). It can recall some past experiments and discussions about self-monitoring (Snyder 1974), self-concept as dynamic process (Markus et al. 1985), and relations between self-censoriousness, behavior and personality consistency (Scheier 1980; Gibbons 1990; Krahe 1992). Principally, Abrams (1999) had shown some comprehensive structures about social identity, social cognition, and the self. His specific attention was conceived to stress on social and group identification integrating with cognitive aspects and social behavior.

8.4.2 Shared Group Belief as Identification

Group beliefs express the extent of identification and committing attitudes for belonged group (Bar-Tal 1998). From viewpoint of group level, diversity of group belief denotes varieties of members' opinions and attitudes as a proof of their group commitments. Group structure and members' identity can alter its intensities of commitment to own belonging group (Levine et al. 1998). Namely, when they are tightening their group belief in each group and increasing their similarity, the extent of commitment and identity among group members will be heightened (Ellemers et al. 1999). These good examples are observed in somewhat religious groups. From standing on this viewpoint, there seems to be related aspects on interpersonal attraction (Hogg and Turner 1985) and assumed similarity.

On the other side, it is certain that stereotypic belief such as illusory correlation will be hard to be reexamined by social experiments. In these regards, both group belief and shared stereotype can be offered for some hypotheses on in-group organizations and state of intergroup conflicts. Given group belief denotes social and group identity, each member makes a contribution to socially establish sharing common representation in their surroundings. Thus, each member shapes own ontological bases and cognitive tendencies as prejudice against the other groups and its belonging members. At least, group belief committed to target group was endorsed by traditional experimental studies. For instance, such topics included such as becoming in-group identity as a part of the self (Smith and Henry 1996), reconstructing cognition and social stereotypes about the world (Monteith 1993), and related matters to social identity (Spears et al. 1997).

Proceeding further, it could contemplate several issues on sharing belief and transition process of group. At first, SIT stands on preconditions that shared belief among group members does not absolutely mean group minds and collective consciousness (Turner et al. 1987). Second point, there are arguments on critical factors for grouping property whether it is categorical or dynamic (Wilder and Simon 1998). The former assumes *entitativitiy* aware of categorical groupness, and it realizes collectively to be organized by clustering sympathized and like-minded members. But the latter, group phenomenon itself can be derived from unrecognized factors as *emergent properties* in dynamics, before clustering somewhat groupness, there is only arbitrary category as group in advance, and then role of each member can be determined by the categorical factors to functionalize both group formation and collective visibility as entitativitiy. By inspecting on group formation and appearances, both categorical belief and actual behaviors will be examined as mutual interdependent factors and its relations.

8.4.3 Social Trust, Group Identification, and Self-Consciousness

Social uncertainty environment has considerable suggestions on some social psychological facets. First of all, it is discrimination between in-group and out-group, and it might mostly derive from intergroup prejudice and conflicts (Spears et al. 1997). Secondly, these social surroundings denote that it is salient to distinguish "groupness" as entitativity (Hamilton et al. 1998). It means a perceivable totality as a group by gathering numerable members wholly. In this concern, Kramer (2001) argued group identification in social uncertainty in terms of social trust and self-consciousness. Public self-consciousness contains the relationships with social unrest and group commitments. Thereby, he discussed a possibility of heightening public self-consciousness in higher social uncertainty.

Further, it is essential to stand on a view of social trust (Yamagishi and Yamagishi 1998; Yamagishi 2011). The lower level of social uncertainty in society makes members' safety, because there are similar members around neighborhoods. However, in the digital age of the global networking, for example, social members must manage the others who hold different cultural backgrounds. Certainly, social trust might become a key factor for understanding the high level of social uncertainty in society. Increasing relational interconnections with different group members inevitably heighten the uncertain degree which is defined as *entropy*. If social trust among the members is insufficient, perhaps such society lacks functional bases to reduce entropy in higher uncertainty.

8.5 Sharing Place and History for Social Identity

Next, it must review such psychological and sociocultural backgrounds mentally committing as a part of identity in own rooted place (i.e., nation, homeland, hometown, community) and history. The author wonders why peoples ought to be shared with *spatiality* and *temporality* accompanied with interconnections with each other.

Psychologically, those mental mechanisms of emotional commitments to the specific place are specified as *place identity* and *place attachment*. Mental bonding of preferential attitudes for own environmental places interacts with own identity. On the other hand, standing on the sociological theory, for example, as Weber's concept on *communities of believers* and his relative discussions, identity has been organized mainly as an identifiable entity by categorical groupings such as race, region, religion, language, and linage. As one of other theories, which is social category theory in social psychology, had specified in social structural conditions, their paradigms were already discussed in this chapter.

For another example, a philosophical theory suggested that the interlinkage with religious value, historical time, and geospatial sites was to socialize and initialize every member committing in own specific community (Nishida 1987). Ancient peoples often organized their distinguishable cluster hardening human relationships in own lands. In this context, their *sacred* place was rather crucially needed as one of the necessary factors for identifying themselves in many religions. They often intensively share common representations about their ancient history, and stay holding sentimental attitudes interweaving with own geospatial areas (Paxton and Moody 2003). Namely each member internalized own social identification with both inheritable constraints of geological space and historical time. After that, lastly, conceptualizations of themselves and attribution consciousness to own community usually ascended to form own national identity and ethnical prides.

Above traditional human relations based on social identity exemplifies blood bonds, linages, and family trees among members. After the end of initialization stage, participation of newcomers in specific community strongly encourages them to commit the rules and customs with each other. But to date, these old custom patterns have different meanings. The digitized world makes it reconsider such bindings of human relationships.

Recently, Whitehouse et al. (2019) published an intrigued result on emergent possibility of "the moralizing God" in the world history of the humankind. Their estimations based on big data could facilitate the globally accumulated demographic evidences to understand the facts. When population size in larger community exceeded somewhat degree of criteria, such "God" had appeared in each history of humankind civilization. Their results meant that each civilization did not require the "God" to be initially organized, but rather growing process of civilization consequentially engendered "the moralizing God" for ruling numerous peoples in community. If so, it was possible that social identification committed in own community and group memberships had been steadily initialized for their morality in such process.

8.6 National Identity and Social Media

As described at the beginning of this chapter, social media vigorously encourages citizens to participate in own like-minded partners, and P2P-based clustering community unifies engaging civic powers to reshape somewhat entitativity as groupness. Social aspect of each identity can be represented in a categorizing manner to roughly distinguish members participating in some opinion groups. Such social category can be characterized as a symbol for identifying "*we-ness*" as same group members (Ellemers et al. 1999; Bastian and Haslam 2008). Then those social situations further call the dividing society among citizens. Given those social categories have much inevitable influences against social aspect of each identity among collective civic engagements, it should pay more attention to public opinion formation and its manipulation caused by the extent of larger exposures by social media and digitized interactive tools.

However, commitments belonged in social group usually mean somewhat unneglectable modus for identification of each individual. Own identity can be socialized by interactive manners, rooted in own communities and committed in own nation. And then, as discussed in succeeding Chap. 10 in detail, the digitized social network provoked new problems to reshape our identity for actual and virtual nation. Those identifications among participatory citizens would be weighted on rather a relationship to the virtual nation than physical commitments for the actual nation.

References

Abrams, D. (1990). *Social identity theory*. London: Harvester.

Abrams, D. (1994). Social self-regulation. *Personality and Social Psychology Bulletin, 20*, 473–483.

Abrams, D. (1998). Context and social self-regulation of stereotyping. In R. Wyer Jr. (Ed.), *Stereotype activation and inhibition*. Mahwah, NJ: LEA.

Abrams, D. (1999). Social identity, social cognition, and the self. In D. Abrams & M. A. Hoggs (Eds.), *Social identity and social cognition*. Oxford: Blackwell.

Bar-Tal, D. (1998). Group beliefs in an expression of social identity. In S. Worchel et al. (Eds.), *Social identity*. London: Sage.

Bastian, B., & Haslam, N. (2008). Immigration from the perspective of hosts and immigrants: Roles of psychological essentialism and social identity. *Asian Journal of Social Psychology, 11*(2), 127–140.

Beiro, M. G., et al. (2016). Predicting human mobility through the assimilation of social media traces into mobility models. *EPJ Data Science, 5*, 30. https://doi.org/10.1140/epjds/s13688-016-0092-2.

Carver, C. S., & Scheier, M. F. (1998). *On the self-regulation of behavior*. Cambridge: Cambridge University Press.

Cederman, L. S. (2005). Computational models of social forms: Advancing generative process theory. *American Journal of Sociology, 110*(4), 864–893.

Davis, J. B. (2007). Akerlof and Kranton on identity in economics: Inverting the analysis. *Cambridge Journal of Economics, 31*(3), 349–362.

Deaux, K., & Philogene, G. (Eds.). (2002). *Representations of the social*. Oxford: Blackwell.

Deville, P., et al. (2016). Scaling identity connects human mobility and social interactions. *PNAS, 113*(26), 7047–7052.

Dolata, U. (Ed.). (2018). *Collectivity and power on the internet: A sociological perspective* (Springer briefs in sociology). Cham: Springer.

Ellemers, N., Spears, R., & Doosje, B. (1999). *Social identity: Context, commitment, content*. Oxford: Blackwell.

Eurasia Group. (2018). *Top 10 risks*. https://www.eurasiagroup.net/issues/top-risks-2018

Festinger, L. (1954). A theory of social comparison processes. *Human Relations, 7*, 117–140.

Fujita, T., & Yoshino, R. (2009). Social values on international relationships in the Asia-Pacific region. *Behaviormetrika, 36*, 149–166.

Fukuyama, F. (2018). *Identity: Contemporary identity politics and the struggle for recognition*. London: Profile Books.

Gibbons, F. C. (1990). Self-attention and behavior: A review and theoretical update. *Advances in Experimental Social Psychology, 23*, 249–303.

Gudykunst, W. B. (1995). Anxiety/uncertainty management theory. In R. Wiseman (Ed.), *Intercultural communication theory*. Thousand Oaks, CA: Sage.

Hamilton, D. L., Sherman, S. J., & Lickel, B. (1998). Perceiving social group: Importance of the entitativity continuum. In C. Sedikides, J. Schopler, & C. A. Insko (Eds.), *Intergroup cognition and intergroup behavior*. Mahwah, NJ: LEA.

Hofstede, G. (1980). *Culture's consequences: International differences in work-related values*. London: Sage.

Hogg, M. A., & Mullin, B.-A. (1999). Joining groups to uncertainty. In D. Abrams et al. (Eds.), *Social identity and social cognition*. Oxford: Blackwell.

Hogg, M. A., & Terry, D. J. (2001). *Social identity processes in organizational contexts*. New York: Psychology Press.

Hogg, M. A., & Turner, J. C. (1985). Interpersonal attraction, social identification and psychological group formation. *European Journal of Social Psychology, 15*, 51–66.

Kanzawa-Kiriyama, H., et al. (2017). A partial nuclear genome of the Jomons who lived 3,000 years ago in Fukushima, Japan. *Journal of Human Genetics, 62*, 213–221.

Kappes, A., Nussberger, A.-M., Siegel, J. Z., Rutledge, R. B., & Crockett, M. J. (2019). Social uncertainty is heterogeneous and sometimes valuable. *Nature Human Behaviour, 3*, 764. https://doi.org/10.1038/s41562-019-0662-y.

Kasperson, J. X., et al. (2003). The social amplification of risk. In N. Pidgeon, R. E. Kasperson, & P. Slovic (Eds.), *The social amplification of risk*. Cambridge: Cambridge University Press.

Krahe, B. (1992). *Personality and social psychology: Towards a synthesis*. London: Sage.

Kramer, R. M. (2001). Identity and trust in organization. In M. A. Hogg & D. J. Terry (Eds.), *Social identity process in organizational contexts*. New York: Psychology Press.

Levine, J. M., Moreland, R. L., & Ryan, C. S. (1998). Group socialization and intergroup relations. In C. Sedikides et al. (Eds.), *Intergroup cognition and intergroup behavior*. Mahwah, NJ: LEA.

Luhman, N. (1984). *Social systems*. Stanford, CA: Stanford University Press.

Luhman, N. (2002). *Risk: A sociological theory*. New Brunswick, NJ: Aldine Transaction.

Maalouf, A. (2012). *In the name of identity: Violence and the need to belong* (original title "Les Identites meurtries", English translated edition). New York: Arcade Publishing

Markus, H., Smith, J., & Moreland, R. L. (1985). Role of self-concept in the perception others. *Journal of Personality and Social Psychology, 49*, 1494–1512.

Monteith, M. J. (1993). Self-regulation of prejudiced responses: Implications for progress in prejudice-reduction efforts. *Journal of Personality and Social Psychology, 65*, 469–485.

Nishida, K. (Ed.). (1987). The logic of the place of nothingness and the religious worldview. In D. A. Dilworth (Trans.), *Nishida Kitaro. Last writings* (pp. 47–123). Honolulu: The University of Hawaii Press.

Paez, D., et al. (1998). Constructing social identity. In S. Worchel et al. (Eds.), *Social identity*. London: Sage.

Papacharissi, Z. (2010). *A networked self: Identity, community, and culture on social network sites*. Routledge: New York.

Paxton, P., & Moody, J. (2003). Structure and sentiment: Exploring emotional attachment to group. *Social Psychology Quarterly, 66*(1), 34–47.

Scheier, M. F. (1980). Effects of public and private self-consciousness on the public expression of personal beliefs. *Journal of Personality and Social Psychology, 39*, 514–521.

Shibuya, K. (2017). Bridging between cyber politics and collective dynamics of social movement (Chapter 307). In M. Khosrow-Pour (Ed.), *Encyclopedia of information science and technology* (4th ed., pp. 3538–3548). Hershey, PA: IGI Global.

Smith, E. T., & Henry, S. (1996). An in-group becomes part of the self: Response time evidence. *Personality and Social Psychological Bulletin, 22*, 635–642.

Snyder, M. (1974). Self-monitoring of expressive behavior. *Journal of Personality and Social Psychology, 30*, 526–537.
Spears, R., Oakes, P. J., Ellemers, N., & Haslam, S. A. (1997). *The social psychology of stereotyping and group life*. Oxford: Blackwell.
Triandis, H. C. (1995). *Individualism and collectivism*. Boulder, CO: Westview Press.
Turner, J. C., et al. (1987). *Rediscovering the social group: A self-categorization theory*. Oxford: Blackwell.
Whitehouse, H., et al. (2019). Complex societies precede moralizing gods throughout world history. *Nature, 568*, 226–229.
Wilder, D., & Simon, A. F. (1998). Categorical and dynamic groups. In C. Sedikes et al. (Eds.), *Intergroup cognition and intergroup behavior*. Mahwah, NJ: LEA.
Yamagishi, T. (1988a). The provision of a sanctioning system in the United States and Japan. *Social Psychology Quarterly, 51*, 265–271.
Yamagishi, T. (1988b). Exit from the group as an individualistic solution to the public good problem in the United States and Japan. *Journal of Experimental Social Psychology, 24*, 530–542.
Yamagishi, T. (2011). *Trust: The evolutionary game of mind and society*. New York: Springer.
Yamagishi, T., & Yamagishi, M. (1998). Trust and commitment as alternative responses to social uncertainty. In M. Frunin (Ed.), *Network and markets: Pacific rim investigations*. Oxford: Oxford University Press.
Yamagishi, T., Cook, K. S., & Watabe, M. (1998a). Uncertainty, trust and commitment formation in the United States and Japan. *American Journal of Sociology, 104*, 165–194.
Yamagishi, T., Kikuchi, M., & Kosugi, M. (1998b). Trust, gullibility, and the social intelligence. *Asian Journal of Social Psychology, 2*, 145–162.
Yoshino, R., & Hayashi, C. (2002). An overview of cultural link analysis of national character. *Behaviormetrika, 29*, 125–142.
Yoshino, R., Nikaido, D., & Fujita, T. (2009). Cultural manifold analysis (CULMAN) of national character: Paradigm of cross-cultural survey. *Behaviormetrika, 36*(2), 89–114.
Yuki, M. (2003). Intergroup comparison versus intragroup relationships. *Social Psychology Quarterly, 66*, 166–183.

Chapter 9
Collective Identity

Even though previous chapter indicates that people has own strong commitments and incentives to stay in their rooted community, some of them may decide to leave to another place. What are crucial factors for such effortful actions? Recently, the problem on collective aspect of identity in global level appears, and their collective mobility can be observed globally. It should turn eyes to social phenomena as *collectiveness* of identity.

The author introduces collective dynamics of citizens using ubiquitous computation and measuring ways to analyze such spatiotemporal phenomena. These technologies and methodologies have been contributed toward investigating social movements and complicated dynamics among citizens in smarter city. In light of this, especially, the author picks up immigration issues. Digital civilization encourages citizens to immigrate to other countries beyond the borders. Those mechanisms and socioeconomic backgrounds can be unveiled by interdisciplinary studies. Finally, by means of demonstrating agent-based simulation, the author examines both collective migration phenomena and acceptance policy in each city in social simulation process.

9.1 Collective Dynamics and Mobility

9.1.1 Collective Dynamics of Human Behavior

Please look around now the large public space in your living town, you can find collectiveness of people. Over-tourism and overcrowding conditions may be observed occasionally in your town. Some of them are now walking with own smartphone. Another person may play one of the online games using geospatial information, and other people perhaps communicate with online friends accompanied with virtual partners. At the time of an earlier morning in Japan, you can find that innumerable

K. Shibuya, *Digital Transformation of Identity in the Age of Artificial Intelligence*, https://doi.org/10.1007/978-981-15-2248-2_9

commuters stampede into the trains at the station. And if you have any Google account, you can easily check your mobility history as geospatial tracking records.

As "*Birds of a feather flock together*" says, it is called a *flocking behavior* and *homophily* as a part of collective dynamics of human behavior in social science. Historically, Physicist Newton said "*I can calculate the motion of heavenly bodies but not the madness of people.*" Newtonian physics explained especially interactive complexity by mutual distance between two entities such as moving asteroids and stars. However, Newtonian physics had crucial limits. Certainly, "people" does not always gently behave for our expectations. His statement suggests that his discovered law and differential equations were impossible to solve the three-body problem. The crucial core of this difficult problem means the complexity of *N-body interactions* (e.g., numerous asteroids and planets more than three). Perhaps, he truly realized it. At the beginning of twentieth century, classical mechanics in physics were critically overturned by Einstein's general relativity theory and quantum mechanics by the Copenhagen School. Namely, "science" and "scientists" always seek universal laws and hidden principles ruling in those phenomena. Ironically, as mentioned later in detail, gravity equations derived from Newton's physics can be applied for collective patterns of human behaviors and other applied mechanisms in social sciences.

9.1.2 Modality on Human Behavior

9.1.2.1 A Possibility of Ubiquitous Computation

Little attention has been given to behavior sciences during the last several decades. It literally aims to analyze our social dynamics of behavior in various social contexts. In a nutshell, the core of this academic study attempts to unveil varieties of either nonlinear dynamics or goal-directed dynamics which organized and interacted by massive peoples. Till now, little is known about both collective phenomena and those dynamics of human behavior. However, at the time, the door of behavior science on human behavior has reopened since the beginning of ubiquitous and mobile computing services. Namely, it might be the time when Apple president Jobs did fantastic presentations for public releases of iPhone and iPad.

Until a few decades ago, before the dawn of ubiquitous and mobile computing (as introduced a brief history of ICT at Chap. 1, it was almost by 2000s), *behavior science* as a disciplinary part of social sciences has only just traditional methodologies such as lab-experimental ways, social survey, computer models, and statistical analysis. Although ergonomics and human engineering had still weak influential contributions to this discipline, many researchers in this field could be trained only by older methodologies.

Since the earlier stage of the internet revolution (i.e., 1995 later), the author has been investigating complexity and collective dynamics of human behavior using ubiquitous computing (before smartphone appeared (Shibuya 2004, 2006, 2011)).

The author intended to explore the possible frontier of behavior sciences for the next future (recently, those academic areas can be included as a part of computational social sciences). These technologies could be founded in interdisciplinary backdrops such as geospatial informatics, sensing devices, measuring metrics on human kinetics, and other necessary engineering. Then, the author repeatedly asserted that human behavior and its dynamical mobility patterns should be carefully examined by real-time analyzation using ubiquitous computations. As readers may realize, such ideas finally also established as a part of MaaS and XaaS in recently. Since that time, we can examine those dynamics in actual environment.

To date, the advent of social media and smartphone has completely achieved the ways to monitor and analyze those social dynamics in daily contexts. Then, those issues on social complexity in computational social sciences aim to enlarge our new understating of social phenomena. Big data will be accumulated by our daily living, and our insight should try to unveil the nature of complicated matters step by step. Collective dynamics of people can be represented as collective moving patterns in geospatial field, and it relates with cyber-physical linkages, as ubiquitous computation enables numerous people handled with mobile devices to coordinate with each other in geospatial environment (Shibuya 2004, 2006; Pentland 2014). This trend of academic progress stands for the journal publications (e.g., *Nature Human Behavior* (Springer-Nature), *Computers in Human Behavior* (Elsevier)).

For example, there was a recent case of "*Pokemon-Go*" fevers. During the summer of 2016, using smartphone, this location-based game application had been launched all over the worlds. Graells-Garrido et al. (2017) reported this fever in Chile case, and they analyzed dynamics of collective population in urban city. Such social fever can be understood as a pattern of collective mobility dynamics using ubiquitous computing. In the past, Abowd and Mynatt (2000) already expected the three directions of ubiquitous computing such as *natural interfaces, context-aware applications,* and *automated capture and access* in cyber-physical environment. Their future directions on ubiquitous computing and mobile devices had lastly established to provide commercialized daily applications and entertainment services which localized reality and embedded in social contexts.

In Japan, some rural governments tied up with Niantic (Pokemon-Go's administrative company), and they intended to attract Pokemon-Go users visiting their city. Eventually, their goals were often achieved successfully, and this social phenomenon further suggests that event-based onsite touring using such entertainment tools can inspire collective dynamics of ordinary citizens to motivate moving toward arbitrary area. It is renewable and repeatable social mechanism on moving patterns and transportation of collective behavior. As using such tools for entertainments, collective dynamics can be operated and induced moving toward designated area; of course, consumer behaviors and socioeconomic influence across localized areas by them will be expected to facilitate better consequences in order to develop rural cities.

These phenomena have significant academic meanings in behavior science (and computational social science). As introducing Newton's statement on collectiveness (madness) of people, according to traditional theories, it was harder intentionally to motivate mass of peoples to the common goal-oriented dynamics. But now, it just gives them only a smartphone. Let them present specific common purposes, their collectiveness can sooner become its characteristic patterns from pseudo-randomness to somewhat vector (direction and weight) in geometrics of the cyber-physical world. Such inducing factors may be either fascinating incentives or charismatic leaders.

As the author was firstly trained as a social scientist, at a glance, it could obtain an insight that the latent core of ubiquitous computation was significant to handle collective dynamics of human behavior (Shibuya 2004). Then, the author had already conceptualized and designed *cyber orienteering* in science museums and natural parks using ubiquitous and mobile computing (Shibuya 2006), and any printed map is now substituted for embedding in local geospatial data and human behaviors using those applications with smartphone. Those collaborative and participatory activities by massive citizens can be self-organized and conducted by the common value among them. Those dynamics and its big data will be optimized by the AI driven operation, and consequently the human-being already becomes IoP (internet-of-people) as one of node in social networking. Those interconnected peoples can be interacted with each other, and their moving dynamics will be synchronized and coordinated with localized contexts and events in the city.

9.1.2.2 Measuring Dynamics in City

To date, social design based on ubiquitous computational services can inspire actual behaviors of citizens, and it led their daily activities to enlarge from offline to online (Shibuya 2017). And it became more important to include geographical and spatial perspectives for modeling and simulation (Gimblett 2002). Spatial informatics and statistics can afford to show critically statistical meaning and significance of collective dynamics. For instance, one of those challengeable concepts is now eager to actualize as the smart city (Jiang and Yao 2010). In this way, citizens handling mobile and wearable devices may engage in ecological-oriented and healthy life (Bécu et al. 2019). As consequences of those social services[1], dynamical patterns of their behaviors can be statistically accumulated as big data (Qi et al. 2019). Thus, both collective and individual data of behaviors can be traced and tracked by the advancement of above technologies (Rossi et al. 2015; Toole et al. 2015). In this point, those patterns of dynamics actually represent their hidden needs, their mobility and aggregation manners can be unveiled in the era of computational social sciences (Abraham and Hassanien 2012).

And recently, on the other hand, Pramanik et al. (2017) reported feasibility for criminal cases using big-data analysis. And Oliveira et al. (2018) has shown dynamical patterns of criminal occurrences in the larger city, as they analyzed big data such as communication logs and geospatial data. Conceptualization of smart city has been discussing such topics, and big data accumulated from those dynamics of

[1] NTT Docomo: mobile geospatial statistics: https://mobaku.jp/

collective behavior with geospatial and time data in city has greater potentials to be further equipped in social contexts for daily and socioeconomic foundations.

On the other examples, collective panics among numerous citizens have been frequently happening in emergency conditions (e.g., fires, natural disasters, human-made accidents, terrible congestions). Those panics could be modeled and simulated in various assumed contexts. Wang et al. (2019) recently revealed a pattern on those panics by mathematical models, and they hypothetically categorized three types of citizens such as *calm, herding,* and *panic peoples* in emergency conditions. By constructing a small-world-like network, they verified how mutual interactions and interconnected weights among three types of peoples will engender panic chaos situations. And then, they found the evidence that there is a possibility to engender a panic situation and get worse if panic degree of herding type excess the threshold.

9.2 Modeling Mobility Flow and Dynamics

Larger mobility and aggregation of human behaviors has already regained the top-level place of the central topics in social science (Azose et al. 2016). Those spatio-temporal dynamics of human behavior calls much attention from social physics (Gonzalez et al. 2008; Helbing 2012; Pentland 2014; Goncalves and Perra 2015), and they often endorse that our daily life could be clarified by physical laws.

In Table 9.1, the author attempts to illustrate that collective human behavior in outdoor conditions distinctively categorizes four quadrants whether participants have the common goal or not. And another axis for categorization is whether some leaders exist or absent (Table 9.1).

1. First Quadrant (leader exists and sharing common intention)

   This case shows that everyone share with social conditions such as disaster, socioeconomic and others, and some leaders can handle total controls for them. Then, those leading mobility can be distinctly realized as collective mobility. In natural condition, zoological species can sometime show herding dynamics led by the leader and it was followed by massive members (e.g., birds, gnus, fishes). *Boid* simulation which was proposed by Reynolds could examine such collective dynamics of biological species in natural environment, and those biological pseudo patterns can be arranged only by several rules such as *separation, alignment,* and *cohesion*. Thus, this pattern indicates stronger commitments to the belonged group among members (i.e., social identification), and stable structure of memberships ordered by the leader.
2. Second Quadrant (leader exists and not sharing common intention)

   This case shows that everyone does not share with any purposes yet, but some leaders announce to discuss for their next political actions, for example. Then, gathering mobility of each individual can be lastly realized as collective mobility, and those patterns often shift to first type.
3. Third Quadrant (leader absents and sharing common intention)

Table 9.1 An example for categorizations on collective human behavior

	Specific leader(s) exists	Specific leader(s) absents
With common intention(s)	1. Political demonstrations with leader(s) 2. Political revolutions and social movements organized by political leader (s) 3. Refugees leading by charisma leader(s) 4. Migration organized by leader(s) 5. Evacuation organized by leader(s)	1. Migrations (as mentioned later in this chapter) 2. Transportations (e.g., commuters, car traffics, international trades) 3. Evacuation at disaster (see at Chaps. 10 and 11) 4. Social movements using social media (see at Chap. 10)
	Models: *Boid* models, ABM, stochastic models, social physics, social network model, etc.	Models: ABM, stochastic models, social physics, social network model, ACO, etc.
Without common intention(s)	1. Incidental and temporal mass event meeting at the public sphere gathering with various opinion groups organized by leader(s) 2. Massive demonstrations by coordinating with participants 3. Earlier discussion stage of political actions	1. Crowds 2. Mobs 3. Collective hysteria panic 4. Congestions 5. Chaotic situations 6. Temporal mass meeting without any purposes
	Models: ABM, social network model, random walk model, stochastic models, social physics, etc.	Models: ABM, random walk model, stochastic models, social physics, fluid dynamics, quantum field theory, ACO, etc.

This case exemplifies above *Pokemon-Go* fever. This case ostensibly shows that everyone share with social conditions such as socioeconomic, emergency, and other purposes. Then, gathering mobility of individuals can be lastly realized as collective mobility. And as mention later, in recently this type of collective behavior has been focused. Migration crisis beyond the nations and social movements using social media have been revealed by computational social sciences. It often assumes to be modeled as agent-based model (Namatame & Sasaki, 1998). On the other hand, ACO (ant colony optimization) can be modeled as a pattern on forging behavior (i.e., exploring something) by large amount of artificial ants (i.e., software agent represented as the AI) in AI studies, and those collective swarm dynamics can be regarded as an adaptive process by experiential learning in uncertainty environment.

4. Fourth Quadrant (leader absents and not sharing common intention)

   This case indeed shows that everyone does not share with specific common purposes and any leaders absent. Then, complicated patterns of mobility by numerous people can be often totally appeared as randomness (e.g., Brownian motion as a kind of random walk phenomena) and chaotic dynamics. Traditionally, there are various patterns such as crowds, mobs, collective panics, congestions, and spontaneous demonstrations. Some of those models could be formalized as an application of physics theories (e.g., fluid dynamics, quantum field). In social sciences, computer simulations are often based on random walk and specific stochastic distributions (e.g., Poisson distribution) in mathematical models and operations researches. Thus, this pattern shows the most unstable structure and disorder state among members, and those mechanisms have stronger potentials to engender social unrest, out of control, and collapse of social justice.

Here, the author especially exemplifies some measuring ways on collective mobility based on above third type. For example, there are varieties of well-known measurements on mobility flows and spatiotemporal patterns of collective dynamics (Renso et al. 2013). Those patterns can be defined as numerical vectors, probabilities, and spatial distributions. Namely, any movers (e.g., humans, cars, animals, moving robots) have specific goals and directions to walk or drive toward the final destination from arbitrary geospatial site, and it means that they do not always intend to behave as literally *random walk* and ad hoc moving.

1. Mobility flow
2. The laws of migration
3. Mathematical models in socioeconomic issues
4. Spatial autocorrelation

9.2.1 Geospatial Informatics on Mobility Flow

To date, mobility flow models based on the mutual distance between two sites have a good reason to be selected for recent studies on human behavior (AIST 2013). For example, there is a *potential model* on moving tendencies among collective people, which can be formalized as the following equations. It is often applied when mobility flow of numerous people seems to be related to the nature of space and physical distance between two sites (S_i and S_j).

$$P_{ij} = \frac{V_{ij}}{\sum_{j-1}^{m} V_{ij}} \tag{9.1}$$

$$V_{ij} = \frac{E_j}{D_{ij}^a} \tag{9.2}$$

where

- S_i, S_j: Site$_i$ and Site$_j$
- P_{ij}: Probability of moving dynamics of people from S_i to S_j
- m: Total numbers adjoining S_i
- V_{ij}: Potentials on moving tendency of people from S_i to S_j
- E_j: Size of population in S_j
- D_{ij}: Geospatial distance (e.g., Cartesian distance)

- a: Constant (it is usually inputted two)

Among above potential model, variable E_j can be actually regarded as the scalability of facility and capacity. And the distance D_{ij} will not be treated as inversely proportional and rather it will be better to be defined as square ($a = 2$). In such way, this mobility flow can be refined as moving patterns between two geospatial sites, and nearer distance can be supposed to be an influential factor for explaining and predicting social phenomena on collective dynamics of behavior.

9.2.2 The Law of Migration

"*The law of migration*" was historically discovered by Ravenstein (1885) during the era of the industrial revolution in the UK. His statistical analysis revealed that migration pattern was strongly committed by physical distance. It means that many workers and immigrants were apt to choose nearer location for their jobs and settlements. This law, to date, has been endorsed by numerous social science studies.

9.2.3 Mathematical Models of Mobility in Socioeconomic Issues

Mathematical model helps to understand the dynamic mobility flows. For other examples, preference index (PI) can be formalized as below (Shryock and Siegel 1973).

$$\mathrm{PI} = \frac{M_{ij}}{\left(\frac{P_i}{P_t} \cdot \frac{P_j}{P_t}\right) \cdot \Sigma M_{ij}} \tag{9.3}$$

where

- M_{ij}: mobility from area$_i$ to area$_j$
- P_i: total population in area$_i$
- P_j: total population in area$_j$
- P_t: total population in whole country
- ΣM_{ij}: gross migrants within whole country

Otherwise, another case formalized mobility patterns in the urban population (Billari and Prskawetz 2003). Ganning and McCall (2012) statistically analyzed geographical weighted regression models, which are formalized as below equation. They calculated their equation-based models, and mutual interactions of immigrations of commuting workers were estimated from each statistical resource.

$$C_{ij} = f\left(M_i, W_{ij}, D_{ij}, H_{ij}, E_{ij}\right) \tag{9.4}$$

where

- C_{ij}: *net number of workers commuting from country i to country j*
- M_i: *net migration into county i in the previous period, normalized by the population in county i in the previous period*
- W_{ij}: *wage in county j minus wage in county i (multiplied by 1000)*
- D_{ij}: *distance between counties i and j, using population-weighted centroids*
- H_{ij}: *standardized housing cost in county j minus county i*
- E_{ij}: *four-year college degree attainment rate in county j minus county i*

In contrary, Heiland (2003) reported their ABM on the mobility patterns from states of the former east-Germany to states of the former west-Germany after the Berlin Wall collapse. Statistically, he described "*from July 1989 to June 1991 about 675,000 East Germans, or roughly 4.1 percent of the population of the former German Democratic Republic, emigrated to West Germany.*" On the bases of statistical analyzed data, he examined an aggregation pattern from east side local city to larger urban city (e.g., Berlin).

$$E_1\left(\sum_{t=1}^{T-1} \delta^{t-1}\left(u\left(c_t\right) + v\left(r_t\right) + \delta U_t\right)\right) \tag{9.5}$$

A destination choice of each agent was defined as some utility functions, where E denotes an expected value function of each agent. Variable δ is discount variable, c implies consuming factor, and r is evaluations of residential amenities in each agent.

Otherwise, Rakowski et al. (2010) analyze stochastic patterns on the commuters' mobility in Poland using ABM. They discussed the complexity of urbanized collective transition patterns, and their ABM could be utilized in social needs.

9.2.4 Gravity Model

Geospatial models had been rearranged by social scientific needs. As one of the representatives, gravity model is adopted in social science studies, and it was originally derived from the universal law of Newtonian classical physics. Obviously, it reminds us of the nature of classical interactive dynamics. Where, F_{ij} denotes an interactive influence between parameter p_i and p_j divided by mutual distance (D) between p_i and p_j. Parameter a and b are constants. For example, both variables P_i and p_j are often inputted actual data from mutual immigrants among cities or GDP in economics (Anderson 2011).

$$F_{ij} = \mathrm{ap}_i \mathrm{p}_j / D^b \tag{9.6}$$

It exactly indicates that gravity model can formulate two interactive entities with geospatial factors of distance. Further Eq. 9.6 can be redefined as a logarithmic estimation model in Eq. 9.7, where ε implies residual error and a is constant.

$$\ln F_{ij} = a + \beta_1 \ln(p_i) + \beta_2 \ln(p_j) + \beta_3 \ln D + \varepsilon \tag{9.7}$$

Of course, gravity model can deal with somewhat interactive patterns between two cities as well as total amounts of intercommunication between two areas. For example, Toole et al. (2015) and Beiro et al. (2016) conducted comprehensive works in geospatial factors and mobility flow using big data. And Pan et al. (2013) examined urban communication patterns in gravity model. Further, Deville et al. (2016) reported new findings on scaling identity compared with gravity model between mobility and social communication interactions. They concluded that their model was more fitted than gravity model, and both geospatial and scaling factors were crucially significant in social interactions.

In this way, social physics and applied mathematics on collective human dynamics have revealed some physical principles (Gonzalez et al. 2008; Helbing 2012; Goncalves and Perra 2015). Their efforts unveiled the universal patterns on power distribution in any social phenomena (e.g., network dynamics, human activities), and it can denote strict formalizations on those patterns.

9.2.5 *Spatial Autocorrelation*

The author articulates a spatial distribution measurement using the spatial autocorrelation. It is essential to study geospatial modeling as well as demographic dynamics. Namely it indicates a correlated coefficient of spatial relationship among adjoining sites. It figures out a stochastic distribution of people and movers within target area and region, and it helps to visualize a degree of geospatial distributions and clustering. If these adjacent areas are related statistically, it can estimate linear correlation. And for statistical examination, for example, there are some formulas such as *Moran's I* (Eq. 9.8) and *Geary's C* (Eq. 9.9) (Haining 1990).

Where, as one of the spatial autocorrelation coefficients, *Moran's I* can range from 0.0 to 1.0. When coefficient I nearly reaches to 1.0, it maximizes value of autocorrelation. Both variables W and w_{ij} indicate weight of matrix. Both variables x_i and x_j are each spatial data. By the way, a model on weight matrix of network (Leenders 2002) was almost regarded as a similar meaning to this geospatial autocorrelation.

$$I = \frac{n}{W} \frac{\sum_{i=1}^{n} \sum_{j=1}^{n} w_{ij} \left(x_i - \overline{x}\right)\left(x_j - \overline{x}\right)}{\sum_{i=1}^{n} \left(x_i - \overline{x}\right)^2} \tag{9.8}$$

Where, as another spatial autocorrelation coefficient is *Geary's C,* when C means positive correlation under 1.0, whereas C means negative correlation over than 1.0. If C equals 0.0, then it considers random condition. Similarly, both W and w_{ij} indicate weight of matrix. Both x_i and x_j are each spatial data.

$$C = \frac{n-1}{2W} \frac{\sum_{i=1}^{n} \sum_{j=1}^{n} w_{ij} \left(x_i - x_j\right)^2}{\sum_{i=1}^{n} \left(x_i - \overline{x}\right)^2} \tag{9.9}$$

9.3 Immigration Dynamics Crossing Nations

Hereafter, the author especially discusses an example case on immigration crisis in the EU. Digital tools and social media could effectively facilitate a large number of migrants to navigate toward the EU. But such larger dynamics of migration engendered larger exclusive reactions by citizens in each country of the EU, and lastly those troubles broke EU's firmer alliance and the international order. Brexit of the UK and "America First" slogan by a candidate for the US president had been induced by those global aftermaths on collective migration crisis. Then, there are a lot of considerable points on those social mechanisms to understand collective identity.

9.3.1 The World History as Collective Immigration

The history of our civilization was coeval with inevitable and strong interlinks between population and geospatial constraints (Forman and Wu 2016). Ever since civilizations, it was too suggestive that total population of humankind repeatedly settled in somewhere and moved to somewhere. However, both long-term settlement and mobility toward unknown places usually require the convincible reasons for each person thoughtfully. And it can expect that identification of each individual has strong interlinkages with own self-esteem grounding on their social and cultural backgrounds. In the past, French government assigned *the Minister of Immigration, Integration, National Identity and Codevelopment*. They ostensibly sought "national identity" for coexistence and integration among citizens. This legal act suggests imminent needs based on serious backgrounds and many troubles against immigrations from overseas.

At first place, why do many people leave from their homeland? All citizens prefer something by each own standard of evaluations, and their motivations for mobility seem to vary widely such as socioeconomic reasons, urbanization, environmental matters, conflicts, disasters, and human-made tragedies.

It is naturally considered as principal bases of the legal rights "*right to live*" in English or "*droit à la vie*" in French. Actually, in 1948, United Nations declared the "*Universal Declaration of Human Rights*" and their ideal is supposed to be shared among member states (United Nations 1948). This unconditional declaration includes "*the right to freedom of movement and residence*."

- Article 13
- *Everyone has the right to freedom of movement and residence within the borders of each state.*
- *Everyone has the right to leave any country, including his own, and to return to his country.*
- Article 14
- *Everyone has the right to seek and to enjoy in other countries asylum from persecution.*

Principally, these rights should be approved for everyone. It includes one of the human rights (Nakanishi 2017). However, self-excessive awareness for these rights and ideologies among stakeholders has grown to become more complicated social dilemmas and contradictions in global orders. The recent collective flow of mobility and aggregations by larger immigrants seems to be realized social phenomena representing the depth of those matters. In social sciences, it further considers the core of this challengeable issue.

9.3.2 What Is the Meaning of "Arab Spring"?

First of all, let me reexamine an actuality. As analyzing big data from social media, we equally became the historical witness of the "*Arab Spring*" and its consequences (Choudhary et al. 2012; Agarwal et al. 2013). This revolution actualized domino effects and the *legitimation crisis* (Habermas 1976; Fukuyama 1992) around middle-east nations. And those massive migration matters suggested the importance of democratic legitimacy for the national regimes among citizens (Shibuya 2017).

We usually notice the visibility for the collectiveness in social situations such as massive social movements, intergroup conflicts, interlinkages in social networking and racial ethnicity. Sociologically, those patterns might be categorized as collective dynamics by numerous citizens (Le Bonn 1931a, b; Smelser 1962). Such movements occasionally show social influence and catastrophic collapse of regimes caused by the crowds. More concretely, the categories of those actions by citizens include opinion expressions, group processes, identification, emotional attitudes, fads, mobs, criminal riots, demonstrations, and politically assertive agitations as well as social change against ancient regimes caused by large groups of citizens in uprisings and revolution (Olson 1971; Brown 2000; Rohlinger and Snow 2003).

As the consequences of those actions, it realized that massive demographic mobility engendered severe controversial problems on not only loss of the common wealth but disruptions of the nation. As the revolution consequences caused by those collective actions, it called further devastations and conflicts. Namely their daily life might be suffered in extreme poverty. Lastly, as the consequences of those revolutions, those severe conditions forced them to leave their hometown and nations.

Furthermore, those mobility patterns of migrations beyond the international borders seemed to be dependent on geopolitical factor. Many of them went straight toward the final destinations (e.g., EU) from their homelands via various routes. And they were navigated to far destinations using smartphones and social media. Those populations estimated as nearly half a million people (Migration policy centre 2012) and approached to five million (UNHCR, Syria Regional Refugee Response 2016).

Hence, with such situations in mind, it became to consider not only unreasonable matters on the disputes of international affairs but systematic models for solutions. On the standpoint from international affairs, those massive migrants across countries consequently called an imbalance of international order and breaking balance of power among stakeholders. Social issues such as ethnicity, religious minority, social unrest, and conflicts among citizens were induced in host country.

9.3.3 Migration Crises as One of the World Affairs

Regarding those backgrounds, it was held in one of the most important congresses on 19th September 2016, the General Assembly of the United Nations called for a summit at the Heads of State and Government level on large movements of refugees and migrants (United Nations 2016; Bauboeck 2019). Needless to say, those who migrated to foreign countries has been increasing, that is because they escaped from the tough problems such as national conflicts, wars, terrorisms, human-made hazards, and natural disasters.

According to UN statistics[2] (United Nations 2015), gross migration was estimated as 65.3 million people in 2015 (e.g., forcibly displaced people worldwide (Castles 2003): refugees, asylum-seekers, and other reasons). Thus the United Nations required each member states to ensure gentle acceptance for them, actually no countries embrace unlimited immigrants. This summit was a first time to worldly discuss those controversial and international affairs on "*migration crises*." Worldwide cooperation and acceptances for large migrants and refugees among member states stepped forward to the next phase.

[2] https://www.unhcr.org/

However, to date, it was certain that those matters has not been resolved completely yet. In a philosophical essay "*Perpetual Peace*" described by Kant (1795), his critics clearly explained "*the law of world citizenship shall be limited to conditions of universal hospitality.*" It subtly means to differentiate between hospitality and immigration policy related with ethnicity and nationalism (Wimmer & Schiller, 2003).

As Buzby (2016) suggested, those immigration policies have been disputed as controversy issues in the USA. Brader et al. (2008) examined triggers' effect to opposition against immigration policy in the USA, and they clarified citizens' attitudes such as anxiety, group stereotypes, and vigilance to immigrants.

In contrast, in EU, standing on the grounds of the Schengen agreement, each ratified member country shall permit citizens to travel and move across these countries (Yuval-Davis et al. 2017). However, those migrants are clearly aimed at surging to loose gates of the EU by the Schengen agreement. Then, calls for opposition demonstrations against immigrants' policy were also made in the EU member states, and antagonists described inequities and aggressions committed against them. The Brexit of the UK and political confusions in Germany were exactly similar to a stalemate condition.

Taken together, it globally requires understanding both the equipment of soundly networking in economic activities and achieving social system design on the bases of reconstruction through demographic change and international immigration (Azose et al. 2016). Rogers (2015) proposed extensively multiregional demography to define those mobility patterns as *flows*. Certainly, there were many cases to institutionalize immigrant's policies for national developments (Hatton 2015). In addition, how is significant between immigration and economic development statistically (Rodiguez-Pose and Berlepsch 2015; Orrenius and Zavodny 2016)? However, to date, aforementioned demographic mobility and its socioeconomic activities were varied dependently by enormous factors such as population dynamics, democracy level among ordinary citizens, and multiple intercommunication networks. Those patterns became too complex.

9.4 Simulation of Migration by Deferred Acceptance Algorithm

Next, the author demonstrates a migration model, which was defined as an agent-based model (Shibuya 2004, 2017, 2018). This Java component based software framework for social simulation has the advantages of analyzing, visualizing, and serving functions for the "*virtual computational social worlds*" (Conte et al. 2012).

Using a deferred acceptance algorithm (DAA: Roth 2008), this migration model searches stable patterns of matching between each individual's preferences and each city' policy. Actual city might have a prioritized acceptance policy for immigrants (Powell 2015; Kahanec and Zimmermann 2016). Namely, each city has somewhat prioritized preference orders. Some cities will offer much opportunity for higher educated skill engineers and well-educated professionals than lower-wage immigrant workers.

Obviously, mobility patterns of immigrants are not necessarily a random diffusion process. As both long-term settlement and mobility toward unknown places usually require the convincible reasons for each person thoughtfully, they do not often behave randomly and unintentionally. All citizens prefer something by each own standard of evaluations, and their motivations for mobility seem to vary widely such as socioeconomic reasons, urbanization, environmental matters, safety, amenity, and human relationships. Each individual holds own desires to improve and maintain daily life, and they preferentially drive to the destination selected by their standard of evaluation. Their preference may be conjoined with geospatial factor for their moving.

9.4.1 Basic Scenario

Each agent plays a role of migrant, and for example, total population of migrants was configured up to 2000. For simulation, the author defines a following scenario to match both preferences between migrants and cities using DAA.

1. Acceptance limitation for migrants in each city

   In each policy as follows, each city has own limitation to accept for migrants.

2. Preference order of migrants:

   It depends on AHP (see below). Each city chooses a policy to prioritize own preference order for acceptance of migrants among first-come-first-served (FCFS), young workers oriented (YWO), or urgency for acceptance (UO). FCFS means availably accepting who applied to cities. YWO policy announces a prioritization for socioeconomic enhancement by active generations as immigrating workforce (workers' age around 15 to 55) in cities. UO is underlying in a humanitarian policy and it prioritizes urgency conditions (from minimum 0 to maximum 5) of migrants. And these policies' evaluations are consisted of socioeconomic budget (cost), socioeconomic merits, humanitarian aids, and latent danger. These initial parameters are configured randomly.

9.4.2 Matching Algorithm

It can be reconsidered as a variation of matching problems in the job market in economics on mechanism design (Roth 2008). Namely, as DAA for matching problems in the labor market has clarified its effectiveness in advance, its strategy-proof was completely endorsed in standard manner. As an honor for his algorithm, Roth obtained a Nobel Prize in economics. Similarly it possibly applies to a matching problem between migrants and cities based on their preference order. Certainly, it must institutionalize the centralized clearing house which offers useful information for migrants.

For example, preference orders to three cities among three migrant agents (m_1, m_2, m_3) are $m_1 = \{c_1, c_3, c_2\}$, $m_2 = \{c_3, c_2, c_1\}$, and $m_3 = \{c_1, c_2, c_3\}$ in order. And conversely, preference orders of cities (c_1, c_2, c_3) are respectively $c_1 = \{m_3, m_1, m_2\}$, $c_2 = \{m_1, m_3, m_2\}$, and $c_3 = \{m_2, m_1, m_3\}$. Here, m_2 matches to c_3, and m_3 matches to c_1. Although his third preference is m_1, as a consequence, m_1 matches to c_2. Those results of matching are finally more stable. Then, matched pairs' set can be lastly completed as $\{m_1, c_2\}$, $\{m_2, c_3\}$, and $\{m_3, c_1\}$, and no blocked pairs remain. Finally, it allocates each migrant for their acceptable city.

9.4.3 AHP

Preference-based computation can be often arranged in computational social choice studies (Brandt et al. 2016; Rothe 2016). For example, AHP (Analytical Hierarchy Process) is characterized with applied mathematical procedures (e.g., Linear Algebra). It can be defined by pairwise comparisons between several alternatives among all combinations. Calculation procedures of canonical AHP in detail can refer to Saaty (2013). Below example case shows to prioritize and evaluate a subjective order of each preference among several alternatives such as *physical closeness from hometown, safety around houses, human relationships, total costs,* and *amenity in new residence area.* Where, n is total sum of factors, $\lambda_{\max}$ is calculated by W (weight matrix) and Eigen values. Each preference and final objectives can be determined by calculating vectors and weight matrix. And finally, *C.I.* (consistency index) for model verifications should be kept under 0.1.

$$\lambda_{\max} = \frac{\Sigma \mathrm{W}_n}{n} \tag{9.10}$$

$$C.I. = \frac{\lambda_{\max} - n}{n - 1} \tag{9.11}$$

As following figures exemplified a result, an agent autonomously evaluated three alternatives by this prioritized ordering: candidate City A (0.460) > City C (0.401) > City B (0.137). Thus it is rational that an individual moves to primary city A. As this evaluation differs among citizens, it engenders different mobility patterns among them. In geospatial constraints, each individual decides destination by preferred order.

Until one of the cities accepts their immigration, all the while each citizen seeks the route for going to final destination on the bases of own preference (if city does not accept an immigration, preference order is reevaluated in next phases) (Figs. 9.1 and 9.2).

Due to the results of AHP, each agent applies to the most available city on own preference order. In DAA, each city also offers finite residence capacities for agents and it accepts them in preferred order up to the limitation. For example, each city is configured to prioritize an acceptance order for migrants such as employability of job seekers and social demands. After that, final matching results between them expect to reach the core and Pareto rational states.

9.4.4 Exploring the Route by Genetic Algorithm

Before running a simulation, genetic algorithm (GA) explores the most costless routes from arbitrary sites to final destination in specific constraints. It optimizes the total moving cost. In simulation, generation of genes is initially randomized in cost allocations. Each gene represented specific costs for transition among sites. And crossover and mutation of genes can be applied in each phase (Goldberg 1989). Thus simulator explores the most adaptive gene (it means the most reasonable routes) in process.

- Sites: 1000, for example
- Sites of stopover in transition (n): from 1 up to 10, for example
- Goal to optimize total cost: below 90, for example
- Total cost between sites (R): 20, for example
- Mutation: from 0.01 up to 1.0
- Each cost of route between n_i and n_j: from 1 up to R

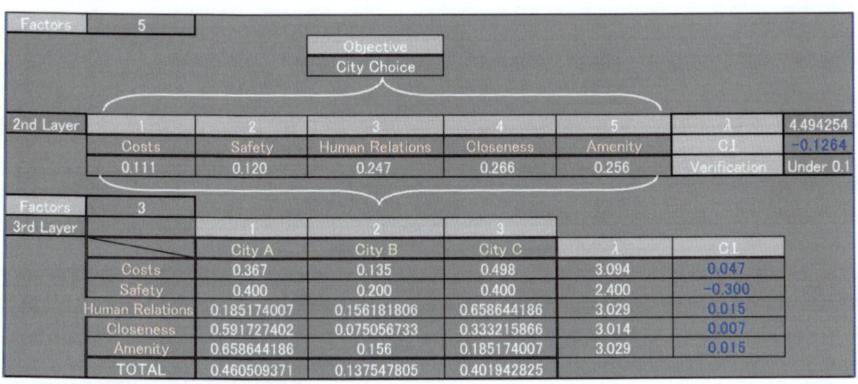

Fig. 9.1 An example of AHP calculation

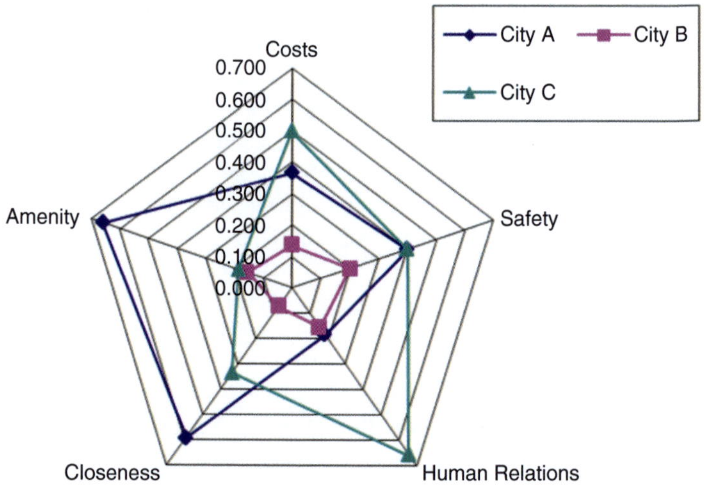

Fig. 9.2 An example of AHP

Table 9.2 An example result on FCFS using DAA

First come-first served	Limit of acceptance	Accepted population	Threshold of migrants' age (15 to 55)	Average of migrants' age	Threshold of urgency (0 to 5)	Average of urgency
Host City 1	450	321	23	42.8	4	2.49
Host City 2	396	360	68	44.7	1	2.56
Host City 3	434	319	31	44.9	1	2.55

9.4.5 *Running Simulation*

Below tables indicate final consequences on limit of acceptance, accepted population, threshold of migrants' age, urgency and those averages in each type such as FCFS, YWO, and UO, respectively. When thresholds are exceeded its value, each city can accept migrants. In only FCFS case, each city is apt to accept migrants in proposed order (first-come, first-served).

Thus, the matching process was almost verified. It endorsed a matching sufficiency between migrants and cities in each scenario. It means to allocate own living space in each city for migrants. Of course, as a list for matching is linearly ordered $O(n)$, DAA has serious troubles to deal with the larger extents of migrants' population, and it directly performs much time consumption in matching process (Tables 9.2, 9.3 and 9.4).

Table 9.3 An example result on YWO using DAA

Young worker oriented	Limit of acceptance	Accepted population	Threshold of migrants' age	Average of migrants' age	Threshold of urgency	Average of urgency
Host City 1	419	180	47	33.3	3	2.36
Host City 2	453	86	29	25.6	1	2.16
Host City 3	479	182	43	31.3	0	2.46

Table 9.4 An example result on UO using DAA

Urgency oriented	Limit of acceptance	Accepted population	Threshold of migrants' age	Average of migrants' age	Threshold of urgency	Average of urgency
Host City 1	372	366	49	42.3	3	3.96
Host City 2	390	390	32	42.9	2	3.57
Host City 3	363	363	48	42.2	2	3.63

In this way, for any governors, social simulation can help in understanding collective dynamics of migrants and their trajectory patterns, albeit it reformed to be simplified. In other words, the conundrum against acceptance of migrations can be considered as allocation of migrants in each city, as it depends on host city' threshold and physical constraints (e.g., geospatial distance, transportation cost and time). When temporal migrants finally become long-term residents and settlers in city, each city should reconsider socioeconomic balances for acceptance of migrations. In this context, it reaches to the migration policy studies in economics (Powell 2015). Increasing active workers' population raises economic benefits for accepting city, but retired generations require their future pensions and social welfare services. Further studies need to assess those scenarios corresponding with socioeconomic and demographic factors.

On the other hand, those matching studies in mechanism design have pioneered to settle unsolved social contradictions which should be coordinating with innumerable stakeholders. Aforementioned DAA by AI driven matching system recently exemplifies many cases such as kindergartens matching, venture companies matched with municipal cities in Japan, and alternative plans can be rearranged immediately in the automatic matching process based on each preference among consumers. Roth originally achieved kidney exchange program, medical interns for their hospital choice based on their preferred orders, and school matching resolutions in local city using his DAA (Roth 2008; Rothe 2016; Hamada et al. 2017).

9.5 Conclusion of Simulation

What this simulation suggested was that migration problems are firstly relevant to national sovereignty and identity among the citizens in the host country. Second problem is how to resolve between human right and acceptance of innumerable migrants. And thirdly, atmosphere of such acceptances in the host country will engender different consequences, whether migrants intend to be hired as longer-term workers or accepted as tentative refugees and travelers.

Collective phenomena such as migration and dynamics of human behavior vividly appear very salient trends of sequential mobility which were self-organized by innumerable peoples. Even if each migrant hesitates to drive toward another country and each of them does not want to be conspicuous for the accepted host countries, their collectiveness itself may threaten against ordinary citizens in host countries. As such collective phenomena of migrants often threaten against the boundaries of own identification for host citizens, their hospitality for travelers may sooner turn into hateful emotion against migrants. Migrants are often employed as lower-wage workers in the host nation, but some of such jobs itself may be negatively prejudiced by host citizens. Their unfavorable status symbolizes the stigma as prejudice and negative stereotypes against them.

In contrary, immigrated countries, from where many migrants leaved, have been often steadily stagnated in socioeconomic development, because necessary investments, workforces, and resources have been shortages during longer-term periods. Although AI and other technologies have possibilities to be fulfilled for workforces, any government must crucially manage to optimize any resources and prioritize planning.

As this chapter demonstrated an example simulation using DAA, DAA-based acceptance of migrants in each city has to be exemplified to plan for future workforce balance. Each city might be considered to activate their planning and evaluate acceptable risks against migration policy itself. Therefore, the digitized age will further reshape our working needs, commuting patterns, and working environment, and the cases of migration policy in each nation will alter both outbound and inbound mobility patterns, and the modality of national identifications among citizens and migrants will be also perhaps transformed.

References

Abowd, G. D., & Mynatt, E. D. (2000). Charting past, present, and future in ubiquitous computing. *ACM Transactions on Computer-Human Interaction, 7*(1), 29–58.

Abraham, A., & Hassanien, A. E. (2012). *Computational social networks: Tools, perspectives and applications*. London: Springer.

Agarwal, N., Wigand, R. T., & Lim, M. (2013). *Online collective action: Dynamics of the crowd in social media*. Vienna: Springer.

AIST. (2013). *Handbook of measuring human behavior*. Tokyo: Asakura Press. (in Japanese, ISBN:978-4-254-20155-0).

Anderson, J. E. (2011). The gravity model. *Annual Review of Economics, 3*(1), 133–160.

Azose, J. J., et al. (2016). Probabilistic population projections with migration uncertainty. *PNAS, 113*(23), 6460–6465.

Bauboeck, R. (2019). *Debating European citizenship*. Cham: Springer.

Bécu, M., et al. (2019). Age-related preference for geometric spatial cues during real-world navigation. *Nature Human Behavior*. https://doi.org/10.1038/s41562-019-0718-zDO.

Beiro, M. G., et al. (2016). Predicting human mobility through the assimilation of social media traces into mobility models. *EPJ Data Science, 5*, 1. https://doi.org/10.1140/epjds/s13688-016-0092-2.

Billari, F. C., & Prskawetz, A. (Eds.). (2003). *Agent-based computational demography*. Heidelberg: Springer.

Brader, T., Valentino, N. A., & Suhay, E. (2008). What triggers public opposition to immigration? Anxiety, group cues, and immigration threat. *American Journal of Political Science, 52*(4), 959–978.

Brandt, F., et al. (2016). *Handbook of computational social choice*. Cambridge: Cambridge University Press.

Brown, R. (2000). *Group processes* (2nd ed.). Malden, MA: Blackwell.

Buzby, A. (2016). Locking the borders: Exclusion in the theory and practice of immigration in America. *International Migration Review, 52*(1), 273–298. https://doi.org/10.1111/imre.12291.

Castles, S. (2003). Towards a sociology of forced migration and social transformation. *Sociology, 77*(1), 13–34.

Choudhary, A., et al. (2012). Social media evolution of the Egyptian revolution. *Communications of the ACM, 55*(Issue 5), 74–80.

Conte, R., et al. (2012). Manifesto of computational social science. *The European Physical Journal, Special Topics, 214*(1), 325–346.

Deville, P., et al. (2016). Scaling identity connects human mobility and social interactions. *PNAS, 113*(26), 7047–7052.

Forman, R. T. T., & Wu, J. (2016). Where to put the next billion people. *Nature, 537*, 608–611.

Fukuyama, F. (1992). *The end of history and the last man*. New York: Free Press.

Ganning, J. P., & McCall, B. D. (2012). The spatial heterogeneity and geographic extent of population deconcentration: Measurement and policy implications. In L. J. Kulcsar & K. J. Curtis (Eds.), *International handbook of rural demography*. New York: Springer.

Gimblett, H. R. (2002). *Integrating geographic information systems and agent-based modeling*. New York: Oxford University Press.

Goldberg, D. E. (1989). *Genetic algorithms in search, optimization, and machine learning*. New York: Addison-Wesley Professional.

Goncalves, B., & Perra, N. (Eds.). (2015). *Social phenomena: From data analysis to models (computational social sciences)*. Cham: Springer.

Gonzalez, M. C., Hidalgo, C. A., & Barabasi, A. (2008). Understanding individual human mobility patterns. *Nature, 453*, 779–782.

Graells-Garrido, E., et al. (2017). The effects of Pokemon go on the pulse of the city: A natural experiment. *EPJ Data Science, 6*, 23. https://doi.org/10.1140/epjds/s13688-017-0119-3.

Habermas, J. (1976). *Legitimation crisis*. London: Heinemann.

Haining, R. (1990). *Spatial data analysis in the social and environmental sciences*. Cambridge: Cambridge University Press.

Hamada, N., Hsu, C.-L., Kurata, R., Suzuki, T., & Yokoo, M. (2017). Strategy-proof school choice mechanisms with minimum quotas and initial endowments. *Artificial Intelligence, 249*, 47–71.

Hatton, T. J. (2015). United States immigration policy: The 1965 act and its consequences. *The Scandinavian Journal of Economics, 117*(2), 347–368.

Heiland, F. (2003). The collapse of the Berlin Wall: Simulating state-level east to west German migration patterns. In F. C. Billari & A. Prskawetz (Eds.), *Agent-based computational demography*. Heidelberg: Springer.
Helbing, D. (2012). *Social self-organization*. Heidelberg: Springer.
Jiang, B., & Yao, X. (2010). *Geospatial analysis and modelling of urban structure and dynamics*. Dordrecht: Springer.
Kahanec, M., & Zimmermann, K. F. (2016). *Labor migration, EU enlargement, and the great recession*. Heidelberg: Springer.
Kant, I. (1795). *Perpetual peace*. Macmillan Company. (English translated edition).
Le Bonn, G. (1931a). *The crowd: Study of the popular mind*. iBook Version (Apple iBook Store).
Le Bonn, G. (1931b). *The psychology of revolution*. iBook Version. (Apple iBook Store).
Leenders, R. T. A. J. (2002). Modeling social influence through network autocorrelation: Constructing the weight matrix. *Social Networks, 24*, 21–47.
Migration Policy Centre. (2012). *The European response to the Syrian refugee crisis, what next?* http://www.migrationpolicycentre.eu/docs/MPC%202012%2014.pdf
Nakanishi, Y. (Ed.). (2017). *Contemporary issues in human rights law: Europe and Asia*. Singapore: Springer.
Namatame, A., & Sasaki, T. (1998). Self-organization of complex adaptive systems as a society of rational agents. *Artificial life and robotics, 2*(4), 189–195.
Oliveira, M., et al. (2018). Spatio-temporal variations in the urban rhythm: The travelling waves of crime. *EPJ Data Science, 7*(29). https://doi.org/10.1140/epjds/s13688-018-0158-4.
Olson, M. (1971). *The logic of collective action: Public goods and the theory of groups*. Cambridge, MA: Harvard University Press.
Orrenius, P. M., & Zavodny, M. (2016). The economics of U.S. immigration policy. *Journal of Policy Analysis and Management, 31*(4), 948–956.
Pan, et al. (2013). Urban characteristics attributable to density-driven tie formation. *Nature Communications, 4*, 1961. https://doi.org/10.1038/ncomms2961.
Pentland, A. (2014). *Social physics*. New York, NY: Penguin Press.
Powell, B. (2015). *The economics of immigration*. Oxford: Oxford University.
Pramanik, M. I., et al. (2017). Big data analytics for security and criminal investigations. *WIREs Data Mining and Knowledge Discovery, 7*, e1208. https://doi.org/10.1002/widm.1208.
Qi, M., Wang, Z., He, Z., & Shao, Z. (2019). User identification across asynchronous mobility trajectories. *Sensors, 19*(9), 2102. https://doi.org/10.3390/s19092102.
Rakowski, F., et al. (2010). Large scale daily contacts and mobility model -an individual-based countrywide simulation study for Poland. *Journal of Artificial Social Simulation, 13*(1), 13. http://jasss.soc.surrey.ac.uk/13/1/13.html.
Ravenstein, E. G. (1885). The Laws of migration. *Journal of the Statistical Society of London, 48*(2), 167–235.
Renso, C., Spaccapietra, S., & Zimányi, E. (2013). *Mobility data*. Cambridge, MA: Cambridge University Press.
Rodiguez-Pose, A., & Berlepsch, V. (2015). European migration, national origin and long-term economic development in the United States. *Economic Geography, 91*(4), 393–424.
Rogers, A. (2015). *Applied multiregional demography: Migration and population redistribution*. New York: Springer.
Rohlinger, D. A., & Snow, D. A. (2003). Social psychological perspectives on crowds and social movement. In J. Delamater (Ed.), *Handbook of social psychology*. New York: Kluwer Academic Publishing.
Rossi, L., et al. (2015). Spatio-temporal techniques for user identification by means of GPS mobility data. *EPJ Data Science, 4*, 11. https://doi.org/10.1140/epjds/s13688-015-0049-x.
Roth, A. E. (2008). Deferred acceptance algorithms: History, theory, practice, and open questions. *International Journal of Game Theory, 36*, 537–569.
Rothe, J. (2016). *Economics and computation: An introduction to algorithmic game theory, computational social choice, and fair division*. Heidelberg: Springer.

Saaty, T. L. (2013). *Theory and applications of the analytic network process: Decision making with benefits, opportunities, costs, and risks*. English Edition: RWS Publications.
Shibuya, K. (2004). A framework of multi-agent based modeling, simulation and computational assistance in an ubiquitous environment. *Simulation, 80*(7-8), 367–380.
Shibuya, K (2006) Collaboration and pervasiveness: Enhancing collaborative learning based on ubiquitous computational services, including as chapter 15 Lytras, M & Naeve, A (Eds.) Intelligent learning infrastructures for knowledge intensive organizations: A semantic web perspective 369-390, IDEA Group Publishing Hershey PA.
Shibuya, K. (2011). *A study on ubiquitous collaborative activities* (pp. 1–110,. *Doctor Dissertation*). The University of Tsukuba (in Japanese).
Shibuya, K. (2017). Bridging between cyber politics and collective dynamics of social movement (Chapter 307). In M. Khosrow-Pour (Ed.), *Encyclopedia of information science and technology* (4th ed., pp. 3538–3548). Hershey, PA: IGI Global.
Shibuya, K. (2018). A design of Fukushima simulation. In *The Society for Risk Analysis: Asia Conference 2018, Japan.*
Shryock, H. S., & Siegel, J. S. (1973). *The methods and materials of demography* (Vol. vol. 2). Washington: U.S. Bureau of Census.
Smelser, N. J. (1962). *Theory of collective behavior*. New York: The Free Press.
Toole, J. L., et al. (2015). Modeling and understanding intrinsic characteristics of human mobility. In B. Goncalves & N. Perra (Eds.), *Social phenomena: From data analysis to models (computational social sciences)*. Cham: Springer.
UNHCR, Syria Regional Refugee Response. (2016). http://data.unhcr.org/syrianrefugees/regional.php
United Nations. (1948). *Universal declaration of human rights*. http://www.ohchr.org/EN/UDHR/Documents/UDHR_Translations/eng.pdf
United Nations. (2015). *Global trends forced displacement in 2015*. http://www.unhcr.org/576408cd7
United Nations. (2016). Addressing large movements of refugees and migrants: Draft for adoption by The President of the General Assembly. United Nations. http://www.un.org/pga/70/wp-content/uploads/sites/10/2015/08/HLM-on-addressing-large-movements-of-refugees-and-migrants-Draft-Declaration-5-August-2016.pdf
Wang, J., et al. (2019). A utility threshold model of herding–panic behavior in evacuation under emergencies based on complex network theory. *Simulation, 93*(2), 123–133.
Wimmer, A., & Schiller, N. G. (2003). Methodological nationalism, the social sciences, and the study of migration: An essay in historical epistemology. *IMR, 37*(3), 576–610.
Yuval-Davis, N., Wemyss, G., & Cassidy, K. (2017). Everyday bordering, belonging and the reorientation of British immigration legislation. *Sociology, 52*(issue 2), 228–244.

Chapter 10
Networked Identity

Here, *Networked identity* indicates interconnected modality on social synchronization and in/exclusiveness among massive global citizens. Those social mechanisms beyond the physical distance should be revealed how and why occurs or not in each condition. Although the advent of social media gained harmonized opinion formation and contagions by through massive influences, the digitalized social world would be coined in sharing emotional experiences among global citizens. Such social dynamics has regenerated influentially fake news, social movements, and responsive support networking. An online virtual nation has been committed and identified among participatory citizens, and their motivations to be recognized as an independent nation could facilitate their progress by online- interconnected citizens.

10.1 Synchronizing Networked Identities

Synchronization (or asynchronization) through social media and digital tools already reshape not only their clustering of social networking but actual social situations. As renowned studies on synchronization of coupled limit-cycle oscillators in nature had been mathematically revealed by Kuramoto model and Strogatz model (2000). Such related phenomena have recently laid much attention in computational social science. As mentioned in Chap. 7, synchronizing patterns could exemplify a dyad interaction model, and further Chap. 9 indeed discussed varieties of collective patterns organized by innumerable peoples. And furthermore, in this chapter, the digitized world exhibits networked patterns such as tweeting online, telecommunication patterns, online participations, and collective dynamics of human behaviors taking with a mobile phone. And without saying, those contents of intercommunications and other necessary factors should be also investigated in computational social science.

In this point, the era of social media has indeed envisaged either synchronous or asynchronous patterns for structural conformity in networking and collective

K. Shibuya, *Digital Transformation of Identity in the Age of Artificial Intelligence*, https://doi.org/10.1007/978-981-15-2248-2_10

dynamics online (Jordan 2001; Zhao et al. 2013; Chan 2014). It is certainly an extended pattern of public sphere (Habermas 1991). Those social trends could be concluded as a certain reason caused by ubiquitous and mobile computing which has been supported among the larger number of global citizens. Then, each individual now stands on own spaces in actual geolocation as well as interconnected with someone in virtual space (Papacharissi 2010; Candeago et al. 2019).

Otherwise, as it turns eyes into identity matter, at the time of 2018, the extent of global users (multiple counts) using social networking sites such as Facebook (globally 2.2 billion users), Twitter (globally 300 million users), and the others exceeds more than approximately 30% of the total population of citizens on the globe (globally 7.5 billion peoples at 2017). Whereas China holds just the total populations of 1.4 billion citizens in their homeland, those online networking are already seemed as one of "*pseudo-nations*." Those pseudo-national borders of larger networking-cluster are definitively independent beyond cultural and racial differences. Many citizens are still apt to only believe in group opinions among like-minded partners online. Such peoples include not only ordinary citizens but hiding antagonists and terrorists (e.g., The Islamic State of Iraq and Syria). Echo-chamber and filter-bubble phenomena may show an opinion divide between citizens, and their networked clustering become to be actualized the bit-nation government using blockchain technology. Which country will we stand on own identity grounds?

10.2 Synchronization in Networking

10.2.1 A Moment of Truth Against Reality Lost

Political propagandas and post-truth assertions by political leaders often call further populisms and nationalism. Such fake information often divides in-group opinions and fastens attitude of extremists in the nation. Recently, Hendricks and Vestergaard (2018) argued "*Reality Lost*." His "reality" here meant what we can understand the truth is. They especially articulated vague information sources such as fake news and post-truth, as well as opinion manipulations using social media at the political campaigns. It did not just discussions on the bases of Searle's meanings (1996). Misinformation and fake news have tremendous latent risks to linkage with malfunctions against democracy and identification of the citizens (Shibuya 2021 (in press)).

Lexically, Oxford dictionary defines "post-truth" as "*relating to or denoting circumstances in which objective facts are less influential in shaping public opinion than appeals to emotion and personal belief*". And Nielsen and Graves (2017) identified types of fake news such as satire, poor journalism, propaganda, some advertising, and false news. Similar concepts can enumerate stereotypes (Lippmann 1922), prejudice, and rumor. Moreover, it definitely claims that it also lacks falsifiability (Popper 1963) and it seems a kind of solipsism of each claimant.

"*The world is the totality of facts, not of things*" (Wittgenstein 1922). Fake news recently reminds us this renowned sentence. Wittgenstein also said that the world consists of a set of facts, and facts are determined by truth or false as a proposition style. Namely, he contemplated questions on our world cognition, provable logics, facts, and knowledge. However, to date, fake news and post-truth are too far from truth, facts, impartiality, and fairness. Any assertions of fake news are never valid and those contents are almost based on malicious intentions or misunderstandings. Moreover, fake news can spread faster than other reliable information through social networks (Doer et al. 2012; Vosoughi et al. 2018). Especially, these mediated information, which are recognized as fake news can be found in the fields such as politics, medical health, socioeconomics, ethnicity, ideology, gender, and other sensitive topics. Regarding populism, some governors demonstrate "self-first" actions and they sometimes use it as political propaganda against their opponents (Bobo 2017; Lamont et al. 2017; Wang et al. 2017).

The problem is the very clear. In the context of finite space, computable and manageable constraints (*Church-Turing thesis*: Copeland and Shagrir 2019), the core is similarly how to prove fact-checking on each counterfactual information and uncomfortable knowledge online (Viviani and Pasi 2017; Shao et al. 2018; Ciamapglia 2018). As through social media, our contemporary society has vulnerability against verification on existential fact and credibility of information sources. And of course, Facebook, Twitter, and other social media sites have already launched eligible regulations and managements in verifying submitted contents and fact-checking. However, there will be stumbling blocks against those objectives. These identifiable evidence verifications on the bases of inductive reasoning and "*evidence of absence*" are still harder for us.

At present 2019, Facebook officially collaborates with international governments (e.g., France), and they announced that their developed AI system can detect 65% of total amount of online problematic contents such as fake news and hate speeches (Johnson et al. 2019), but 35% of the residuals still remain not to be automatically enough detected yet (they also said that problematic image files can be nearly perfect detected by the AI). Subtle nuances of online contents can be harder interpreted by the AI, and then it should manually check them by the human experts.

Revealing fake news is very similar to somewhat detective story as well as scientific verifications (i.e., reviewing papers and reexaminations on published papers) in academics. For example, it enumerates *gathering facts, examining by empirical ways, estimating stochastic variables, confirming by crowd intelligence (e.g., witness), interacting with the suspicious persons, endorsing the evidences by the third parity authority,* and *detecting falseness*. Hence, in digitized society, there are greater needs to verify evidences and entrust quality management of information by the authority (please see Chaps. 5, 6 and 7 in this book).

In actual cases, there were empirical observation data obtained from the USA president election campaign at 2016. Before this voting day, according to a report conducted by Pew Research Center (2017), two-third of Americans got some news from social media including fake news. Further, for example, one of fake news "*Pope Francis shocks world, endorses Donald Trump for president, releases statement*" was supported by three-fourth (75%) of the total of Trump supporters, whereas his opponents were only 46%. It really exhibits in-group bias for interpretation consolidated with their political ideologies. If still yet, those who eagerly accept fake news such as delusional, anti-intellectual and anti-scientific believers may be still confused.

Democracy has been underlying in sound judgment and opinions among sovereigns (i.e., citizens). Machiavelli ironically said, "*Though the populace may be ignorant, it is capable of grasping the truth.*" However, this sentence should be ignored at once. When all the citizens do not have any methods to validate mediated vague information and defend their opinions, they are easily influenced by other influencers. Political propaganda and opinion manipulations with exposing malice neglect those process itself, and fake news and post-truth are actually vanguards to perturb our sound understanding of entire knowledge (DiResta et al. 2019; Howard et al. 2019; Shi et al. 2019).

10.2.2 Social Movement and Collective Protests All Over the Worlds

Synchronization by few opinion leaders with massive followers indicates not only their clustering pattern of social networking but actual social modality. Their patterns of inclusiveness and exclusiveness in social networking may reflect actual intergroup conflicts and prejudicial antagonisms against opponents (Golbeck and Hansen 2014; Huckfeldt et al. 2014; Bakshy et al. 2015; Bauer et al. 2018). Those are actually regarded as a style of social identification by in-group commitments and supporting specific opinion leaders. As results of these patterns, there is still anxiety that those mechanisms of social synchronization can potentially minimize varieties for our world views among citizens.

Certainly many reports have described that social media easily raise social movements or anti-autocratic revolutions by ordinary citizens. Those media have eagerly inspired ordinary citizens to participate in vigorous discussions as well as actual social movements (Casilli and Tubaro 2012; Choudhary et al. 2012). It seems readily apparent that participants indeed have shared further motivations and common goals.

In the below table, although the total population of participants could not be estimated accurately in each case, the entries exemplify serial movements (e.g., "Arab Spring" Jasmine and Egyptian revolutions, "Occupy Wall Street" movement, the umbrella revolution in Hong Kong, a series of protests against the Charlie Hebdo shooting at France, a series of protests against policy-making on the national security related bills in Japan, and other similar cases in various nations). Those who participated expressed their desires for democratic institutions, freedom of opinion expression, socioeconomic chances, and opposition against controversial political issues. They also made other demonstrative appeals. In details, on the other hands, it was obvious that many of participants had simultaneously different purposes, and each of participants eagerly appealed own discomforts against the government.

Especially, outcomes of the "Arab Spring" became the trigger of similar actions. Furthermore, in Arab cases, each political regime was overturned by numerous participating citizens using social media. Subsequently, the results spurred regional and global strain (Boening 2014; Sadiki 2015). Democratic progress in Tunisia was awarded a Nobel Prize in Peace at 2015, but their activities require more efforts.

As mentioned before, those collective phenomena might be categorized as the collective dynamics of behavior caused by numerous citizens (Le Bonn 1931a, b; Smelser 1962). The movements occasionally realized social influence and catastrophic collapse of regimes caused by the crowds. Here, the categories of those actions by citizens include opinion expressions, group processes, identification, emotional attitudes, fads, mobs, criminal riots, demonstrations, and politically assertive agitations as well as social change against ancient regimes caused by large groups of citizens in uprisings and revolution (Table 10.1) (Olson 1971; Brown 2000; Rohlinger and Snow 2003).

As it may notice soon, it should differentiate at least four patterns of participants. Namely, these are *hard core, only site participants, only SNS participants*, and *bystanders* crossed by two categories (Onsite Participation or SNS Participation). Ordinary participants can demonstrate with the collectives and tweets their opinions real time in social movement. As many of them can participate in social demonstrations taking with smartphone, these patterns cannot be often separated distinctively whether they use only SNS or not (Table 10.2).

1. Hard core: They engage in both participations to SNS and actual onsite demonstrations.
2. Only SNS participation: They are just eager to communicate with other sympathizers in online communities.
3. Only onsite participation: They are eagerly to attend the large meeting and demonstrations in public spaces, but they don't attend in social networking services.
4. Bystanders: They are indifferences against political and social movements, and they are not apt to attend mass meetings whether online or not.

In addition, these following reasons usually make it difficult to estimate accurately the modus of participants.

- Sock puppet: Creating multiple accounts of SNS. These are online camouflage identities, typically created by a person or group in order to propagate their own opinions.
- VPN (virtual private network) and other network specifications: These networking can make it difficult to be traced by the governmental authority. For example, these are interconnection tools using Tor (The Onion Router) and P2P-based direct messaging (e.g., fire-chat application).
- Geospatial constraints: Geospatial constraints sometimes make it difficult to estimate the modality of participants and their locations.

Table 10.1 A part of recent movements

Events	Date	Nations	Estimated max of participants	Main intention and goal of participants	Final results
Jasmine Revolution	Dec. 17, 2010–Jan. 14, 2011	Tunisia	100,000?	Political claims?	Revolution
Egyptian Revolution	Jan. 25, 2011–Feb.11, 2011	Egypt	More than 2 million people? (Tahrir square, Cairo)	Political claims?	Revolution
"Occupy wall street"	Sep. 17, 2011	USA	15,000? (Foley Square, New York)	Socioeconomic claims and occupying area	Failure and dissolution
The Umbrella Revolution	Sep. 26–Dec. 15, 2014	Hong Kong (China)	100,000? (nearby the admiralty)	Political claims on freedom for democracy and occupying area	Failure and dissolution
Protests against the Charlie Hebdo shooting	Jan. 7, 2015	France	35,000? (Paris)	Protests for public opinion	Temporal mass meeting
Greece Economic Crisis 2015	Jun.–Aug. 2015	Greece	30,000? (opposition side, Atene)	Socioeconomic reasons	Temporal mass meeting
Opposition against the policy making on the national security related bills	Aug. 30, 2015	Japan	120,000? (Around the National Diet, Tokyo)	Political claims and oppositions against controversial issues	Temporal mass meeting and oppositions
Catalan Nationalism for Independence	Sep. 11, 2015	Spain	More than a million people? (Català)	Political claims for independence	Ongoing matter. Participants repeatedly appeal the independence
Impeachment of Park Geun-hye, President of the South Korea (Lee and Oh 2018)	26 October 2016–11 March 2017	South Korea	More than 2.3 million people?	Political claims against president	Impeachment of president
Impeachment of Brazil presidents	2016–2017	Brazil	More than 3 million people?	Political claims against presidents	Impeachment of presidents (two cases)
Mouvement des Gilets jaunes	Nov. 17, 2018–?	France	More than hundred thousand people?	Political claims against fuel tax, labor issues, and other purposes.	Decreasing the total numbers of participants and dissolution?
Oppositions against the parliament of Hong Kong and Government of the China (mainland)	June. 2019–	Hong Kong	More than 2 million people?	Political claims and oppositions against the parliament of Hong Kong and the government of China (mainland)	Ongoing matter

Table 10.2 Four types of participating styles

	SNS participation	Not SNS participation
Onsite participation	Hard core of participants	Only onsite participation
Not onsite participation	Only SNS participation	Bystanders or never participation

- Usage of non-digitized tools: Many participants in larger demonstrations recently wear on facial masks preventing leakages of own identifiable information. They use non-digitized money and transportation ways, and further they often commit fraud of own identifications.

Recently, Ertugrul et al. (2019) attempted to endorse co-occurrence phenomena with both online and offline participations by statistical and spatiotemporal model, and they had been fitting data which obtained from actual movements to forecast events and movements. By observing behaviors of citizens at the smaller size, further study will clarify that machine learning discerns a possibility to forecast exponentially growing larger size of such participations.

10.2.3 Greater Disaster

The author exemplified another issue on disaster study using big data (Shibuya 2012, 2017b, 2018, 2021 (in press)). On 11 March 2011 (Japan time), many people had become witness for the Tohoku Earthquake Disaster and consequential accident of nuclear power plants at Fukushima in Japan. The magnitude of this Tohoku quakes was estimated as 9.0, and it ranked the fourth level among the world records (first case was magnitude 9.5 at Chile (1960), second case was magnitude 9.3 at Sumatra (2004), and third case was magnitude 9.2 at Alaska (1964)).

Still now, it presents a number of challenges indeed to overcome. As details are discussed at a later chapter, one of the hardest problems from the inventory of the Fukushima case was characterized as intensive purification of nuclear pollutions around Fukushima and collective migration of evacuees caused by both natural and human-made disasters (Science Council of Japan 2011; OECD/NEA 2016).

Since the aftermath of actual shocking events, it happened to synchronize with online users immediately. Especially, log data of Twitter had shown such synchronizing patterns among collective users during dynamic timeline (Li et al. 2014). Their services displayed an analyzed visualization of pulse patterns proposed by Twitter cooperation.[1] Global citizens cared about safety of Japanese citizens and sought to know the latest information by through Twitter. Recently, Mendoza et al. (2019) attempted to build nowcasting system using twitter log data at earthquake disaster, and their motivation might be rooted in such dynamics of networking.

[1] http://blog.twitter.com/2011/06/global-pulse.html (*currently, original page cannot be accessed).

Real-time analyzation based on social media can reveal people's tendencies with geospatial-tagged information and personal characteristics.

The most severe event is the accident of several nuclear power plants in Fukushima damaged by serial tsunami attacks, and it had completely lost both electric energy supply and total controls of vent and inner reactor. And lastly those systems had not prevented from meltdown and serious system failures.

After that, the author clarified online information seeking behavior sooner after the disaster. Below figure indicates a result of query a keyword "*Fukushima*" into Google Insights for Search (∗at the present, this service name alters to Google Trends; Ginsberg et al. 2009) at the end of 2011. It was clear that many of the citizens had googled a keyword "*Fukushima*," and their top level domains (i.e., it means national domain such as .uk, .jp, .cn, etc.) had suggested home countries of each viewer (but this domain information cannot perfectly match with accessed users who live in actual nation). Consequentially, this crisis affected the European countries such as Germany, Italy, Switzerland, and France (Topçu 2013). Consequentially, these countries abandoned to build nuclear power plants in their homeland.

As a result, it appears that the disaster of nuclear power plants caught much attention all over the world. It was too critical that these trend patterns by citizens' behaviors could be detected easier by national and municipal government at the Tohoku Disaster. Of course, Google, Yahoo! and other search engines could promptly provide their web-queried data to governments and rescue teams for their decision-making and practical solutions for victims and survivors. But actually these ideas were unrecognizable for them at the Tohoku Disaster (Fig. 10.1).

On the other hands, the most important was that collective civic engagements for helping victims by ordinary citizens were voluntarily observed in both online and offline (Shibuya 2012). And it should evaluate numerous donations, charities of gifts, information, and substances (Wilson 2000; Wang and Graddy 2008). At the end of June 2011, Japan government estimated that totally participations of Japanese volunteers were more than a half million. Thanks a lot to global volunteers and charities, Japanese already launched reconstructions from terrible ruins around Tohoku area (included Fukushima prefecture). As standing on the views of foreigners, those altruistic, gentle, and voluntary grassroot networking by Japanese citizens were often too miraculous to understand. As these voluntary networking represents "*Kiduna*," this invisible tie basically offered entire community resilience, identifications, commitments, and civic engagement of voluntary participations as well as firmly bonds.

This phenomenon exactly had shown some implications of the utilities on reciprocity and social capital among Japanese. As often said in rural communities, reciprocity custom has strong bonds tendency, which grounds on a variety of give-and-taking relationships. Reciprocal helping had been clarifying by past social surveys. For instance, Yoshino et al. (2009) reported that the answer rate of Yes was up to 56.8% in the item of "*Repaying people who have helped you in the past*" among Japanese. It is clear that reciprocity custom would be apt to bind the mind of recipients to reward for donor. These mutual reciprocal activities often engender the base of interweaving with members of communities. Taken together, as standing on

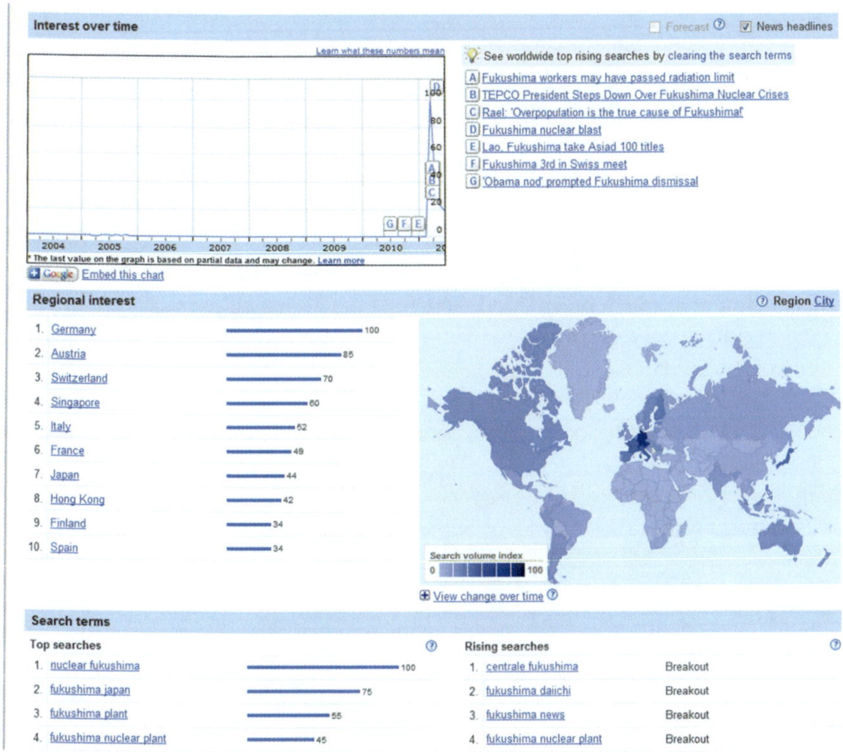

Fig. 10.1 A result of query by a keyword “Fukushima" into Google insight for search (a captured screen photo at the time of the end of 2011)

such social factors, spontaneous networking by voluntary participants was actualized by social capital and cooperative relationships among local community members. And these activities would undertake the social commons characterized as priceless value and standards in local community.

In this concern, biologists Nowak and colleagues already categorized five types on reciprocity such as *Direct, Indirect, Kin selection, Network*, and *Group Selection* (Nowak and Sigmund 2005; Nowak 2006). Those activities by many of Japanese spontaneously launched and organized social support networking, and it could be fitted for indirect and network reciprocity. It was certain that kinship reciprocity had influences for the community members, but online activities for volunteering have rich meanings on reciprocity in depth among participants. Rewarding incentives and motivations had not been triggered for their actions, and those social capitals have been equipped in larger citizens.

Here, the author summarizes the lessons on big data for the disaster research in digital social science (Shibuya 2012, 2017b, 2018, 2021 (in press)).

1. Statistics

   It should analyze correlational (and causal) relationships between actual and online activities by both ordinary citizens and survivors (as discussed later in Chap. 11).
2. Data Provisions for Rescue

   Google, Yahoo! and other search engines could promptly provide their web-queried data to governments and rescue teams for their decision-making and practical solutions for victims and survivors. The needs of survivors for their living became signaling for supports. How did survivors around Tohoku Area obtain their necessary substances? And what did they require since the end of the quake? Fortunately, according to Yahoo!'s data, it can easily make sense what the needs of survivors were.
3. Data Management and Archiving

   Data management and enduring archives should be properly considered (as described later in Chap. 12).
4. Timing for Support

It was too important that timing of supports for survivors could be matched for them.

Below next figure illustrates a concept of disaster management in the era of social media. It should entirely conduct disaster studies *before, after,* and *real time* as necessary. At each stage of the disaster, the quantity of needs is quite different. Before disaster stage, various case studies on scenarios for evacuees and governors' risk management should be examined. At real-time disaster stage, governor will be required to conducting immediately and simultaneously multiple rescue plans for ordinary citizens and evacuees being fronted disasters. AI-driven systems, ICT tools, and social media can be useful for information provisions if available at the disaster, and collective dynamics of evacuees should be arranged by nowcasting analytics (Widener et al. 2013). And after disaster stage, as soon as possible, there will be greater needs to decide the comprehensive plans to support victims and evacuees, and governors shall devote to steadily operating multitasks in order to recover citizens' daily life and the whole community. After that, it should analyze big-data social media usage patterns and posted data, and those historical trends will be operated by the clear goals to manage for the future disaster (Fig. 10.2).

In Japan, as innumerable applications of smart phone are freely available for ordinary citizens, when quakes tremendously occur, those who hold smart phone motivates to be localized and navigated toward the nearest evacuation area. Those big data have been accumulated and quantified to conquer against the future disasters.

However, governors should realize a necessity of the effective ways to inform necessary directions for the citizens. At recent disasters (e.g., quakes, typhoon disasters, and other natural hazards), information provision ways and timing for evacuations among citizens in community have been criticized. Such information provision and warning announcements for immediate evacuation, which directed by the authority cannot be often obeyed by many of the ordinary citizens. One of the

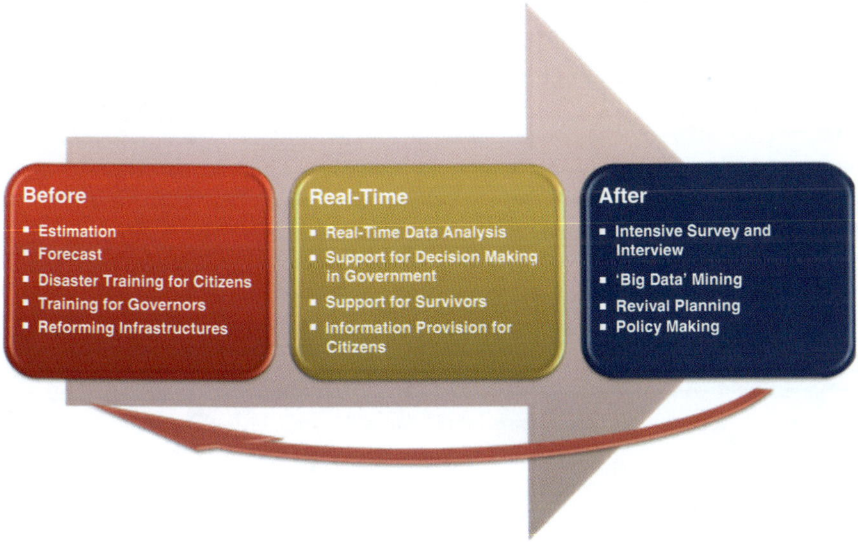

Fig. 10.2 A concept on disaster management process

reasons was to consider as cognitive psychological bias in each person, and many of us sought to believe own normal daily surrounding will be continued nevertheless they faced the critical danger against their life at the disaster. As Japanese has rich experiences in natural disasters, many of them are usually apt to behave by own judgment. As results, such self-judgment might lose their crucial chances to evacuate in advance. In addition, some of the people (especially, elder persons and disabilities) felt something troublesome to evacuate, and intended to stay their home. Actually, there were massive citizens who did not evacuate in each community, and consequently there were many deceased persons in multiple cases.

10.3 Influencers' Power in Social Networking

It cannot neglect much social influence by the influencer and opinion leader online (Watts and Dodds 2007; Nash et al. 2013). As such social influence in social networking sites has been traced by massive studies on social media usages in terms of computational social sciences and social psychology (MacCoun 2012; Malinick et al. 2013; Dolata 2018). For example, Weinberger (2011) reported that computational social science studies successfully revealed latent risks on networking patterns among terrorists, and consequently the most influential leaders who intended to conduct severe terrorisms had been arrested by the authority of the USA before committing terrorism.

10.3.1 *Social Influences of Power*

Generally speaking, a concept on social power consists of vaguer conceptual meanings in social psychology and relative fields (Cook et al. 1983, 1988; Bozzo and Franceschet 2016). Social powers for the purpose of obtaining social resources can implicitly and explicitly affect enormous social activities. In social psychology, Claire and Fiske (1998) discussed about social power in social interaction based on stereotypic belief among dyad or more members.

Similarly, sociological explorations had been investigating social power and its determinants for social capitals and social relationships. For example, Coleman (1974, 1990) and Blau (1998) attained their original ways along with establishing a part of social foundations. Especially, their activities were based on mathematical ways and network analysis in sociology. Their formalization on the social power was to be articulated and sophisticated by game theory, network analysis, and rational choice theory. Reciprocity and inequality in social justification and legitimatizing process could be regarded as more significant in social relationships (Cook et al. 1983, 1988). These studies could find the importance of centrality of network to distribute various social resources. And they discussed that social power and its dominant relationships are crucial factor for the sake of organizing social structural relationships in group.

According to Deutsch and Coleman (2000), power can be enumerated such as *coercive power*, *cooperative power*, *powerless and dependent*, *empowerment,* and *independent*. Their mentions aimed to investigate on social conflicts and its resolution in various social contexts. These factors of social power and powerless are an indicator in relation to individual condition in social status and social group. On the other hand, such networking has certainly potentials to engender dominance relationships, and socioeconomic emergent structure which can be defined as *network externality* (Katz and Shapiro 1985) and *scale-free network* (Barabási 2016) would become more influential power among stakeholders.

10.3.2 *Power Holder and Followers*

To date, by through intercommunications and social media, followers are massively eager to commit and support for opinion leaders. Here it should discern between preference and prejudice. Assistance for opinion and activity of such leader may be favorably depended on each preference of followers. However, social phenomena gradually accumulated by such many followers' preferences will further engender the evidence which is mostly supported by many like-minded citizens, and lastly it excludes present and latent opponents and antagonists apart (Bauer et al. 2018). And then such social condition should be realized as prejudice and stereotypes. Networked identification namely should be reconsidered in the context of influential power, preference, and prejudicial process.

The power of social influence was already clarified by social psychological studies (Claire and Fiske 1998). Their studies indicate significance of social power as influence under interpersonal communication and its confirmation process including alternative standpoint of third person. And social cognitive aspects can help contagious influences from power holders. This social power has a potential to intensify or reduce dominance for both social relationship and shared stereotypic belief. Operario et al. (1998) examined that social power has social and personal control for either activating or inhibiting stereotypic cognition and information processing. And Goodwin et al. (2000) exemplified that power holders are apt to pay attention to opponents' stereotypic attributes and prediction. Then, it is certain that social power can determine to activate or inhibit stereotypic understanding in human relationships.

Social power enacts power holders to obtain the legitimatizing conditions for dominance and regulation of social recourses. When power holders have persistent stereotypic cognition against some members, they often require them to become dependent followers and maintain own dominant relationships. Stereotype and prejudice are almost always considered as cognitive processes for interpretation and understanding of reference frameworks of the perceived social facts. In this concern, social power cannot only coerce cognitive functions to adjust in-group norms (Christensen et al. 2004), but also it encourages group members to aspirate gaining higher status of socioeconomic success. Group status and position of relationships are critically significant meanings for belonged members, and then they would be inspired by the possibility of obtaining powerful relationships. If they wish to attain higher success, they will also follow more power holders. Likewise, they hope to calibrate and reconnect with other nodes of people for the purpose of getting acquaintanceships in order to fulfill what they seek in another chance.

In addition, past studies on social power in social psychology sometimes approached to social influence, which consists of social persuasion and conformity. The social cognitive approach had yet to analyze social influence effectively in 1990s (Mackie and Skelly 1994), but the important notice is that interpersonal communication and influential media have ubiquitously much potential to influence enormous online members. Social networks in acquaintanceships and social community can cultivate social reality and cognitive construal, and then social consensual meaning of the reality will be growing. This implicit power as social influence cannot be neglected, and we should pay attention to this latent social structure.

On the other hand, as mentioned before, it should be noticed that social relationships and communication among members will be certainly underlying in reduction of uncertainty (Gudykunst 1995). Aforementioned stereotypic belief against unfamiliar others will be depended on uncertainty which is impossibility to predict their behaviors in social contexts. In other words, social power for in-group members would define a way of understanding to reduce social uncertainty in such social situations.

10.3.3 Algorithm

10.3.3.1 Networked Clusters

There are some algorithms which detect one of the most influential persons have been investigating in social network studies (Wasserman and Faust 1994). Especially, Newman and Kleinberg's works have established one of breakthrough for further understandings (Newman 2010; Easley and Kleinberg 2010). Those patterns are usually measured by network centrality, network betweenness, and other necessary indices (Bozzo and Franceschet 2016; Skibski et al. 2019).

Further, exploring social structure in social networking sites, there is a need to divide between opinion groups clustering with massive members. Girvan–Newman method is one of the representative methods, their algorithm's steps for community detection are summarized below (Girvan and Newman 2002; Newman 2010).

1. The betweenness of all existing edges in the network is firstly calculated.
2. The edge with the highest betweenness is removed.
3. The betweenness of all edges affected by the removal is recalculated.
4. Steps 2 and 3 are repeated until no edges remain.

10.3.3.2 Homophily Model

Bakshy et al. (2015) articulated that ordinary people have the strong dispositions based on biased approaching for specific opinion groups. They intentionally prefer to access website and online resources based on their ideologies and attitudes. Hence, at last, their opinion group clustered by intensification and motivation among participants may distinguish from other groups. Regarding those backgrounds, Stovel and Fountain (2009) modeled one of the clustered networks among two groups.

$$P_{ij} = (1-2p)\left((1-\theta)H_{ij} + \theta\left(1-H_{ij}\right)\right) + p \tag{10.1}$$

Here, it considers some social organizations such as employers and employees. Here, θ represents the strength against network weight, H_{ij} denotes some special characteristics shared between i and j. Variable p is the probability (0–1) to be connected to each other. As θ approaches 1.0, the mutual distances become isolated, and when they are 0.5, they become closest to each other and approach the Bernoulli random graph structure. Of course, this property is applicable as long as the graph structure can be taken.

10.3.3.3 Convergence of Opinions

In social sciences, there are computational models on opinions dynamics of influence among heterogamous members. There are, for example, threshold models (Granovetter 1978; MacCoun 2012), averaging models, and networked models using mathematical or agent-based simulation.

Principally, Helbing (2012) reported a mathematical model on opinion dynamics, which underlies in the *theory of integrating forces* proposed by sociologist Durkheim. In this simulation, each agent has incentives to adapt in specific opinion which weighted means ($O_{j(t)}$) at the time$_t$ interacting with other agent$_j$. Lastly, in this way, this social progress converges to a single opinion among agents, and it organizes homophily clustering. When an opinion of agent$_i$ is more influential than agent$_j$, opinion distance (d_{ij}) will minimize its value as the following formula suggests.

$$\mathrm{W}_{ij}(t) = e^{-d_{ij}(t)/A} = e^{-|o_j(t)-o_i(t)|/A} \tag{10.2}$$

where, variable A denotes a range of influential power of each agent. When this value is smaller, and it means that this agent strongly commits to specific opinion. But if not so, as Durkheim's theory presumes, each agent can alter to another opinion. And $\zeta_{i(t)}$ defines white-noise, and it plays a role of disintegrating disturbance factor in opinion dynamics.

$$\Delta \mathrm{o}_i = \frac{\sum_{\substack{j=1 \\ j\neq i}}^{N} \left(o_j(t) - o_i(t)\right) w_{ij}(t)}{\sum_{\substack{j=1 \\ j\neq i}}^{N} w_{ij}(t)} + \zeta_i(\mathrm{t}) \tag{10.3}$$

Thus, as above configured model, Helbing concluded that it successfully found a process on opinion formation and consensual integrations among heterogeneous members. Apparently these equations seem to be complicated, but it identifies a minimization of distances between heterogeneous opinions. But actually this model was similar to previous studies. For example, Kozma and Barrat (2008) critically examined a mathematical physics model on opinion formation process defined by Deffuant et al. (2000). Otherwise, as Amblard and Deffuant (2004) also simulated the opinion dynamics including *Extremists* agent as a representation of extreme opinion claimers, and they verified disturbance and converging differences in opinion dynamics. Therefore, there are several differences in each model, but their models seek consensual opinion and minimize differences of attitudes among members.

On the other hands, by social simulation of ABM, Hegselmann and Krause (2002) approached opinion dynamics in terms of averaging opinions among interconnected agents. By through collectively conforming to each opinion, their total direction of the opinion gradually nears to the consensual state in society.

$$\overline{x_0} \triangleq \frac{1}{N}\sum_{k=1}^{N} x_0(k) \tag{10.4}$$

Those models are suggestive for understanding in-group opinion formation and internetworked opinion converging process. In actual reasons, because clustering process with like-minded members within opinion group has been observed online through social media, and it can be managed as representatives of bottom-up process using ABM. And it examines how localized optimum of the opinion, which interacted among adjoined interconnected agents, can possibly enlarge to the wholly optimum as consensus in the networking society.

10.3.3.4 Percolation in Spatial Graph

Percolation phenomena can be widely observed as a variety of contagion, collapse, and self-organized clustering patterns in spatiotemporal geometric field (Broadbent and Hammersley 1957; Baggag et al. 2018; Shibuya 2012, 2017a). So far, percolation had been studying in mathematics, statistical physics, and chemistry as well as computational complexity sciences (Hoshen and Kopelman 1976; Newman and Park 2003; Newman and Ziff 2001; Malarz and Galam 2005). As a matter of fact, in the fields of social sciences, many of the social patterns are expectedly obeyed in this natural law of percolation. Social influential contagion of opinions spreading through SNS networking is a good example in recently, and its mathematical characteristics have potentials to become better adaptive models for understanding.

In the most fundamental case of a two-dimensional square lattice and a vicinity of graph structure, there are two cases such as *bond* percolation (interconnected edges or not) and *site* percolation (spatial adjacent nodes are occupied or not). In addition, each network pattern has own specific critical point (threshold) for the *phase transition*, and it indicates the borderline between clustered organization (or disruption) of networked clustering in spatiotemporal field. At below table, it exemplified some patterns of percolation and its criteria for phase transitions.

It is too critical to examine spatiotemporal patterns of clustering and geometric conditions. The square lattice is characterized as a pattern of grid matrix (e.g., chessboard). Except for the Bethe and binary lattice, percolation criterion of many networks including the Square lattice could be calculated and estimated approximately by only statistical physics and computational manners (Table 10.3) (Malarz and Galam 2005).

Otherwise the Bethe lattice identifies an infinite radial spreading pattern and depicts a hierarchical tree graph structure (more than z = 3: At the conditions of z = 2, this case can be treated as another model (i.e., binary tree model)). Each edge restricts not to interlink across edges stemmed from other branches, and there is not any bypass and closed loops. The Bethe lattice is defined by the formula P_c

Table 10.3 Some types of percolation and its criterion

Example types	z: Edges of each vertex	Site percolation criterion	Bond percolation criterion
Binary	Root is 3, branch is 2	0.50	0.50
Bethe lattice	Arbitrary number more than 3	1/(z – 1)	1/(z – 1)
Square lattice	4 (Neumann)	0.592	0.50
Square lattice	8 (Moore)	0.407	*

*Value was not specified (Malarz and Galam 2005)

(Percolation criterion) = $1/(z - 1)$, and z indicates edge numbers (branches) deriving from each node. For example, if z is 3, its percolation criterion equals 0.5. And in this case of the Bethe lattice, both bond and site types of percolation can be calculated strictly by this formula. For example, for the purpose of applying disaster education, the author recommended a Pay-It-Forward networking model on distributing relief supports and collaborations in widely area because such models can be formalized by the Bethe graphs (Shibuya 2012). Below figure displays a contagion phenomenon on the surface of visualized sphere by drawing patterns of the Bethe lattice (Fig. 10.3).

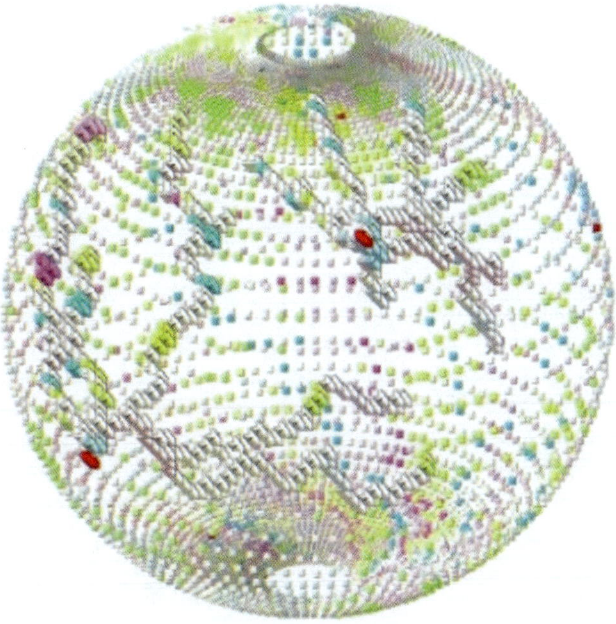

Fig. 10.3 This example of simulation represents information contagion phenomena on the Bethe lattice

Therefore, percolation in social network can be applied for collective clustering process on opinion dynamics. In the context of this, the phase transition in percolation principle can be regarded as an emergence of the majority as clustering group in finite networks, and after that, it expects that this cluster exponentially accelerates its social influence within whole society. Social processes ruled by percolation principles can exemplify such as contagion patterns of fake news and rumors on social media, clustering patterns of distributions, innovation progress (Helbing 2012), systemic risks (e.g., financial collapse, disease infections) and diffusing process of flows in spatial modeling (O'Sullivan and Perry 2013).

10.4 Estimating Activities Online

Measurement for the large amount of online collective participations has been required in various social situations. As it estimates clustering patterns, and there are mathematical and physical models such as social network theory (Granovetter 1978; Newman 2010; Easley and Kleinberg 2010; Nash et al. 2013), social physics (Pentland 2014; Goncalves and Perra 2015), and spatial informatics (O'Sullivan and Perry 2013).

However, as mentioned before, recent social movements have been observed, but the total amount of online participations and their modal activities could not be accurately estimated. For pursuing such needs, it enumerates mathematical models, statistical estimations, AI technologies (i.e., pattern recognition), geospatial information system (GIS), remote sensing, and other ICT-based technologies, for instances. Then, hereafter, the author picked up some measuring models and methodologies.

10.4.1 Wireless Sensing System

Measuring collective dynamics of people indoor has been possibly implemented and worked by wireless sensor networking (e.g., Wifi, Bluetooth, RFID, etc.) and IoT-based infrastructures (Alberto et al. 2010). It can be coordinated with GIS of user's taking smartphone, and context-based computational services have been now applicable anywhere. A stochastic model of the Bayesian network may be used to predict human needs and behavior (Castro et al. 2001; Shibuya 2004). These functions could also be used for practical services, especially when there are more physical constraints in indoor situations than in outdoor situations.

In contrast, population estimation methods using wireless communication technology have been developed (Botta et al. 2015). For example, there is a technique to estimate how many people are on the other side of a wall by irradiating Wifi across a wall (Barbera et al. 2013). They estimated the number of people by scrutinizing the data that bounced back by irradiating the electromagnetic waves of Wifi. However, since Wifi's electromagnetic waves are close to the frequency of infrared rays emitted by microwave ovens, there is another risk of physical interference (i.e., 2.4 GHz). Even if there are physical obstacles, there is a possibility to open up the availability of estimating the number of people.

10.4.2 Image Analysis by Artificial Intelligence

As mentioned before, thanks to dramatic improvements in the AI technology (e.g., image and video analysis), these techniques mechanically allow us to dynamically monitor and detect each person in a specific area (Tistarelli et al. 2009; Massimo and Christophe 2017). Therefore, it is possible to dynamically determine the number of people in target images and video clips.

10.4.3 Remote Sensing

According to Yang (2011), *remote sensing* can dynamically monitor whole lands from the artificial space satellites, and it estimates total residents in specific areas, which is based on observed data of the number of dwellings. They reexamined seven population estimation models as "*persons per household*" around the target area through satellite remote sensing. If it assumes that each household has own family members, it can lastly expect total population within community. But the precisions of estimations would be fluctuated and varied by assumed parameters.

Further, another type of remote sensing should be considered. It means the horizontal way via networking rather than the vertical way by artificial space satellites. XaaS-based interconnections in social platform such as the smart city and digitized smart house can be traced by external observation not limited to satellite-based sensing. Those data can be assumed to estimate total population in each household by through revealing their daily rhythms and energy consumptions of private activities. Connected vehicles can be similarly offered to notice for the purpose of analyzation by such ways. Those canonical external usages among interconnected houses and vehicles would intend to be principally operated and optimized electric energy generations and allocations within smarter city by smart-grid, and sequentially such big data within smart city has possibility to be measured for forecasting needs and activities in each household and auto-driving car. However, the extent of energy consumptions and interconnected data of wireless networking in each household has another vulnerability to unintentionally leak privacy information to external malicious purposes. Thus their goals must be cared about higher security.

10.4.4 Estimation of Online Participants

In the case of Wifi wireless communication, by examining the number of connections managed by a specific router or a relaying base station, it may be possible to estimate how many people exist within the reach of radio waves. Besides, ad hoc mode and mesh structure by Wifi wireless networking have advantages for direct interconnecting with each user (Johansson and Jäntti 2011).

Otherwise, He and Lin (2017) have measured the degree of collective attention such as burst-like increase in tweet numbers on shocking events by online data distilled from social media patterns. Analyzing those patterns, total participants online will be realized corresponding with specific contexts.

10.5 An Aspect of National Identity

10.5.1 Emerging Bit-Nation

On the other hand, it picks up another problem. There are some cases to socially reconsider own identity based on nationality and conflict resolution mapping between the actual nation and bit-nation (i.e., online virtual nation using blockchain technology). For example, Estonia (total population: 1.31 million people at 2017) and Catalonia (a state of the Spain: total population: 7.44 million at 2017) have plans to achieve their independent virtual nation using blockchains and advanced information managements. They firstly intend to apply blockchain technology for finance, tax, and government digital documentation. And they lastly establish their virtual nation defined as the style of bit-nation. Then, will they differentiate their identity between actual nation and no-bordered virtual nation? Or will they vigorously still hold on their identity which based on the actual regime and bordered nation?

10.5.2 Digital Transformation Against the National Border

The emergence of the bit-nation possibly forces to reshape our actual worlds. There are some cases to realize ambiguous identity conditions mapping between the actual nation and aforementioned bit-nation. As noted above, Estonia and Catalonia have effective policies to achieve their electric governances using blockchains and advanced information managements.

On the other hands, global companies will erode the actual borders and frameworks of ancient regime. Actually, at June 2019, Facebook announced that they launch own cryptocurrency (*Libra*: a kind of electric currency using blockchain technology, refer to Chap. 5). But they are facing at the stumbling blocks offended by the established nations and international institutions. The reasons are not only presumptions against cracking possibility of global financial trades. Their project has latently both meaning of cutting-edge and land-breaking matters against global and domestic governance, because Facebook user accounts already exceed more than 2 billion and its larger online economic zone will appear beyond the actual international borders. Comparatively, at the present 2019, TPP (trans-pacific partnership) economic zone within participated 11 nations has only total populations as 500 million people and GDP based economics as approximately 11 trillion US dollars. In EU case, at the present 2018, euro currency zone within 19 nations was estimated as only totally 300 million citizens and a billion euros (approximately 13 trillion US dollars). Therefore, probably, other GAFA companies follow its trends. Two billion populations of Facebook can be indeed said as an actual nation (or more than a nation?). Namely, they globally hold citizens, own online networking, and they will handle own electric currency for financial transaction online. When Facebook will further register citizenship, offer daily life supports, manage own taxation, and legislate own laws and institutions, traditional nation regimes and its governances will decay sooner or later. Then, their activities steadily approach to the *genuine* nation (not virtual nation) in actually.

As a result of this, autonomous, interconnected, self-regarding, and virtual nation regime might be self-organized by larger dynamics of participatory citizens exiting from older establishments (ancient-regimes) and untangling the existing national borders. Such dynamism will further engender new enclaves clustering by each race, ethnicity, cultural backgrounds, socioeconomics, ideology, and other reasons in each possible world (Hayek 1945). Those interconnected nexus among sympathizers in each cluster will be apt to gain the extent of legitimacy of their assertions and activities.

To keep those contexts in mind, as standing on the present and future perspective, it should step further reconsidering what identity is once more. Namely, it should not be managed with only ideology matters. Rather it refines as a problem on how to redefine an identity including complicated ontological issues in digitized society. Given AI and big data enhances further possibilities for our living styles, those who entirely immersed into the virtual worlds might ambiguously alter own boundary between self and the others sharing interconnected reality and experiences. Further, it catalyzes the patterns what their existence will reform to be composited of belonging awareness of the citizenships, race, self-definition, and other social aspects of identity.

10.5.3 Rewiring International Orders

National border might be already contextualized to reshape and transform to human-made clustering networks crossing globally like-minded partners rather than imaginary geospatial borderline. It calls opinion-divide among committing citizens. In these conditions, citizens have blindly tendencies to believe in information, which apparently accepted by innumerable followers online. Then vulnerabilities still remains to confuse any evidences. Those who support mediated fake news desire to believe only self-regarding "fact" and see "fact" as they want are likely to organize their networking with like-minded partners ("*echo chamber*": Sunstein 2001).

Consequently, as social media empower vigorous networking by ordinary citizens and their numerous opportunities against political and ethnic opponents (Corner 2017; Jost et al. 2018), those cyber politics opened the door for emerging another types of public opinion formation (Hill and Hughes 1998; Choucri 2012; Shibuya 2017a). They have dispositions to neglect both objective verification and sufficient fact checks on fake news, and further they will interconnect with like-minded partners and repel their opponents (Bakshy et al. 2015). Gradually, this polarization between opinion groups certainly organizes their ideology as a part of the real world. In this way, when majority who eagerly desire to support extremist's demagogy gains total legitimacy in whole democratic society, it will be time to encompass the political power by the name of the democracy. As a result of this, their nation may lose a compass for governing direction.

During the age of the cold war, Morgenthau (1978) had forecasted our future of world conditions. When ordinary citizens harden their attitudes for nationalism and exclusionism by globally intermediated communication tools (global opinion or world opinion), with enthusiasm, it will be inevitably to next world war in the future. His telling future is our present time, and political demagogues in his era can be regarded as fake news in our era. His foretelling clearly said the negative facet of social media?

References

Alberto, B., Maurice, H., & Mikael, V. J. (2010). *Networked control systems*. London: Springer.

Amblard, F., & Deffuant, G. (2004). The role of network topology on extremism propagation with the relative agreement opinion dynamics. *Physica A, 343*, 725–738.

Baggag, A., et al. (2018). Resilience analytics: Coverage and robustness in multi-modal transportation networks. *EPJ Data Science, 7*, 14. https://doi.org/10.1140/epjds/s13688-018-0139-7.

Bakshy, E., Messing, S., & Adamic, L. A. (2015). Exposure to ideologically diverse news and opinion on Facebook. *Science, 348*(6239), 1130–1132. https://doi.org/10.1126/science.aaa1160.

Barabási, A.-L. (2016). *Network science*. Cambridge, UK: Cambridge University Press.

Barbera, M. V., et al. (2013) Signals from the crowd: Uncovering social relationships through smartphone probes. http://conferences.sigcomm.org/imc/2013/papers/imc148-barberaSP106.pdf

Bauer, M., et al. (2018). Social contagion of ethnic hostility. *PNAS, 115*(19), 4881–4886.

Blau, P. M. (1998). *Exchange & power in social life*. Piscataway, NJ: Transaction Publishers.

Bobo, L. D. (2017). Racism in Trump's America reflections on culture, sociology, and the 2016 US presidential election. *The British Journal of Sociology, 68*(S1), S85–S104. https://doi.org/10.1111/1468-4446.12324.

Boening, A. (2014). *The Arab Spring: Re-balancing the greater Euro-Mediterranean?* Cham: Springer.

Botta, F., et al. (2015) Quantifying crowd size with mobile phone and Twitter data. *Royal Society Open Science*. http://rsos.royalsocietypublishing.org/content/2/5/150162

Bozzo, E., & Franceschet, M. (2016). A theory on power in networks. *Communication of the ACM, 59*(11), 75–83.

Broadbent, S. R., & Hammersley, J. M. (1957). Percolation processes. *Mathematical Proceedings of the Cambridge Philosophical Society*. https://doi.org/10.1017/S0305004100032680.

Brown, R. (2000). *Group processes* (2nd ed.). Oxford, UK: Blackwell.

Candeago, L., Bertagnolli, G., Bosetti, P., Vescovi, M., Sacco, F., & Lepri, B. (2019). Cities of a feather flock together: A study on the synchronization of communication between Italian cities. *EPJ Data Science, 8*, 19. https://doi.org/10.1140/epjds/s13688-019-0198-4.

Casilli, A. A., & Tubaro, R. (2012). Social media censorship in times of political unrest - A social simulation experiment with the UK riots. *Bulletin de Methodologie Sociologique, 115*(1), 5–20. https://doi.org/10.1177/0759106312445697.

Castro, P., Chiu, P., Kremenek, T., & Muntz, R. (2001). *A probabilistic room location service for wireless networked environments*. Atlanta, GA: Ubiquitous Computing. http://godfather.cs.ucla.edu/publications/pdf/ubicomp01.zip.

Chan, M. (2014). Social identity gratifications of social network sites and their impact on collective action participation. *Asian Journal of Social Psychology, 17*(3), 229–235.

Choucri, N. (2012). *Cyberpolitics in international relations*. Cambridge, MA: The MIT Press.

Choudhary, A., et al. (2012). Social media evolution of the Egyptian revolution. *Communications of the ACM, 55*(5), 74–80.

Christensen, P. N., Rothgerber, H., Wood, W., & Matz, D. C. (2004). Social norms and identity relevance: A motivational approach to normative behavior. *Personality and Social Psychology Bulletin, 30*(10), 1295–1309.

Ciamapglia, G. L. (2018). Fighting fake news: A role for computational social science in the fight against digital misinformation. *Journal of Computational Social Science, 1*, 147–153.

Claire, T., & Fiske, S. T. (1998). A systemic view of behavioral confirmation. In C. Sedikides et al. (Eds.), *Intergroup cognition and intergroup behavior*. Boca Raton, FL: LEA.

Coleman, J. S. (1974). *Power and the structure of society*. New York: Norton.

Coleman, J. S. (1990). *Foundations of social theory*. Cambridge, MA: Berknap Press of University of Harvard Press.

Cook, K. S., Emerson, R. M., Gillmore, M. R., & Yamagishi, T. (1983). The distribution of power in exchange networks: Theory and experimental results. *American Journal of Sociology, 89*, 275–305.

Cook, K. S., Hegtvedt, K. A., & Yamagishi, T. (1988). Structural inequality, legitimation and reactions to inequity in exchange networks. In M. Webster & M. Foschi (Eds.), *Status generalization: New theory and research*. Palo Alto, CA: Stanford University Press.

Copeland, B. J., & Shagrir, O. (2019). The church-turing thesis: Logical limit or breachable barrier? *Communications of the ACM, 62*(1), 66–74.

Corner, J. (2017). Fake news, post-truth and media–political change. *Media, Culture & Society, 39*(7), 1100–1107.

Deffuant, G., Neau, D., Amblard, F., & Weisbuch, G. (2000). Mixing beliefs among interacting agents. *Advances in Complex Systems, 3*, 87–98.

Deutsch, M., & Coleman, P. T. (Eds.). (2000). *The handbook of conflict resolution theory and practice*. San Francisco, CA: Jossey-Bass Publisher.

DiResta, R., et al. (2019). The tactics & tropes of the internet research agency. https://disinformationreport.blob.core.windows.net/disinformation-report/NewKnowledge-Disinformation-Report-Whitepaper.pdf

Doer, B., Fouz, M., & Friedrich, T. (2012). Why rumors spread so quickly in social networks. *Communications of the ACM, 55*(6), 70–75.

Dolata, U. (Ed.). (2018). *Collectivity and power on the internet: A sociological perspective (Springer briefs in sociology)*. Cham: Springer.

Easley, D., & Kleinberg, J. (2010). *Networks, crowds, and markets: Reasoning about a highly connected world*. Cambridge, UK: Cambridge University Press.

Ertugrul, A. M., Lin, Y.-R., Chung, M. T., Yan, M., & Li, A. (2019). Activism via attention: Interpretable spatiotemporal learning to forecast protest activities. *EPJ Data Science, 8*, 5. https://doi.org/10.1140/epjds/s13688-019-0183-y.

Ginsberg, J., et al. (2009). Detecting influenza epidemics using search engine query data. *Nature, 457*, 1012–1014.

Girvan, M., & Newman, M. E. J. (2002). Community structure in social and biological networks. *Proceedings of National Academy Science, 99*(12), 7821–7826.

Golbeck, J., & Hansen, D. (2014). A method for computing political preference among Twitter followers. *Social Networks, 36*, 177–184.

Goncalves, B., & Perra, N. (Eds.). (2015). *Social phenomena: From data analysis to models (Computational social sciences)*. Cham: Springer.

Goodwin, S. A., Gubin, A., Fiske, S. T., & Yzerbyt, V. Y. (2000). Power can bias impression processes: Stereotyping: Subordinates by default and by design. *Group Processes Intergroup Relations, 3*, 227–256.

Granovetter, M. S. (1978). Threshold models of collective behavior. *American Journal of Sociology, 78*(6), 1420–1443.

Gudykunst, W. B. (1995). Anxiety/uncertainty management theory. In R. Wiseman (Ed.), *Intercultural communication theory*. Thousand Oaks, CA: SAGE.

Habermas, J. (1991). *The structural transformation of the public sphere*. Cambridge, MA: MIT Press.

Hayek, F. A. (1945). The use of knowledge in society. *The American Economic Review, 35*(4), 519–530.

He, X., & Lin, Y.-R. (2017). Measuring and monitoring collective attention during shocking events. *EPJ Data Science, 6*, 30. https://doi.org/10.1140/epjds/s13688-017-0126-4.

Hegselmann, R., & Krause, U. (2002). Opinion dynamics and bounded confidence: Models, analysis and simulation. *Journal of Artificial Social Simulation, 5*, 3. http://jasss.soc.surrey.ac.uk/5/3/2.html.

Helbing, D. (2012). *Social self-organization*. Cham: Springer.

Hendricks, V. F., & Vestergaard, M. (Eds.). (2018). *Reality lost: Markets of attention, misinformation and manipulation*. Cham: Springer.

Hill, K. A., & Hughes, J. E. (1998). *Cyberpolitics: Citizen activism in the age of the internet*. Lanham, ML: Rowman & Littlefield Publishers.

Hoshen, J., & Kopelman, R. (1976). Percolation and cluster distribution. I. Cluster multiple labeling technique and critical concentration algorithm. *Physical Review B, 14*, 3438.

Howard, D., et al. (2019). Evolving embodied intelligence from materials to machines. *Nature Machine Intelligence, 1*, 12–19.

Huckfeldt, R., et al. (2014). Noise, bias, and expertise in political communication networks. *Social Networks, 36*, 110–121.

Johansson, M., & Jäntti, R. (2011). Wireless networking for control: Technologies and models. In *Networked control systems* (Vol. 406, pp. 31–74). London: Springer.

Johnson, N. F., et al. (2019). Hidden resilience and adaptive dynamics of the global online hate ecology. *Nature*. https://doi.org/10.1038/s41586-019-1494-7.

Jordan, T. (2001). Language and libertarianism: The politics of cyberculture and the culture of cyberpolitics. *The Sociological Review, 49*(1), 1–17.

Jost, J. T., et al. (2018). How social media facilitates political protest: Information, motivation, and social network. *Advances in Political Psychology, 39*(1), 85–118.

Katz, M. L., & Shapiro, C. (1985). Network externalities, competition and compatibility. *The American Economic Review, 75*(3), 424–444.

Kozma, B., & Barrat, A. (2008). Consensus formation on adaptive networks. *Physical Review, E, 77*, 016102.

Lamont, M., et al. (2017). Trump's electoral speeches and his appeal to the American white working class. *The British Journal of Sociology, 68*(S1), S153–S180. https://doi.org/10.1111/1468-4446.12315.

Le Bonn, G. (1931a). *The crowd: Study of the popular mind*. iBook version. (Apple iBook Store).

Le Bonn, G. (1931b). *The psychology of revolution*. iBook version. (Apple iBook Store).

Lee, J., & Oh, J. J. (2018). What motivates a citizen to take the initiative in e-participation?: The case of a south Korean parliamentary hearing. *Communications of the ACM, 61*(12), 56–61.

Li, J., Vishwanath, A., & Rao, H. R. (2014). Retweeting the Fukushima nuclear radiation disaster. *Communication of the ACM, 57*(1), 78–85.

Lippmann, W. (1922). *Public opinion*. San Diego, CA: Harcourt Brace and Company.

MacCoun, R. J. (2012). The burden of social proof: Shared thresholds and social influence. *Psychological Review, 119*(2), 345–372.

Mackie, D. M., & Skelly, J. J. (1994). The social cognition analysis of social influence: Contributions to the understanding of persuasion and conformity. In P. G. Devine, D. L. Hamilton, & T. M. Ostrom (Eds.), *Social cognition: Impact on social psychology*. Cambridge, MA: Academic.

Malarz, K., & Galam, M. (2005). Square-lattice site percolation at increasing ranges of neighbor bonds. *Physical Review E, 71*, 016125.

Malinick, T. E., et al. (2013). Network centrality and social movement media coverage: A two-mode network analytic approach. *Social Networks, 35*, 148–158.

Massimo, T., & Christophe, C. (Eds.). (2017). *Handbook of biometrics for forensic science*. Cham: Springer.

Mendoza, M., Poblete, B., & Valderrama, I. (2019). Nowcasting earthquake damages with Twitter. *EPJ Data Science, 8*, 3. https://epjdatascience.springeropen.com/track/pdf/10.1140/epjds/s13688-019-0181-0.

Morgenthau, H. J. (1978). *Politics among nations: The struggle for power and peace*. New York: Knopf.

Nash, R., et al. (2013). Investigating in people: The role of social networks in the diffusion of a large-scale fraud. *Social Networks, 35*(4), 686–698.

Newman, M. E. J. (2010). *Networks: An introduction*. Oxford, UK: Oxford University Press.

Newman, M. E. J., & Park, J. (2003). Why social networks are different from other types of networks. *Physical Review E, 68*, 036122.

Newman, M. E. J., & Ziff, R. M. (2001). Fast Monte Carlo algorithm for site or bond percolation. *Physical Review E, 64*, 016706.

Nielsen, R. K., & Graves, L. (2017). *"News you don't believe": Audience perspective on fake news*. Reuters Institute Fact Sheet (October 2017).

Nowak, M. A. (2006). Five rules for the evolution of cooperation. *Science, 314*(5805), 1560–1563.

Nowak, M. A., & Sigmund, K. (2005). Evolution of indirect reciprocity. *Nature, 43*(7), 1291–1298.

O'Sullivan, D., & Perry, G. L. W. (2013). *Spatial simulation: Exploring pattern and process*. Hoboken, NJ: Wiley-Blackwell.

OECD/NEA. (2016). Five years after the Fukushima Daiichi accident: Nuclear safety improvements and lessons learnt. https://www.oecd-nea.org/nsd/pubs/2016/7284-five-years-fukushima.pdf

Olson, M. (1971). *The logic of collective action: Public goods and the theory of groups*. Cambridge, MA: Harvard University Press.

Operario, D., Goodwin, S. A., & Fiske, S. T. (1998). Power is everywhere. In R. S. Wyer Jr. (Ed.), *Stereotype activation and inhibition*. Boca Raton, FL: LEA.

Papacharissi, Z. (2010). *A networked self: Identity, community, and culture on social network sites*. Abingdon, UK: Routledge.

Pentland, A. (2014). *Social physics*. London: Penguin Press.

Pew Research Center. (2017). News use across social media platforms 2017. http://assets.pewresearch.org/wp-content/uploads/sites/13/2017/09/13163032/PJ_17.08.23_socialMediaUpdate_FINAL.pdf

Popper, K. (1963). *Conjectures and refutations*. Abingdon, UK: Routledge.

Rohlinger, D. A., & Snow, D. A. (2003). Social psychological perspectives on crowds and social movement. In J. Delamater (Ed.), *Handbook of social psychology*. Dordrecht: Kluwer Academic.

Sadiki, L. (2015). *Routledge handbook of the Arab Spring: Rethinking democratization.* Abingdon, UK: Routledge.

Science Council of Japan (SCJ). (2011). Report to the foreign academies from science council of Japan on the Fukushima Daiichi Nuclear Power Plant accident. http://www.scj.go.jp/en/report/houkoku-110502-7.pdf

Searle, J. R. (1996). *The construction of social reality*. London: Penguin Press.

Shao, C., Ciampaglia, G. L., Varol, O., Yang, K.-C., Flammini, A., & Menczer, F. (2018). The spread of low-credibility content by social bots. *Nature Communications, 9*, 4787.

Shi, F., Teplitskiy, M., Duede, E., & Evans, J. A. (2019). The wisdom of polarized crowds. *Nature Human Behaviour, 3*, 329–336.

Shibuya, K. (2004). A framework of multi-agent based modeling, simulation and computational assistance in an ubiquitous environment. *Simulation, 80*(7–8), 367–380.

Shibuya, K. (2012). A study on participatory support networking by voluntary citizens-the lessons from the Tohoku earthquake disaster. *Oukan, 6*(2), 79–86. (in Japanese).

Shibuya, K. (2017a). Bridging between cyber politics and collective dynamics of social movement. In M. Khosrow-Pour (Ed.), *Encyclopedia of information science and technology* (4th ed., pp. 3538–3548). (Chapter 307), IGI Global.

Shibuya, K. (2017b). An exploring study on networked market disruption and resilience. *KAKENHI report (no. 26590105)*, pp. 1–200 (in Japanese).

Shibuya, K. (2018). A design of Fukushima simulation. *The society for risk analysis: Asia conference 2018*, Japan.

Shibuya, K. (2021). Breaking fake news and verifying truth. In Mehdi Khosrow-Pour (Ed.) *Encyclopedia of organizational knowledge, administration, and technologies* (1st Ed.), IGI Global (in press).

Skibski, O., Rahwan, T., Michalak, T. P., & Yokoo, M. (2019). Attachment centrality: Measure for connectivity in networks. *Artificial Intelligence, 274*, 151–179.

Smelser, N. J. (1962). *Theory of collective behavior*. New York: The Free Press.

Stovel, K., & Fountain, C. (2009). Matching. In P. Hedström & P. Bearman (Eds.), *Oxford handbook of analytical sociology*. Oxford, UK: Oxford University Press.

Strogatz, S. (2000). From Kuramoto to Crawford: Exploring the onset of synchronization in populations of coupled oscillators. *Physica D, 143*, 1–20.

Sunstein, C. R. (2001). *Republic.com*. Princeton, NJ: Princeton University Press.

Tistarelli, M., Li, S. Z., & Chellappa, R. (Eds.). (2009). *Handbook of remote biometrics: For surveillance and security*. London: Springer.

Topçu, S. (2013). *La France nucléaire: L'art de gouverner une technologie contestée*. Paris: Le Seuil.

Viviani, P., & Pasi, G. (2017). Credibility in social media: Opinions, news, and health information-a survey. *WIREs Data Mining and Knowledge Discovery, 7*, e1209. https://doi.org/10.1002/widm.1209.

Vosoughi, S., et al. (2018). The spread of true and false online. *Science, 359*, 1146–1151.

Wang, L., & Graddy, E. (2008). Social capital, volunteering, and charitable giving. *Voluntas: International Journal of Voluntary and Nonprofit Organizations, 19*(1), 23–42.

Wang, Y., et al. (2017). To follow or not to follow: Analyzing the growth patterns of the Trumpists on Twitter. https://arxiv.org/pdf/1603.08174

Wasserman, S., & Faust, K. (1994). *Social network analysis*. Cambridge, UK: Cambridge University Press.

Watts, D. J., & Dodds, P. S. (2007). Influentials, networks, and public opinion formation. *Journal of Consumer Research, 34*(4), 441–458. http://www.uvm.edu/pdodds/teaching/courses/2009-08UVM-300/docs/others/2007/watts2007a.pdf.

Weinberger, S. (2011). Web of war. *Nature, 471*, 566–568.

Widener, W. J., et al. (2013). Simulating the effects of social networks on a population's hurricane evacuation participation. *Journal of Geographical Systems, 15*(2), 193–209.

Wilson, J. (2000). Volunteering. *Annual Review of Sociology, 26*, 215–240.

Wittgenstein, L. (1922). *Tractatus Logico-philosophicus*. San Diego, CA: Harcourt, Brace.
Yang, X. (2011). *Urban remote sensing: Monitoring, synthesis and modeling in the urban environment*. Hoboken, NJ: Wiley.
Yoshino, R., Nikaido, D., & Fujita, T. (2009). Cultural manifold analysis (CULMAN) of national character: Paradigm of cross-cultural survey. *Behaviormetrika, 36*(2), 89–114.
Zhao, S., Grasmuck, S., & Martin, J. (2013). Identity construction on Facebook: Digital empowerment in anchored relationships. *Computers in Human Behavior, 24*(5), 1816–1836.

Chapter 11
Identity Health

Identity health has especially specific meanings for social relationships in contemporary digital age. First, computerized digital communication makes many citizens in severe maladaptation. The WHO[1] often warns mental addictions of internet usages and online gaming among the youth. The advent of social media and online networking has endangered them in ambiguous situations which are not stabilizing in those basic grounds for human relationships. Further, because social networking sites and social gaming frequently enforce each member to interconnect with the others, many of participating members often hold harder mental debts to respond and maintain their interconnections. In this situation, in other words, it can say that all of users simultaneously might share common conditions under mental illness.

Secondly, the big-data era enables us to analyze massive online data accompanied with actual phenomenon. Especially, medical health data can be arranged to coordinate with any solutions for predictions and forecasting in digitized society. Infectious diseases such as influenza and other pandemic events often statistically appear significant and correlational correspondences with actual events and web query among citizens.

Thirdly, subjective well-being and social capital will be still effective concepts for understanding our daily life despite the age of digitized social transformation. It often represents a part of self-definitions embedded in rural community and human relations. Especially, at disaster conditions and emergency situations, it is quite necessary to monitor a patient's longer-term diagnosis. Their identifications surrounded in illness conditions should be cared about intensively.

And fourthly, AI and data science have been equipped to support for our health management and cares in daily life. Those mechanical tools and robots will be eligible to give amenity for various needs among children and elder persons.

[1] https://www.who.int/features/qa/gaming-disorder/en/

K. Shibuya, *Digital Transformation of Identity in the Age of Artificial Intelligence*, https://doi.org/10.1007/978-981-15-2248-2_11

11.1 Digital Society and Health

11.1.1 Addiction Online

WHO (the World Health Organization) has already published their warning reports on gaming disorder and its mental addiction (WHO 2015). And their site also says: "*Gaming disorder is defined in the 11th Revision of the International Classification of Diseases (ICD-11) as a pattern of gaming behavior ("digital-gaming" or "video-gaming") characterized by impaired control over gaming, increasing priority given to gaming over other activities to the extent that gaming takes precedence over other interests and daily activities, and continuation or escalation of gaming despite the occurrence of negative consequences*". Those who can be identified as ICD-11 case are not globally well known, but suspicious cases are roughly estimated as adults (approximately 4.2 million) and youths (approximately a million) in Japan (data at 2017).

Since the beginning of the internet revolution, internet communication and online activities among the youth had been frequently accused by serious concerns in mental health. The first is violent behavior induced by playing violence game. Until recently, video game has been argued by media psychologists (Xu 2018). They often verified violent behavior of adolescents related to experiences and the extent of game playing. The causal relationships across those patterns have not been clarified enough yet. Some reasons endorsed by experimental designs are still difficult to interpret clearly (Bruner and Bruner 2006).

Secondly, there are still controversial issues on online addiction cases (Griffiths 2013). Turel and colleagues (2012, 2016) have been publishing serial reports on online addictions. The number of people diagnosed with the condition of addiction for internet usage has been increasing in the era of social media. Especially, adolescents and young adults are eager to immerse in online cyber-world activities. They consume their time through browsing web, watching online video, participating in online games, and internet communication.

In Japan, MIC[2] (Ministry of Internal Affairs and Communications) at 2018, reported the latest statistics that the total average of internet usage time of young citizens (ages vary from 10 to 17) was 2 h and 49 min every day. Their motivations for internet usages diversified categories such as watching online movie (79%), playing games (76%), and communicating by SNS and emails (66%) (Multiple answers).

The most important is how we think about such online additions and youth's sound development. These are a kind of mental illnesses and conditions as a maladaptation of gaming and social withdrawals from actual society, or they are over-adaptation in somewhat online communities rather than physical environment. The former is to step further grounding in social living, and the latter may suggest that

[2] https://www8.cao.go.jp/youth/youth-harm/chousa/h30/net-jittai/pdf/sokuhou.pdf

they extraordinarily prefer to online human relationships. It should be clinically observed in each case.

Thus, online gaming and social networking sites are indeed based on somewhat human relations and such online communities organized by providers often offer to share some comforts, cooperative achievements, and entertainments among active participants. And simultaneously those services require much engagement among participants, and then such mental obligations enforce each participant to keep playing online game and committing with the other online partners during longer times (Oberst et al. 2017). Here, it experientially indicates that group commitments reduce anxiety of members and enhance comfort and mental bonding among members (Leary and Baumeister 2000; Baumeister et al. 2005). Their group life intends to maintain such conditions in in-group memberships, and commitment belonged in group whether offline or not has crucially important meaning for them.

11.1.2 Medical Data Analysis

To date, advanced data analysis on our PHR (Personal Health Record) and personality dispositions has been conducting in medics (Pol and Thomas 2013). Further, the advancement of big data and the AI driven medical services is to rush into the daily life contexts (Marin et al. 2016; King et al. 2017). Namely these services can offer to assess and promote both mental and physical health of each individual.

First, as mentioned valuation by the AI at Chap. 4, those services already contain some questionnaires on psychological assessments related to personality characteristics, health attitudes, and social adaptation in daily life. Those assessed data might intend to statistically reveal our strength of mental health and degree of adaptation in social relations, and then automatic prediction for those who answered personality tests enables to trustfully measure financial limitations for loans and transactions in actual contexts. Therefore, our mental conditions and its social adaptations, to date, have been unveiled by such ways, and those applied services could be built for somewhat vigilance system on mutual trust among citizens. Mental disorder and maladaptation of each individual have possibilities to further pervade unsound influences among the others, and vice versa. Financial bad-debt by personality dysfunctions of individual will also engender chain bankruptcy among stakeholders, but those services would intend to predict such consequences in advance through checking personality maladaptation in daily life. Hence, our digital life has been already founded in those mechanisms, and the AI and big-data operations indeed interlude into our mentality.

Secondly, wearable devices and sensing tools for human behavior can help monitoring and analyzing latent patterns of physical and mental conditions in daily life (Morahan-Martin and Schumacher 2003; Clifton 2016; Zhu et al. 2019). And telephone communication patterns using smart phone can be interacted by identifiable *chronotype* of each user in daily activities (Aledavood et al. 2018). Medical cares in each country has the demands to organize national health information systems, and it includes big data in the relations to national assurances for health cares, medical quality, demographic statistics, financial investment, quality of life (QOL), quantifications for modeling (Shibuya 2006), and other social welfares (Brady et al. 2012). Those information systems may be governing own centered database, but it will be replaced by distributed blockchain database in the future.

During the previous era, there were some troubles yet to bridge between clinical psychology and sociological studies as well as computational models and experimental cases. Sympathy interpretations for client's latent mental process and their needs should be taken carefully. But case studies often mean that there are no effective ways to explore future patterns and expectations based on past clinical cases in the daily interactions (Leary 1983; Kircher and Leube 2003). Traditionally, the clinical psychological way is usually beneficial to manage mental dynamics that are impossible to be generalized and formalized. Clinical psychology and its fundamental assumptions usually hesitate to do generalizations from clinical case studies. That is because each personal condition and mental distress may be too individualized, and rather researchers in this field recommend qualitative and intensive caring ways for understanding each personal experiences embedded in actual conditions.

11.1.3 A Case of the World Pandemic Flu at 2009

11.1.3.1 The Humankind and Diseases

Historically, emerging diseases have been suffering us ever since our civilization (Roeser et al. 2012). Human history could be said as somehow survival process from lethal diseases. For example, there were smallpox, pest, dysentery, tuberculosis, and other diseases. Otherwise recent outbreaks of emerging diseases such as HIV, Ebola, SARS (severe acute respiratory syndrome: Shibuya 2006), Zika, and others are still ongoing matters, and those medical ways such as drugs and examination tools have not completed enough yet.

In contrary, smallpox can be exemplified as a success case. The humankind had finally achieved the extinction of this disease threat in natural conditions using *vaccine* as a land-breaking medical way. Parts of above diseases could be cured by specific medicines, but known well, there is another problem on resistant bacteria against those medicines. To date, in those fields, immunological and medical investigations have been accelerated by biotechnological and gene-technological advancements. Those who obtained the Nobel Prize in physiology or medicine, and Nobel Prize in chemistry contributed toward enhancing medical progress for well-being of the humankind. In fact, there were contributions by Japanese scientists in the medical field (e.g., Tonegawa, S, Yamanaka, S, Ohmura, S, Ohsumi, Y, etc.).

But, the humankind cannot completely repel both disease and death. According to The WHO report "*Top 10 causes of death globally 2016*[3]," infectious diseases

[3] https://www.who.int/news-room/fact-sheets/detail/the-top-10-causes-of-death

such as lower respiratory infections (over 2 million deaths), diarrheal diseases (nearly 2 million deaths), and tuberculosis (nearly 2 million deaths) still remain within list of top 10. Otherwise, the worst three cases were ischemic heart disease (nearly 10 million deaths), stroke (nearly 6 million deaths), and chronic obstructive pulmonary disease (approximately 3 million deaths). Additionally, total deaths caused by Alzheimer disease and other dementias can be estimated to approximately 2 million per year globally.

In those areas, as mentioned before, computational searches by the AI driven system and big data analysis will boost enhancing diagnosis of subtle symptoms, image processing on medical data, pharmacologic utility discovery, and statistical precisions for future risks of each patient. Precision of diagnosis by the AI's pattern recognition systems on medical images has outperformed more accurately than human doctors (Zhang et al. 2019). As larger growing size of knowledgebase on medical science increasingly requires much experience for them, and it will be impossible to operate any clinical cases unless the AI's supports can be provided.

And other computational contributions to epidemics can enumerate such as computer simulations of mathematical models on spreading infectious diseases (e.g., SIR (susceptible-infected-recovered), small-world networking model (Moore and Newman 2000; Newman 2002)), gene analysis of virus in bioinformatics (Ksiazek et al. 2003), and risk management on health data of patients and medical policies for future controls of emerging diseases (Shibuya 2006).

11.1.3.2 Coincidences Between Online and Offline Trends?

In terms of medical cares, epidemiological actions should lay weights on governmental policy, because there are great needs to control against secondary contagions and predict precisely dynamic trends on diseases. Traditionally, those fields must be accumulated from onsite clinical data on diagnosis of patients and analyze statistical trends which localized in each region. Regarding these concerns, to date, online query results mostly reflect citizens' intentions and latent needs for specific actual events in society. Recently, Ginsberg and his colleagues (2009) had unveiled such facts by their big data, and their findings by statistical analyzations on logistic positive correlations between actual trends of data from CDC[4] (Centers for Disease Control and Prevention, USA) and web-query data among citizens had become a pioneer for big-data age. Namely this study clearly suggested that citizens were usually apt to seek more accurate and necessary information in uncertainty conditions such as disaster (see Chap. 10), unwelcome infectious disease, terrorism, and other fascinated events. Their study seemed to be a first breakthrough for researches using web data analysis. As an implication of this finding, many researchers realized significant meanings on synchronizing and corresponding evidence between web trends and offline events. Namely "big data" can be analyzed by computational

[4] https://www.cdc.gov/

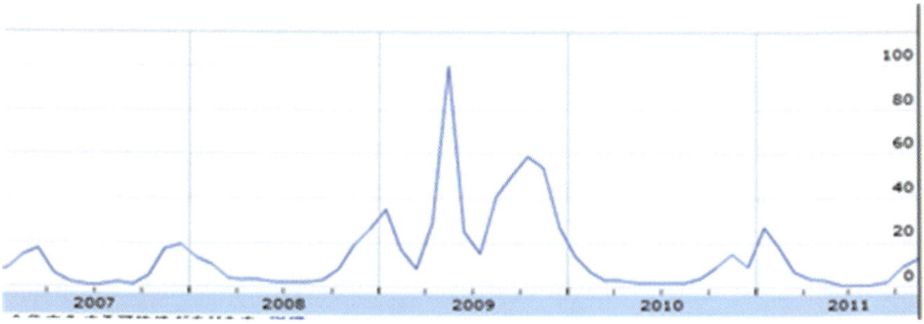

Fig. 11.1 Trends of Google query result which inputted into a keyword "*Influenza*" in Japanese. *Y* axis means frequencies of a query on a specific keyword and *X* axis shows each year in this case

engineering methodologies such as artificial intelligence, statistical machine learning techniques, and natural languages processing, and it can open the gate to investigate novel findings automatically.

Let me exemplify an actual case. According to data from the WHO,[5,6] from the spring of 2009 to 2010, global pandemic caused by new type of influenza (H1N1) had suffered global citizens. The total amount of the death was estimated as 14,286 in globally (the present data at 2010).

At the peak of this pandemic, the author investigated those trends using Google Insights for Search services. Japanese patients were roughly estimated as totally 15.6 million (it finally includes at least 203 death cases), nevertheless many citizens have traditional customs encouraged to treat and keep their hygiene in daily living. Below Fig. 11.1 indicates a trend of Google query result inputted into keyword "*Influenza*" in Japanese. At 2009, it was certainly that there were mostly three peaks during this year. And Fig. 11.2, in contrary, shows only seasonal trends of ordinary influenza (except for data of pandemic patients), and both peaks (earlier weeks of this year and the late of year) can be identified during 2009 year. Seasonal trends on influenza can be also recognized in each year, and only bizarre peak around the middle of this year can be specified. Namely, because the pandemic caused by new type of influenza virus occurred around the beginning of May 2009, it can understand that the middle peak in Fig. 11.1 was underlying in above pandemic influences. And then, in this Japan case, Google trends could entirely indicate correlational patterns between information needs among citizens and actual influenza trends, and each peak corresponded with seasonal or pandemic ones.

As an alternative of Google query, using Twitter as one of microblogging tools, Signorini et al. (2011) revealed synchronizing phenomena on online tweets about influenza corresponded with actual trends of influenza given from CDC data, and

[5] https://www.who.int/csr/disease/swineflu/en/
https://www.who.int/wer/2009/wer8421.pdf?ua=1

[6] Implementation of the International Health Regulations (2005) http://apps.who.int/gb/ebwha/pdf_files/WHA64/A64_10-en.pdf?ua=1

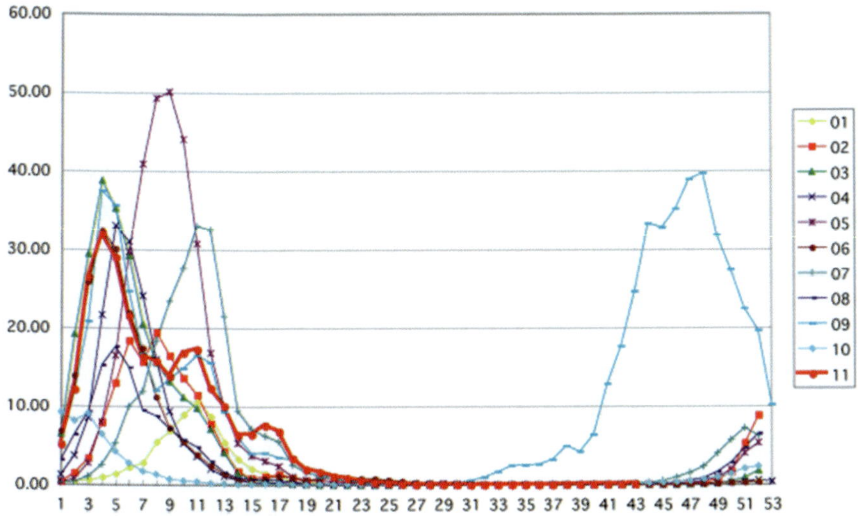

Fig. 11.2 Trends of seasonal influenza patients in Japan at 2009 (except for pandemic data). *Y* axis denotes reported new patients, and *X* axis periodically shows serial weeks. Each line (from 01 to 11) means each site given data from medical hospital. This figure was cited from IDSC (Infectious Disease Surveillance Center, Japan) (http://idsc.nih.go.jp/idwr/kanja/weeklygraph/01flu.html)

they could find efficient results. Those metrics have stronger merits being qualified for online and real-time analyzation than trend data obtained from Google query. Similarity, Broniatowski et al. (2013) reported significant correlation between normalized prevalence data on influenza filtered from Twitter's real-time tweets data and CDC actual trends of influenza. They further attempted to forecast influenza trends using data from Twitter (Paul et al. 2014), and then social media consequently enables to do real-time sensing among citizens.

11.2 Identity Crisis

11.2.1 Disasters in the Big-Data Age

As succeeding previous Chap. 10, in the digitized society, disaster, environment, and climate data have also become a target for big-data analyzation. Larger natural disasters and human-made hazards have globally potentials to corner to the crisis of humanity. Because our global society has been endlessly threatened by various disasters (UNISDR 2018), more than 160 million citizens have been globally harming by natural disasters every year. And this data contains more than 100,000 deaths per year. Natural disasters are almost interrelated to numerous factors such as climate, demography, environment, and anthropogenic events. Further, it seemed obvious that complicated factors related to climate changes in global level have been influencing those meteorological disasters (e.g., hurricanes, drought, flood, etc.). Of course, the anthropogenic factors (e.g., industrial damages to the environment, carbon gas emissions) should be occupied in system models in order to examine detail mechanisms (Meadows et al. 2004).

Especially, at 2015, after serial tragedies of the Tohoku Quakes and Fukushima nuclear disasters, the UNISDR as a part of the United Nations (at 2019, UNISDR was renamed to UNDRR[7]: The UN Office for Disaster Risk Reduction) held the global conference on disaster management at both Tokyo and Sendai city of Japan. As consequences of much discussion, the committee finally proposed following the four priorities for global actions which entitled "*Sendai Framework for Disaster Reduction*[8]."

1. Priority 1: Understanding disaster risk
2. Priority 2: Strengthening disaster risk governance to manage disaster risk
3. Priority 3: Investing in disaster risk reduction for resilience
4. Priority 4: Enhancing disaster preparedness for effective response and to "Build Back Better" in recovery, rehabilitation, and reconstruction

Disaster management means just our preparedness against disasters. Certainly, the digitized global world can offer artificial space satellites, wireless internet, mobile computing, social networking services, and other mechanical relief robots. And the AI driven systems and big-data analyzations on the disaster can be useful for us. Recently, NOAA (The National Oceanic and Atmospheric Administration, USA) has released their online services, which is named as the Coastal Inundation Dashboard.[9] It visually enables people to know and prepare for floods. In this way, nevertheless digitized systems for global monitoring the earth, collaboration with each nation and analyzation of vast necessary data had been pervasively equipped, but our acquired technologies for forecasting and preventing disasters have not been achieved enough yet. As above four priorities said, the UNDRR as a part of the United Nations lays heavier weights on rather mitigations and resilience from disasters. Because of such inevitable reasons why our survival efforts from the tremendous disasters are never diminishing, there are still greater needs to consider the human existential issues in disaster management.

Here, as introduced a bit at Chap. 10, the author had a chance to conduct own researches on the Tohoku quake and nuclear disasters in Fukushima. And this time, a part of findings and evidences can be exhibited in health topic (Shibuya 2012, 2015, 2017, 2018, 2021 (in press)).

[7] https://www.unisdr.org/we/inform/disaster-statistics

[8] https://www.unisdr.org/we/coordinate/sendai-framework

[9] https://tidesandcurrents.noaa.gov/inundationdb/

11.2.2 A Case of Fukushima Disaster

After 2011, the Fukushima case[10] of nuclear power plant accident had incubated another problem (OECD/NEA INES (international nuclear event scale) ranked level 7 (the worst)). This Fukushima case can be called as a nuclear power plants' crisis (NRC 2011; Roeser et al. 2012; OECD/NEA 2016; Shibuya 2017).

At the time, many of Fukushima citizens lost not only hometown, but background of their identities. In this point, they forced to be laid in identity crisis. According to theoretical sociologists Berger et al. (1973) and Giddens (1991), they commonly argued that western post-modernizations could reconstruct mindsets on reality and social identification ways among citizens during achieving industrial progresses, if above severe incidents of nuclear power plants and those systems failures could be regarded as malfunctions as a symbol of modernity, above consequences of nuclear crisis on the Fukushima case (and other human-made disasters) might be contextualized to reexamine social adaptation and consciousness among Fukushima citizens by sociological verifications. How did daily belief systems among Fukushima naïve citizens against the safety surrounding in nuclear power plants deal with? Their attitudes had been rather steadily stabilizing among many of them before the crisis. However, risk cognition, social constructive senses of reality, and meaningful understanding against nuclear disasters among citizens would be collapsed in those conditions. They were betrayed by advanced technologies, government, and optimistic beliefs shared among them. It is probably that their theoretical discussions have few suggestions for any policies on Fukushima case more than awkward theoretical bases, and rather there are quite needs to tackle grounding in actuality, resilience of identification, purifying environment, and rebuilding community for the Fukushima citizens (Science Council of Japan (SCJ) 2011; Shibuya 2021 (in press)).

Actually, this Fukushima case was a reluctantly controversial issue on an accident of nuclear power plant by academic researchers in Japan. Since the Tohoku Quake, many natural scientists in Japan were very criticized by ordinary citizens. Especially these were academic scholars such as nuclear physicists, government-side natural science researchers and engineers at nuclear power plant, and of course politicians must be also confronted with such serious criticisms (Funabashi and Kitazawa 2012). With deep reflections, the Science Council of Japan (SCJ) (2011) published globally their investigated documents by belonging scientists and researchers in various academic fields. This report laid stress on the fact description of the Fukushima Nuclear Power Plant Accidents and the statement of actual conditions for researchers in foreign countries.

Here, the author conducted to investigate this SCJ's statement (2011) in depth using text mining. Table 11.1 shows a result by text mining analyzation on the whole contents of SCJ's statement, and it depicted a part of frequent and important words (it extracted around top 10 words among the most frequent words) and its total

---

[10] http://www.pref.fukushima.lg.jp/site/portal-english/

Table 11.1 This table shows a part of frequent words and its total counts

Words	Frequencies
Water	49
Power	47
Unit	46
Plant	43
Accident	40
Fuel	39
March	38
Reactor	38
Government	35
Radiation	35
Material	33
Area	27
Nuclear	25
Pool	23
Resident	22
Fukushima	22
Result	21
Japan	20

counts within above SCJ's statement text. At a glance, it appeared that their motivations were implicated by some words such as nuclear, radioactive, cooling, accident, safety, emergency, evacuation, and so on. TEPCO means an abbreviated name of administrative company for electric power plants in Fukushima. Figure 11.3 shows an example of network structure of words' co-occurrence. In this case, it configured mathematically to color each separated subgraph structure that limited to important co-occurrence words. The author found some clusters of words on Radioactive materials, Fukushima Nuclear Plant, Accident inside-out and others. These patterns were weighted and frequently articulated by document writers. Otherwise, Fig. 11.4 depicts a result of three-dimensional visualization which analyzed by MDS (Multi-Dimensional Scaling: this time was configured by Kruskal and Jaccard models). This method located statistically each word in cubic dimensions, and it appeared some clusters such as Power Plants (e.g., power, plant, nuclear, and Fukushima), Quake (e.g., tsunami, earthquake, situation, and operation) and others. As result of these malfunctions, confidence for ruling party critically had been fallen down.

What text mining analysis made clear was that this report concluded the negative consensus against the nuclear hazard among scientific community in Japan, and their statements explicitly described that mythical beliefs among stakeholders were no avail in the case of Fukushima nuclear power plant accident. Obviously, they intended to publish the truths for globally foreign academicians and citizens in terms of mainly nuclear physics and energy engineering after the Fukushima crisis. As mentioned at Chap. 10, without doubts, many of global citizens were eager to know more accurate and immediate information in detail at that time of moment.

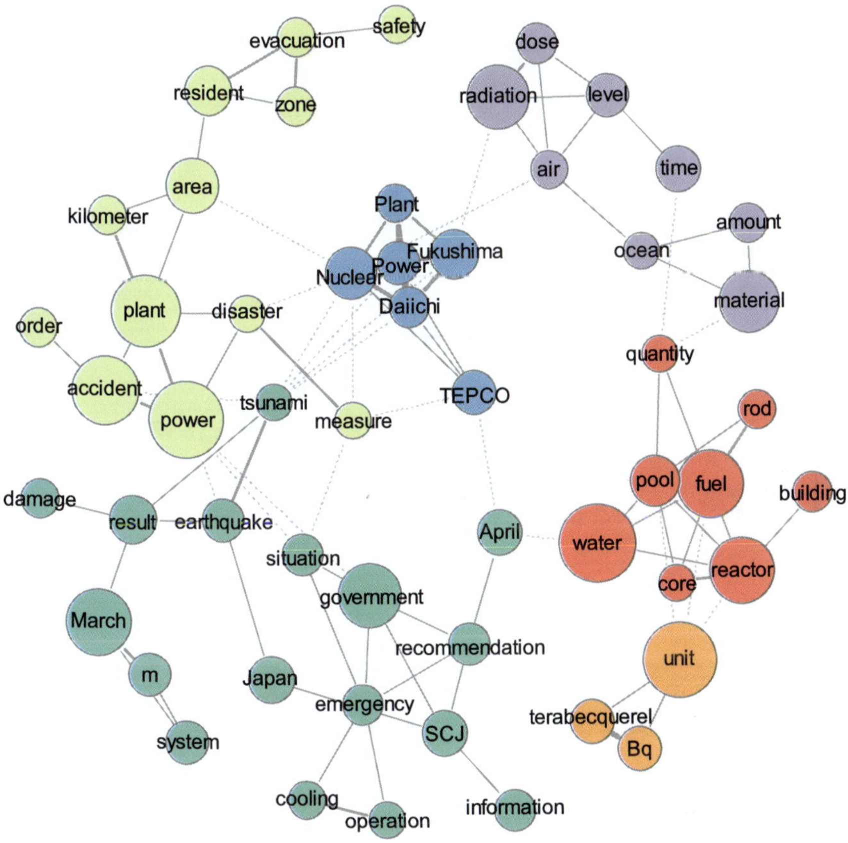

Fig. 11.3 It depicts a part of network structure of co-occurrence words by text mining. This configuration was to separately color each subgraph structure which limited to important co-occurrence words

Those results by text mining could exhibit Japanese governance of risk management against nuclear power plants and energy policies before the Fukushima case. There were no rational reasons for excuses that risk communication and consensual discussion had not been openly organized among stakeholders in Fukushima, as it was differently the Canada's case (Johnson 2008). Rather, TEPCO and governmental ministry had oppressed to scientifically contemplate and examine nuclear risks and their published data by ordinary citizens and external professionals. With these backgrounds, the focal point of disputes on the compensations has been accused by plaintiff (i.e., citizens, evacuees, and victims) under trials in courts (OECD/NEA 2016).

For the digitized society, this Fukushima case indicated further unneglectable facts. IoT and XaaS will be deployed anywhere (Geng 2017). Many of those systems such as computational controls and sensing networking related to power plants and sensitive artifacts will be impossible to keep under controls unless electric

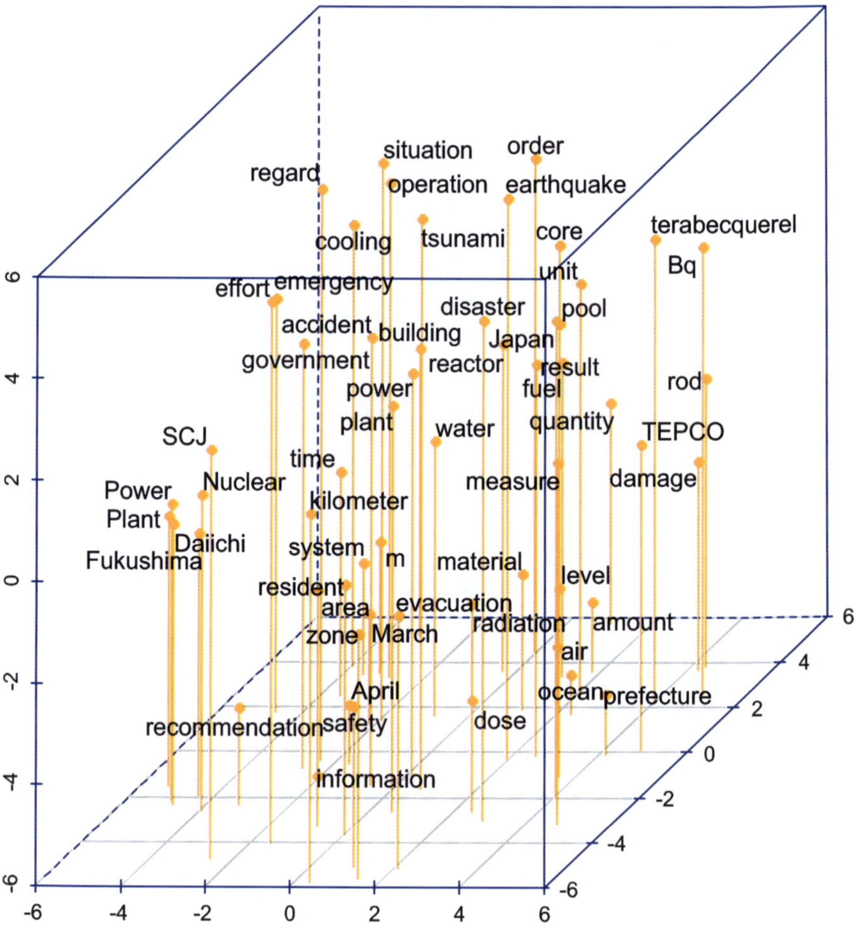

Fig. 11.4 This three-dimension picture depicts a result analyzed by MDS

power can be continually provided. Nuclear power plants and its control systems per se always also require independent electric power for controlling those mechanisms. Serial incidents of the Fukushima nuclear power plants can be determined by the serious factors on both the loss of external electric power supply and vent malfunctions for refrigerating systems caused by tremendous tsunami attacks. Besides, until now, the AI-driven robots sensing inner-damaged power plants cannot be activated over the physical limitations because inner-damaged power plants still remain with higher radiation dose (i.e., the human dies immediately and computational mechanisms will be disabled by radiations sooner). Then, the lessons for the digitized future must be intensively deduced and learned by convincible investigations. In the USA, after this disaster, a taskforce team was immediately assembled for investigations to report the cause of the nuclear power plants, reconstruction of the nuclear power plants for safety, and future policy on energy management in the

USA (NRC 2011). Their 12 recommendations within this report should be carefully read for future engineers.

11.2.3 Victims of the Fukushima

After that, it should turn eyes to persevering purification from the radiation damage caused by the nuclear accidents, its environmental restoration, health monitoring, and socioeconomic reconstruction in community (NRC 2011; IAEA 2011). Consequently, the Fukushima case definitely needs to solve nuclear accidents and future design for long-term reconstruction on their devastated communities and hometown. At least, the following points should be tackled.

1. Decommissioning work for the wrecked nuclear power plants in Fukushima
2. Removal of nuclear fuel and substances
3. Chemical management on nuclear substances
4. Purifications in polluted places
5. Temporary storage and final disposal of contaminated soil, water, and garbage
6. Medical health survey and care for victims
7. Risk assessment against environment and people exposed by nuclear substances
8. City and socioeconomic reconstructions

As consequences of serial incidents of the nuclear power plants in Fukushima, nuclear pollution provoked the severe disputes on human rights, health, radiation contaminations to foods and environmental restorations. Both environments (soil, waters and air) as well as ordinary peoples who lived in Fukushima were polluted and exposed by both nuclear substance and radiation (Gibney 2015; Merz et al. 2015; OECD/NEA 2016). Sampling data accumulated by agricultural scientists endorsed that many of crops absorbed radioactive substances such as ^{90}Sr, ^{134}Cs, ^{137}Cs, and others (Takahashi 2016). And excessive intake and exposure of radioactive pollutants will endanger citizens' and workers' health (Hiraoka et al. 2015).

Consequentially, enacted provisions often reflect social actualities. For conquests against those hardships of citizens and evacuees, the "*Basic Act on Reconstruction in response to the Great East Japan Earthquake (24th, June 2011)*" enacted the basic policies for reconstructions. For example, a part of provisions clearly stipulated as follows.

- Article 2. The reconstruction in response to Great East Japan Earthquake will be implemented based on the following
  - The unprecedented disaster resulted in enormous damage, where countless lives were lost, numerous people were deprived of their basic living infrastructures and have been forced to evacuate in and out of the disaster-affected regions. Also, the disaster's influence extends over the entire nation; the

economic stagnation in the disaster-afflicted areas is affecting business activities and peoples' lives nationwide…(hereafter omitted)

At the 2014FY, the total budget of Fukushima Prefecture for reconstruction after the disaster was approximately 1714.5 billion yen (including both quakes and nuclear disaster countermeasure portion of 870.5 billion yen). And it includes population declining and aging countermeasures as well as restoration of birth number (142.3 billion yen). In addition, other items were 245.5 billion yen for environmental restorations, 144.3 billion yen for living reconstruction assistance, 36 billion yen for expenses to protect medical health of the citizens, 31.6 billion yen for expenses for children and youths who will be responsible for the future, and other expenses. Further, in addition to above costs, the Fukushima case requires unprovoked compensations and its litigation disputes are ongoing matters in courts (Shibuya 2017). Namely, there are still requirements to solve future designing for reconstruction from devastation in their hometown.

Next, the total amount of casualties in Japan was more than 20,000 at the time of 2012. Moreover, as aftermath of the Fukushima disaster, one of the hardest matters was collective immigration of evacuees from their hometowns to other places in Japan (Akabayashi and Hayashi 2012; Library of Congress 2013). At the peak (May 2012), gross migrants from Fukushima (e.g., total population of evacuees) were estimated over 160,000. And including this, gross migrants (e.g., total evacuees in Japan) were estimated over 340,000 at the time of 2012. This estimation was not too low. Please recall similar past cases, for example, the case of Chernobyl in 1986 reported that total population of evacuees was approximately 116,000 around 30 km (INES level 7). And, in 1979, the case of the Three Mile Island accident was estimated over 240,000 around 24 km (INES level 5).

Table 11.2 shows a part of outflow data on migrants across major cities (it includes mobility data within same city). It queried into big data of MIC (e.g., demographic data of migration and population) and the Geospatial Information Authority of Japan (e.g., geospatial data and distance information). The numbers of citizens lived in Fukushima prefecture has been notably decreasing from 2.1 million (2011) to 1.9 million (2015). And this area is statistically 13,780 km^2 (the third widest area in Japan). And it compares Fukushima with people lived in major metropolis such as Tokyo area (total population is approximately 13 million people and

Table 11.2 Outflow data on mobility across cities in Japan (Compared with 2011 and 2015) (Data from Ministry of Internal Affairs and Communications, Japan)

		To								
		Fukushima		Tokyo Area		Nagoya Area		Osaka Area		
From		Sum	Per.	Sum	Per.	Sum	Per.	Sum	Per.	Sum
Fukushima	2011	27,613	**0.526**	21,112	**0.402**	1,605	**0.031**	2,167	**0.041**	52,497
	2015	1,671	**0.686**	664	**0.273**	42	**0.017**	59	**0.024**	2,436
Tokyo Area	2011	9,087	**0.026**	236,446	**0.677**	37,071	**0.106**	66,834	**0.191**	349,438
	2015	581	**0.021**	22,317	**0.800**	1,921	**0.069**	3,077	**0.110**	27,896
Nagoya Area	2011	588	**0.005**	40,809	**0.355**	50,034	**0.435**	23,535	**0.205**	114,966
	2015	58	**0.008**	2,123	**0.279**	4,272	**0.562**	1,150	**0.151**	7,603
Osaka Area	2011	619	**0.004**	74,869	**0.480**	22,575	**0.145**	57,913	**0.371**	155,976
	2015	83	**0.008**	3,849	**0.368**	1,289	**0.123**	5,235	**0.501**	10,456

within 9 million people in 23 special districts), Nagoya area (approximately 2.2 million people within central city), and Osaka area (approximately 2.6 million people within central city) in Japan. It namely denotes moving flows within Fukushima cities, moving toward one of metropolises from Fukushima and moving patterns between metropolises.

Actually, there is still another problem in residential data. In year 2015, a national census every 5 years was carried out in Japan, and it has achieved to unveil many data discrepancies and inconsistencies of population in each local area of Fukushima. Comparing with statistics on resident data holding municipal government office and actual population from census, the latter cases were almost too lower than estimated populations in many cases of Fukushima. To date, these missing populations have not been traced properly and many evacuees did not intentionally apply immigration cards to municipal government office. It means that many of them still have strong intentions to go back to Fukushima in the near future. However, the stumbling blocks still remain against their returns, even though the government purifies radioactive substances and pollutions around their towns.

According to the general surveys by Tokyo capital government for evacuees and interviews for evacuees by the author (Shibuya 2017), they found that evacuees' motivations which choose the destination and refuge were relying on following critical factors: (1) to tie with any kindred relationships (a factor of human relationship), (2) rich opportunities for jobs in the new address (a factor of new job opportunity), (3) conveniences to manage their own real estates, farms, livestock, and factories in their hometown (a factor of holding estates).

First factor implies mutual cooperation and helping among local acquaintances, and second answer clearly reflects their needs for jobs in new dwelling. And third factor relates to geospatial location, and parts of them have been living in two places of both Fukushima and refuges. Namely, their conditions were back and forth between hometown and refuge. Thus they could not leave to far refuges, and geographical area around 300 km within Fukushima and refuges satisfied their above motivations using transportations such as the bullet train, highways, and other land transportations (it drives toward the destination within 2 h). For those who required living needs, Tokyo and nearby area of Tokyo could properly offer new job opportunities and dwelling availabilities for them.

11.2.4 Caring for Victims and Their Community

Our identifications are often determined by not only own cognitive factors but human relational and spatiotemporal factors, and their daily conditions of mental health would be interlinked with those internal and external surroundings. The Fukushima case similarly indicated those mental malfunctions of evacuees caused by the human-made disaster.

When such individuals lose all (or a part) of the linkages with both human relationships and living place, their mental foundation for identification will be seriously damaged in severe situations. Such accidents and events caused by both natural and human-made disasters will be easy to engender secondary damages against social adaptation and subjective well-being of each individual. It is too important to care for each, but there are resilient needs to wholly repair and revive the community among them and their human relationships such as social capital (Putnum 2000; Kawachi and Berkman 2003; Kawachi et al. 2008; OECD/NEA 2016) and family bondages. Even though digitized communication styles renewed our daily commitment for online community, physical-contact based commitments in onsite community have still special meanings for their well-being.

11.3 Monitoring Health in Daily Life

11.3.1 Working in Digital Era

Frey and Osborne (2013) reported how will contemporary industries and jobs be changed and replaced by computerization and the AI-driven robotics, and they simulated socioeconomic trend patterns of jobs fitted by Gaussian stochastic model. Their estimations have indeed shown that the AI society will engender emerging job market and require other skills for citizens. But it will be clearly understood whether correct or not in the future.

But, will a meaning on working life be steeply altered by such innovation? Some theorists said that both living and working are indispensable relations each other. As one of the renown episodes, psychoanalyst Freud answered that it requires "*Lieben und Arbeiten*" for becoming sound and independent adult. The former "*Lieben*" means the accepting and loving for the others as a partner. And the latter "*Arbeiten*" devotes to achieving the goals for maintaining daily living by own works and pursuing enhancement of own intellectual abilities and skills. Both are still certainly the fundamental necessary factors for people.

And a life-course approach in developmental psychology has much suggestion to wholly understand our mental development and health promotion during the life process (Erikson 1950, 1959, 1982). In each development stage of identity, each individual closely faces the problems and what should be conquered by each of them. Such tasks can be furnished for own rich experiences of each, and each can be reorganized to adapt in own life process (e.g., self-actualization). Working, learning, and other daily activities will be achieved by undertaking self-development and adaptation in social surroundings. It motivates to enhance each quality of life (QOL) through own working experiment.

11.3.2 *Quality of Life*

It is certainly that the AI and big-data-based society enables to change our QOL and working life. Such innovation progress in our working styles has already become the cascading to collapse larger barriers by the big wave of digital transformation. Working environment has been crucially invested in the contexts of either employees or employer. But quality in working environment cannot be determined by factors of material and physical surroundings, and it should be cared about human factors such as enhancement of human relationships and well-being (OECD 1976; Strack et al. 1991; Buunk and Gibbons 1997). For example, OECD[11] proposed total framework ("*Measuring Well-being and Progress: Well-being Research*"), and they enumerated necessary factors related to measuring both economic and well-being value in working and daily life.

Quality of Life
1. Health status
2. Work–life balance
3. Education and skills
4. Social connections
5. Civic engagement and governance
6. Personal security
7. Subjective well-being

Material Conditions
1. Income and wealth
2. Jobs and earnings
3. Housing

The digital transforming society will enlarge our working from actual physical space to virtual online space through tele-existence and online collaboration tools. And our working skills and abilities will be required conquering the harder roads of uphill progress of the AI and data sciences (Boyd and Holton 2018). Further, traditional stressful working environments can be attempted to quantify and coordinate with each parameter of employees such as personality characteristics, demanded skill levels, abilities, chemistry among members, and other necessary factors. Now, such HR (human resource) technology (i.e., a case using business microscope[12]) enhances our working styles and improves productivities in various situations (Khartri and Samuel 2019). Analytics teams have vividly rushed to dive into the ocean of big data, but matching between the needs and their analyzed solutions can be improved by further efforts.

Active workers are usually facing issues at the marriage and family in their life courses. In some developed countries, the reasons, due to which unmarried rate

[11] http://www.oecd.org/statistics/measuring-well-being-and-progress.htm

[12] http://www.hitachi.com/New/cnews/121019.html

during the lifespan of the youth generation has been increasing, may be understandable in work–life balance context. In Japan, statistical data of IPSS (National Institute of Population and Social Security Research, Japan)[13] show such facts: men's case of unmarried rate excessed 20%, and women's case was 10.6% at 2010. Many adolescents frequently hesitate to lose their free time and conformity, and it simultaneously means that they hate any interruptions by others and physical contacts. As necessary, they can choose tentative friends online (Su and Hu 2019). Using smartphone, many matching service applications for marriage among future partners have been launched in Japan. And then, the matching needs can be bridged with the youths for their marriages. Those matching services might be applied by stable matching problem in economics of mechanism design (Roth 1982). Such algorithm can be formalized for matching pairs of stable marriage.

Otherwise, daily living enriches its big data (Ganchev et al. 2019). Especially, there are strong requirements for children and elder people in their local community and living environment. First, smart sensing and ubiquitous technologies aim to enhancing our daily life (Shibuya 2004), and smart cities and smart house have cutting-edges for improving health services for us (Lee 2012). For example, the Digital human research center in Japan proposed an autonomous caring system and simulators[14] for toddlers and little infants. Due to their sudden and unpredictable manner of behaviors, serious accidents such as injuries and death at home often happen. This system intends to monitor and analyze daily patterns for improving their safety. Of course, those systems which are equipped in smart houses are also applicable for elder people to watch their daily cares and health monitoring.

On the other hands, secondly, Airbnb and similar sharing house services have been launched in many nations, and big data on those paring patterns between house-owners and visitors will be arranged to analyze such trip purposes, sharing durations, other preferences by the AI-driven services (Koh et al. 2019). In these regards, digitization has already reshaped our quality of living in those contexts.

11.4 Therapy for Human by AI

11.4.1 Therapy for Identity in Both Personal and Social Level

Using smartphone, mental health monitoring can be possible recently (Ben-Zeev et al. 2015; Bakker et al. 2016). Especially, using social media data, there were innumerable examinations to analyze the relations with mental health and diagnosis of discourses on Twitter and SNS services. For example, there were enumerable cases on depression trends in community corresponding with data of geospatial

[13] http://www.ipss.go.jp/syoushika/tohkei/Data/Popular2013/T06-23.xls (in Japanese)

[14] http://www.dh.aist.go.jp/jp/research/enabling/index.html
http://www.dh.aist.go.jp/jp/research/enabling/InfantBehaviorSimulator/

location (Yang and Mu 2015), depression detection on Twitter (Guntuku et al. 2017), ADHD (Attention Deficit Hyperactivity Disorder) diagnosis using discourses on Twitter (Coppersmith et al. 2015), and other borderline cases in clinical psychology (e.g., hopeless, loneliness, social withdrawal). As social networking services clearly indicate a part of human relationships online (Lazakidou 2012), it can consider that their relations itself still have sharing illness personalities and depressed mental health. Namely, there is a possibility that latent patients were apt to be participating in such social media, and some of them flocked online each other. Inclusive cares within communal and relationship level can be also very effective for each person.

Traditionally, in the studies of communication and communal health, the concept on illness identity in interactive communication has potentials to recognize identification process for caring illness and health (Hecht et al. 2004). Illness identity could be regarded as interactive processes from personal to communal layers in terms of the communication theory of identity. These mechanisms should be inclusively cared by various viewpoints from personal to social communal level. That is because, for example, social anxiety, depression, and other mental (and physical) illness could not be easily emancipated from not only individuals but also more diverse interactions and social groups.

According to Wegner et al., they discussed mental control and relationships with the others. And their findings can be understood only by including perspectives of social contexts. That is, it should pay attention to not only individual experiences of depression and mental distress but social relationship and interaction process with standpoints from the others (Aneshensel et al. 2012).

11.4.2 Theory of Mind: The Nature of Understanding Myself and Others

As implied earlier, rapport interaction between clinical psychologists and clients has been better focused in empathy oriented understanding as client-centered therapy (Rogers 1995). Clinical psychological cases such as autism, psychopathy and other diagnoses usually display specific patterns of behaviors and assertions.

Especially, those who have specific disabilities against soundly interactions with the others are managed by ToM (Theory of Mind), and this study intends to reveal our mental manners to recognize and coordinate with the others in dyad models (Semeijn 2019; Freitas et al. 2019). Such client cannot understand any intentions asserted by the others, and they often confuse meanings pretending and deceptions by the others. Namely, understanding for the others flexibly requires more imaginable coordination in social context. And the loss of such basic intellectual skills becomes difficult for them to behave appropriately against troublesomeness with the others. In other words, some evolutionary psychologists and neuroscientists told that the humankind could be evolved both to lie against the others and detect deceptive intentions. They further said that acquisitions for those neural mechanisms of social intelligence took advantages of beating against other wild animals during surviving history of ancient peoples.

11.4.3 Assistance by the AI and Robots

According to attachment theory by developmental psychologist Bowlby, it authorized that physical attachment between mother and children offers strong and comfort foundation during child development (Bretherton 1992). Suggestively, serial experiments also shown that an alternative of "mother" could be sufficiently replaced for children. For example, a fluffy *doll* as alternative of mothership could fulfill having comfort emotions for child (in their experiments, they tried to use a child of monkey).

In this concern, personal intelligent robots have potentials to assist our daily life (Coeckelbergh 2010). As an example case, PARO already achieved improving many patients of mental illness and Alzheimer diseases. It calls *robot therapy* assisted by AI and adorable doll-like robots. Such new services can be adapted in the social welfare institutes, daily cares in home and hospitals (Wada et al. 2007; Yu et al. 2015). Their attachments with physically autonomous entities may offer them some reliefs and comforts.

In such ways, as mentioned earlier, there are certainty to automatically diagnose clients by the AI using telecommunications and smartphone. Sensing data accumulated by wearable devices in daily life has strong potential to detect mental illness and bad mental-physical conditions in earlier stages.

References

Akabayashi, A., & Hayashi, Y. (2012). Mandatory evacuation of residents during the Fukushima nuclear disaster: An ethical analysis. *Journal of Public Health, 34*(3), 348–351.

Aledavood, T., Lehmann, S., & Saramäki, J. (2018). Social network differences of chronotypes identified from mobile phone data. *EPJ Data Science, 7*, 46. https://doi.org/10.1140/epjds/s13688-018-0174-4.

Aneshensel, C. S., Phelan, J. C., & Bierman, A. (2012). *Handbook of the sociology of mental health*. Dordrecht: Springer.

Bakker, D., Kazantzis, N., Rickard, D., & Rickard, N. (2016). Mental health smartphone apps: Review and evidence-based recommendations for future developments. *JMIR Mental Health, 3*(1), e7. https://doi.org/10.2196/mental.4984.

Baumeister, R. F., et al. (2005). Social exclusion impairs self-regulation. *Journal of Personality and Social Psychology, 88*(4), 589–720.

Ben-Zeev, D., Scherer, E. A., Wang, R., Xie, H., & Campbell, A. T. (2015). Next-generation psychiatric assessment: Using smartphone sensors to monitor behavior and mental health. *Psychiatric Rehabilitation Journal, 38*(3), 218–226. https://doi.org/10.1037/prj0000130.

Berger, P. L., Berger, B., & Kellner, H. (1973). *The homeless mind: Modernization and consciousness*. New York: Irvington Publication.

Boyd, R., & Holton, R. J. (2018). Technology, innovation, employment and power: Does robotics and artificial intelligence really mean social transformation? *Journal of Sociology, 54*(3), 331–345.

Brady, K., Sriram, R. D., Lide, B., & Roberts, K. (2012). Testing the nation's healthcare information infrastructure: NIST perspective. *IEEE Computer, 45*(11), 50–57.

Bretherton, I. (1992). The origins of attachment theory: John Bowlby and Mary Ainsworth. *Developmental Psychology, 28*(5), 759–775.

Broniatowski, D. A., Paul, M. J., & Dredze, M. (2013). National and local influenza surveillance through twitter: An analysis of the 2012-2013 influenza epidemic. *PLoS One, 8*(12), e83672. https://doi.org/10.1371/journal.pone.0083672.

Bruner, C., & Bruner, K. (2006). *Play station nation: Protect your child from video game addiction*. New York: Center Street.

Buunk, B. P., & Gibbons, F. X. (1997). *Health, coping, and well-being: Perspectives from social comparison theory*. Boca Raton, FL: LEA.

Clifton, D. A. (Ed.). (2016). *Machine learning for healthcare technologies*. Oxford, UK: Oxford University Press.

Coeckelbergh, M. (2010). Health care, capabilities, and AI assistive technologies. *Ethical Theory and Moral Practice, 13*(2), 181–190.

Coppersmith, G., et al. (2015). From ADHD to SAD: Analyzing the language of mental health on Twitter through self-reported diagnoses. In *Proceedings of the 2nd workshop on computational linguistics and clinical psychology: From linguistic signal to clinical reality* (pp. 1–10).

Erikson, H. (1950). *Childhood and society*. New York: W. W. Norton.

Erikson, H. (1959). *Identity and the life cycle*. New York: International Universities Press.

Erikson, H. (1982). *The life cycle completed*. New York: W. W. Norton.

Freitas, J., Thomas, K., DeScioli, P., & Pinker, S. (2019). Common knowledge, coordination, and strategic mentalizing in human social life. *PNAS, 116*(28), 13751–13758. https://doi.org/10.1073/pnas.1905518116.

Frey, C. B., & Osborne, M. A. (2013). The future of employment: How susceptible are jobs to computerisation? https://www.oxfordmartin.ox.ac.uk/downloads/academic/The_Future_of_Employment.pdf

Funabashi, Y., & Kitazawa, K. (2012). Fukushima in review: A complex disaster, a disastrous response. *Bulletin of the Atomic Scientists, 68*(2), 9–21. http://bos.sagepub.com/content/early/2012/02/29/0096340212440359.

Ganchev, I., et al. (Eds.). (2019). *Enhanced living environments: Algorithms, architectures, platforms, and systems*. Cham: Springer.

Geng, H. (2017). *Internet of things and data analytics handbook*. Hoboken, NJ: Wiley.

Gibney, E. (2015). Fukushima data show rise and fall in food radioactivity: Giant database captures fluctuating radioactivity levels in vegetables, fruit, meat and tea. *Nature*. https://doi.org/10.1038/nature.2015.17016.

Giddens, A. (1991). *Modernity and self-identity: Self and society in the late modern age*. Cambridge, UK: Polity.

Ginsberg, J., et al. (2009). Detecting influenza epidemics using search engine query data. *Nature, 457*, 1012–1014.

Griffiths, M. D. (2013). Social networking addiction: Emerging themes and issues. *Journal of Addiction Research & Therapy, 4*(5). https://doi.org/10.4172/2155-6105.1000e118.

Guntuku, C. S., et al. (2017). Detecting depression and mental illness on social media: An integrative review. *Current Opinion in Behavioral Sciences, 18*, 43–49.

Hecht, M. L., Warren, J. R., Jung, E., & Krieger, J. L. (2004). The communication theory of identity. In W. B. Gudyunst (Ed.), *Theorizing about intercultural communication*. Thousand Oaks, CA: SAGE.

Hiraoka, K., et al. (2015). Review of health issues of workers engaged in operations related to the accident at the Fukushima Daiichi Nuclear Power Plant. *Journal of Occupational Health, 57*(6), 497–512.

IAEA. (2011). Additional report of Japanese Government to IAEA - Accident at TEPCO's Fukushima Nuclear Power Stations. http://www.iaea.org/newscenter/focus/fukushima/japan-report2

Johnson, G. F. (2008). *Deliberative democracy for the future: The case of nuclear waste management in Canada*. Toronto: University of Toronto Press.

Kawachi, I., & Berkman, L. F. (Eds.). (2003). *Neighborhoods and health*. Oxford, UK: Oxford University Press.

Kawachi, I., et al. (Eds.). (2008). *Social capital and health*. Cham: Springer.

Khartri, V., & Samuel, B. M. (2019). Analytics for managerial work. *Communications of the ACM, 62*(4), 100–108.

King, S., Mills, A., Kadirkamanathan, V., & Clifton, D. A. (Eds.). (2017). *Equipment health monitoring in complex systems*. Oxford, UK: Oxford University Press.

Kircher, T. T. J., & Leube, D. T. (2003). Self-consciousness, self-agency, and schizophrenia. *Consciousness and Cognition, 12*, 656–669.

Koh, V., Li, W., Livan, G., & Capra, L. (2019). Offline biases in online platforms: A study of diversity and homophily in Airbnb. *EPJ Data Science, 8*, 11. https://doi.org/10.1140/epjds/s13688-019-0189-5.

Ksiazek, T. G., et al. (2003). A novel coronavirus associated with severe acute respiratory syndrome. *The New England Journal of Medicine, 348*(20), 1953–1966.

Lazakidou, A. A. (2012). *Virtual communities, social networks and collaboration*. New York: Springer.

Leary, M. R. (1983). *Understanding social anxiety*. Thousand Oaks, CA: SAGE.

Leary, M. R., & Baumeister, R. F. (2000). The nature and function of self-esteem: Sociometer theory. In M. P. Zanna (Ed.), *Advances in experimental social psychology* (Vol. 32). Cambridge, MA: Academic.

Lee, G. (Ed.). (2012). *Advances in computational environment science: Selected papers from 2012 international conference on environment*. New York: Springer.

Library of Congress, USA. (2013). Japan: Legal responses to the Great East Japan earthquake of 2011. http://www.loc.gov/law/help/japan-earthquake/Great-East-Japan-Earthquake.pdf

Marin, H., et al. (Eds.). (2016). *Global health informatics (1st edition): How information technology can change our lives in a globalized world*. Oxford, UK: Oxford University Press.

Meadows, D. H., et al. (2004). *The limits to growth*. Hartford, VT: Chelsea Green.

Merz, M., et al. (2015). Analysis of Japanese radionuclide monitoring data of food before and after the Fukushima nuclear accident. *Environmental Science Technology, 49*(5), 2875–2885. https://doi.org/10.1021/es5057648.

Moore, C., & Newman, M. E. J. (2000). Epidemics and percolation in small-world networks. http://www.santafe.edu/sfi/publications/Working-Papers/00-01-002.pdf

Morahan-Martin, J., & Schumacher, P. (2003). Loneliness and social uses of the internet. *Computers in Human Behavior, 19*, 659–671.

Newman, M. E. J. (2002). The spread of epidemic disease on networks. http://www.santafe.edu/sfi/publications/Working-Papers/02-04-020.pdf

NRC (Nuclear Regulatory Commission, U.S.A.). (2011). The near-term task force review of insights from the Fukushima Dai-Ichi accident. http://pbadupws.nrc.gov/docs/ML1118/ML111861807.pdf

Oberst, U., et al. (2017). Negative consequences from heavy social networking in adolescents: The mediating role of fear of missing out. *Journal of Adolescence, 55*, 51–60.

OECD. (1976). *Measuring social well-being: A progress report on the development of social indicators*. Paris: OECD Publication Center.

OECD/NEA. (2016). Five years after the Fukushima Daiichi accident: Nuclear safety improvements and lessons learnt. https://www.oecd-nea.org/nsd/pubs/2016/7284-five-years-fukushima.pdf

Paul, M. J., Dredze, M., & Broniatowski, D. (2014). Twitter improves influenza forecasting. *PLoS Current, 28*(6). ecurrents.outbreaks.90b9ed0f59bae4ccaa683a39865d9117.
Pol, L. G., & Thomas, R. K. (Eds.). (2013). *The demography of health and healthcare* (3rd ed.). Dordrecht: Springer.
Putnum, R. (2000). *Bowling alone: The collapse and revival of American community*. New York: Simon & Schuster.
Roeser, S., Hillerbrand, R., Sandin, P., & Peterson, M. (Eds.). (2012). *Handbook of risk theory: Epistemology, decision theory, ethics, and social implications of risk*. Dordrecht: Springer.
Rogers, C. R. (1995). *Client-centered therapy: Its current practice, implications and theory*. London: Constable & Robinson Ltd.
Roth, A. E. (1982). The economics of matching: Stability and incentives. *Mathematics of Operations Research, 7*(4), 617–628.
Science Council of Japan (SCJ). (2011). Report to the foreign academies from science council of Japan on the Fukushima Daiichi Nuclear Power Plant accident. http://www.scj.go.jp/en/report/houkoku-110502-7.pdf
Semeijn, M. (2019). Interacting with fictions: The role of pretend play in theory of mind acquisition. *Review of Philosophical Psychology, 10*, 113–132.
Shibuya, K. (2004). A framework of multi-agent based modeling, simulation and computational assistance in an ubiquitous environment. *Simulation, 80*(7–8), 367–380.
Shibuya, K. (2006). Actualities of social representation: Simulation on diffusion processes of SARS representation. In C. Dijkum, J. Blasius, & C. Durand (Eds.), *Recent developments and applications in social research methodology, Proceedings of the RC33 sixth international conference on social science methodology*. Leverkusen: CD-ROM Format, Budrich.
Shibuya, K. (2012). A study on participatory support networking by voluntary citizens - The lessons from the Tohoku earthquake disaster. *Oukan, 6*(2), 79–86. (in Japanese).
Shibuya, K. (2015). A simulation on networked market disruptions and resilience from "Fuhyo-Higai". In *Tokyo conference on international study for disaster risk reduction and resilience*. United Nations UNISDR, University of Tokyo.
Shibuya, K. (2017). An exploring study on networked market disruption and resilience. *KAKENHI Report (no. 26590105)*, pp. 1–200 (in Japanese).
Shibuya, K. (2018). A design of Fukushima simulation. *The society for risk analysis: Asia conference 2018*, Japan.
Shibuya, K. (2021 (in press)). A risk management on demographic mobility of evacuees in disaster. In M. Khosrow-Pour (Ed.), *Encyclopedia of organizational knowledge, administration, and technologies* (1st ed.). Hershey, PA: IGI Global.
Signorini, A., Segre, A. M., & Polgreen, P. M. (2011). The use of Twitter to track levels of disease activity and public concern in the U.S. during the Influenza A H1N1 Pandemic. *PLoS ONE, 6*(5), e19467. https://doi.org/10.1371/journal.pone.0019467.
Strack, F., Argyle, M., & Schwarz, N. (1991). *Subjective well-being: An interdisciplinary perspective*. Oxford, UK: Pergamon Press.
Su, X., & Hu, H. (2019). Gender-specific preference in online dating. *EPJ Data Science, 8*, 12. https://doi.org/10.1140/epjds/s13688-019-0192-x.
Takahashi, T. (Ed.). (2016). *Radiological issues for Fukushima's revitalized future*. Japan: Springer.
Turel, O., & Qahri-Saremi, H. (2016). Problematic use of social networking sites: Antecedents and consequence from a dual system theory perspective. *Journal of Management Information Systems, 33*(4), 1087–1116.
Turel, O., & Serenko, A. (2012). The benefits and dangers of enjoyment with social networking websites. *European Journal of Information Systems, 21*, 512–528.
UNISDR. (2018). Disaster displacement: How to reduce risk, address impacts and strengthen resilience. https://www.unisdr.org/files/58821_disasterdisplacement05a.pdf
Wada, K., et al. (2007). Robot therapy for prevention of dementia at home–results of preliminary experiment. *Journal of Robotics and Mechatronics, 19*(6), 691–692.

WHO. (2015). Public health implications of excessive use of the internet, computers, smartphones and similar electronic devices. *Meeting report*. http://apps.who.int/iris/bitstream/10665/184264/1/9789241509367_eng.pdf?ua=1&ua=1

Xu, L. (2018). Exploiting psychology and social behavior for game stickiness. *Communications of the ACM, 61*(11), 52–53.

Yang, W., & Mu, L. (2015). GIS analysis of depression among Twitter users. *Applied Geography, 60*, 217–223.

Yu, R., et al. (2015). Use of a therapeutic, socially assistive pet robot (PARO) in improving mood and stimulating social interaction and communication for people with dementia: Study protocol for a randomized controlled trial. *JMIR Research Protocol, 4*(2), e45. https://doi.org/10.2196/resprot.4189.

Zhang, Z., et al. (2019). Pathologist-level interpretable whole-slide cancer diagnosis with deep learning. *Nature Machine Intelligence, 1*, 236–245.

Zhu, T. T., Pimentel, M. A. F., Clifford, G. D., & Clifton, D. A. (2019). Unsupervised Bayesian inference to fuse biosignal sensory estimates for personalizing care. *IEEE Journal of Biomedical and Health Informatics, 23*(1), 47–58.

Chapter 12
Identity History

Accumulating digitalized data of each person since birth will become lastly a kind of big data. Those digitized personal data represents entirely each personal history. Here, *identity history* denotes sequential information of lifelog and digital footprint to envisage each person in longer-term perspective. Such data of narratives and life history can be converted to a style of digitized folklores in the heritages by socially sharing (Abello et al. 2012).

Furthermore, by the next century, our average of lifespan may nearly approach to 100 years in some developed countries. And the progress of regenerative medicine using iPS cells and genome editing will propose to reconsider our life. Then, it is time that everyone should contemplate these matters in own life course (Erikson 1950, 1959, 1982) and living value.

12.1 An Insight History on Life and Death

As ethics, philosophy and thanatology said (McCorristine 2017), we occasionally recall a term "*Memento mori*." Many of us can realize own lifespan and its value by facing at the death.

For example, during the oldest civilization in ancient Mesopotamia (BC 3500 to BC 3000), ancient peoples scribed their considerations on life and death by the specific letters (*cuneiform*) to large amount of old tablets. According to the "*Epic of Gilgamesh*," Gilgamesh sought immortality by his conducting the tough adventure to obtain an advice from Utnapishtim. Lastly he encountered Utnapishtim, and he was given an advice, and he obtained a panacea for rejuvenation and immortality. But he lost it at front of the big river returning to his homeland (a snake steeled and drunk it).

Here, there are many questions for him. Why did he drink it sooner at that time? Why did he go back to obtain it again? Perhaps, his enlightenment in depth on the life did undertake his limited lifespan.

K. Shibuya, *Digital Transformation of Identity in the Age of Artificial Intelligence*, https://doi.org/10.1007/978-981-15-2248-2_12

At present, everyone will inevitably face at the end of own life, but our identity will be redefined standing on the prolonged lifespan and its perspective. In some developed nations, each citizen may face at necessity for planning their own one hundred lifespan. Developmental psychological studies will be referable for each adaptation in all stages of own life course (Erikson 1950, 1959, 1982).

12.2 Reconsidering Life on Historical Perspective

12.2.1 Two Meanings of Time

In Greeks, the time concept had two different nuances such as *Καιρός* ("Caerus" in Latin) and χρόνος ("Chronus" in Latin). The former emphasizes formally both subjective meaning of the time and internal mental senses of time. In contrary, the latter means the irreversible, unilateral, continual, and equal velocity process of the time from the past to the future. Namely, the differences between them are understood as mental or physical aspect of the time.

Philosopher Heidegger (2008) said that we are equally "*Sein zum Ende*." He cited Hegelian historical philosophy on concepts of time and history, and he further notated "*History essentially depends on each mental, and it passes within internal time of each*." He distinguished clearly aforementioned two types of the time, and he weighed on internal senses of time. Hence, he cast insightful considerations for digitized social science.

The digitized society conjoining with both nuances has appeared. In those contexts, our identity has already been synchronized by both meanings of the time. As synchronizing subjective experience of time with each citizen via vast interconnected networks, social phenomena raises online contexts in these real-time progress. By semi-sharing each experience among interconnected users, those massive and historical lifelogs of their activities are now available for life-long inscriptions in the chronical order. Will it engender some new forms rather than sharing each time perspective among the citizens?

12.2.2 Evaluating Present Time and Future

In brief, subjective quality of time and well-being in finite lifespan are more important than quantity (Oyserman et al. 2004). Because each has only limited lifespan, an individual recognizes gratefully value of time, and it should be rational to behave in social situation. In behavior economics, ordinary citizens have commonly vulnerabilities to evaluate present and future value based on time discount principle (Camerer 2003). If social networking activities have much present value, many of the users rush into investment for gaining an interest of financial value in the future.

But many of them do not so. Fehr-Duda and Fehr (2016) recently revealed six factors as our cognitive deficits on decision-making for the future valuations. Namely it enumerates (1) other-regarding preferences, (2) self-regarding preferences, (3) delay-dependent risk tolerance, (4) feedback-dependent risk aversion, (5) short-term impatience, and (6) failure to carry through on intentions. Thus those cognitive dispositions will possibly make a stumbling block against preference-based judgments.

Both disposable income and "*disposable time*" have frequently higher value in the digitized networking economy (Easley and Kleinberg 2010). Service providers become more competitive among rivals because each user's physical time has a strict limitation per day. Social media, online games, and other online services do not often permit to perform concurrent tasks among users (Brynjolfsson et al. 2019).

Further, online human relationships have associated with priceless value of social capitals (Kawachi and Berkman 2003; Kawachi et al. 2008). As interconnection per se has a meaning of "*consumer surplus*" for each user in economics, valuation on sharing time among online participants will be weighted on each subjective standard. Such utility of economic value can be evaluated by the stated preferences methodologies such as WTP (will-to-pay: Breidert 2005), CVM (continual value method), AHP (Saaty 2013), and conjoint analysis. In Japan, according to a survey, total average of valuation for daily usage of Line (one of the most popular communication tool of the SNS among Japanese) was estimated as 3 million yen.[1] Such amenity among users can be regarded as more valuable than other economic value in daily life (Whillans et al. 2019). So then, each user prioritizes rates of own disposable time (maximum 24 h per a day) in preferred order. In such ways, at the age of social media, "*Verfallenheit*" (it may nearly mean mentality of inertia condition and eagerness) (Heidegger 2008) by intercommunication among users suggests a symbol of invariance in daily life.

12.3 Big Data of "I"

12.3.1 How to Estimate the Data Volume of "I"?

Please imagine how is the total data volume of "you" at the present? As you noticed, the author intends to estimate all data volumes related to "you." Namely, each individual has a set of own genome that continuously inherited from ancients, and further it can include massive data which are created by yourself such as CV, thesis, dissertations, academic papers, mails, address data, visual data, opinions, fictions, poems, and programming codes within personal computers, mobile devices, and other storages through daily life. And if someone is an Adobe Photoshop creator,

[1] Nikkei Press: https://www.nikkei.com/article/DGXMZO41697360V20C19A2SHA000/ (in Japanese).

total file sizes of his designs and art works should be included. And for other worker as a YouTuber, he often uploads larger size data of his movie clips. Then, the author posted a question to estimate the total data volume above all of them.

At first, a set of genome of the human species had been completely parsed in bioinformatics studies (International Human Genome Sequencing Consortium 2004). Then, a set of genome of the human was estimated totally as 6.4×10^9 bits (approximately less than 800 MB), where total 3.2 billion bases of DNA within a set of genome exists and it assumes one base of DNA as 2 bits. Here, the author has a CD-R on the desktop. This CD-R's capacity labels a data limit up to 800 MB. If the author wants to write all of above genome data into this CD-R, it will be complete (time consumes much to complete it).

Secondly, as the author's case, personal pages of the author on social networking sites occupy only data volume of several megabytes. Even if the author further appends all data within USB memories (totally 20 GB), cloud data (5 GB), a personal computer (max 600 GB), and an iPad (max 16 GB) into the list of "mine," those data includes all of my dissertations, academic papers, programming source codes (i.e., Java, C/C++, Visual Basic, R, etc.), lecture files, data files, CV, photographs, and other necessary files for researches without any data compressions.

Next, there is a case of the *Fermi estimation problems* on the total amount of memorized data within the human brain. The author thought in this way. On the desk of the author, there is a CD-ROM which was installed one of the most popular and largest dictionaries in Japan (i.e., *Kojien, version.4*). This dictionary has more 200,000 articles written in Japanese, and it has approximately 500 MB data (but it was compressed data; plain text data maybe excess over than 1 GB) and visual data within this single CD-ROM. The author has already memorized and recognized nearly all of the articles, and then it nearly implies that the authors' brain memorizes total amount of knowledge obtained from this dictionary more than 500 MB at least.[2] Further, given there are N dictionaries and encyclopedias in professional and academic fields and each of them has 500 MB text-based data respectively, the total amount of knowledge can be similarly estimated 500 MB $\times N$. If inputting 500 to N, the brain of the author may hold approximately 0.25 TB as memorized knowledge at least. Personal and private knowledge should be also appended to the list, and it might be more memorized data. And then those memorized data within the brain has vast capacity. Such estimated data would be implicated a mystery on the knowledge base of the inner brain. Then, our brain has much capacity to memorize and recognize knowledge, and those functions of brain have been revealed by the AI-based contexts (Iqbal et al. 2019). Of course, above estimation was roughly assumed, and there seem to be fluctuated the differences among persons.

In contrary, Landauer (1986) examined empirically to estimate *human-memory size* in cognitive science, and he concluded that an adult (35 years old) can memorize 1.5×10^9 bits data (it nearly equals 187.5 MB, and it is roughly 200 MB). But

[2] Plain text based file size of this book what you are now reading occupies only around 1.5 MB within PC storage (more than total 110,000 English words).

his estimations could be a lower status by experimentally forced memorization and recalling methodologies. And his *memory* meant memorization size of experimental data in each subject person (i.e., working memory or short-term memory), and then it might not be an entire set of knowledge in each person (i.e., longer-term memory).

Therefore, the author thought that a total of "my" data could not exceed more than 3 TB ever since the beginning of the research career of the author (i.e., a set of genome, total digital data in computers including backup data, temporary files and not digitized data, and total knowledge in brain). It is similar to a fictional scene described by Doyle that Dr Watson examined total knowledge and its specifications of his partner Holmes.

12.3.2 A Case of the Three Watsons

However, the memorizing size of the AI-driven system has been already overwhelmed the human-being. At 2011, when an AI system Watson[3] developed by IBM wined human champions of the TV Quiz show, Watson memorized the extent of vast knowledge which inputted from more than *a million books* in its own database. Given each book (it may be likely an encyclopedia) was defined as 1 GB text based data, and then Watson memorized total data as 1 GB × million = 1 PB (it equals 1000 TB). The Watson system further required natural language processing for the AI, and it might include meta-knowledge, ontological knowledge, annotations among linkages of information, and other necessary data. Namely, Watson had memorized vaster knowledge than ordinary human citizen and scholars, and then Watson can be identified as an autonomous and mechanical *library* comparing with the Library of Congress,[4] USA.

As mentioned above, when global 7.5 billion peoples have own 200 MB working memory, the total working memory among all global peoples can be estimated as 1.5 EB (it equals 1500 PB). Otherwise, each given person holds 0.25 TB memorized-knowledge in each brain, and the calculated result is 1.875 ZB. Such social brains can interweave eagerly with each other to actualize the global interconnected worlds in daily life through synchronizing actuality and knowledge using own smart devices. Such globally interconnecting knowledgebase-library with social brains has a potential to reach to a level of the AI Watson. Thus, living system of these mechanisms can be impressively for further possibility of computing (Aimone 2019).

By the way, the author also found the *third* Watson. Although it is trivial, there are already renowned persons who published own genome data online. For example, Dr Watson who obtained a Nobel Prize in Physiology or Medicine at 1962 uploaded his digitized data of whole genome to the online site (National Center for Biotechnology Information, USA). As he contributed to firstly finding on the

[3] https://www.ibm.com/watson/jp-ja/quiz/index.html

[4] https://blogs.loc.gov/thesignal/2011/07/transferring-libraries-of-congress-of-data/

structure of DNA double helix, and he perhaps wished that his genome data should be available for advancement of medical sciences (Wheeler et al. 2008).

12.3.3 Sharing Each Lifelog as Human-History

According to Facebook,[5] at year 2014, they have a total volume of hard disk spaces up to 2.4×10^{18} bits (approximately 300 PB). These digitized volumes certainly consist of all activities of users in daily life such as communication logs, posting data, and other information. Similarly, Shapiro and Aneja (2019) who investigated critically the big data holding by GAFA described as follows: "*For a clear sense of how much personal information the large internet platforms collect and analyze… we discovered that Google holds 3.51 gigabytes of personal data on the Millennial, and Facebook holds 631 MB; for the Boomer, Google has 51 MB of personal data, and Facebook has 71 MB. One MB of information is the equivalent of 583 Word pages, and one gigabyte is the equivalent of 583,038 Word pages. So, Google's current files of personal information on the two of us would fill the equivalent of 2,09467,463 Word pages on the Millennial and 29,735 Word pages on the Boomer, and Facebook's personal data files on us would fill the equivalent of 367,897 Word pages on the Millennial and 41,396 Word pages on the Boomer*".

On the other hands, blockchain-based database, which can share perpetually with global citizens, has potentials to achieve semi-world history logging. Namely, when all of each user post own public and private data to blockchain-based distributed database in global networking in every day, lastly it means global logging to cocreate and share a history in the world.

In this way, blockchain should be designed to endorse an identity assurance of each individual, and everyone share with trustworthy data among all of the citizens. Further blockchain-based database system can be built by the *secure by design*, and then it can prevent deceptive attacks, frauds, and tampering. After that, everyone can access and query into log data in such database if necessary, and it can be verified any evidences on those who should be offered such as education history, academic degrees, job careers, acquired skill levels, and other achievements. And these data can be analyzed corresponding with external evidences on the others' histories and the world records.

Of course, it should manage carefully such data and enact "forgotten rights" in GDPR. Further, progressive engagements by all citizens should be necessary for system managements and mutual verifications on the database. Likely "bit-nation," global citizens will prioritize higher networked identity than traditional nation, and many of them will keep going further steps. In those communities based on blockchain interconnection networking, someone may propose for the legislature of necessary laws and regulations on their databases ruled by themselves.

[5] https://code.fb.com/core-data/scaling-the-facebook-data-warehouse-to-300-pb/

12.4 Histories in Simulations

Turning eyes to more statistical concerns in the human history, one of the oldest demographic statistics was "*the Numbers*" as a part of the Bible. Similarly, ancient peoples managed intentionally rough demographic data among clans in each community. There were many estimation results on global population sizes by demographic studies (Durand 1977). By recent studies in historical demography,[6] ever since homo sapiens (and ancient human-being) emerged on the globe, total accumulated population of the humankind was estimated approximately as 100 billion. In Japan, some studies similarly estimated that the total accumulated population of Japanese was approximately as 600 million.

With such data, archeology (Feinman and Price 2007) and Clio-dynamics have possibly unveiled a part of historical patterns by traditional ways, computer simulations and systems science studies (Grow and Jan 2017). As discussed above, such big data have already stored much lifelogs accumulated from many individuals. If someone can analyze to find specific patterns and dynamics in cultural diversities and demographic changes, those historical dynamics can be defined as statistical and mathematical models to run computer simulation. For example, using ABM, each agent is determined by mocked one of someone from the big data, and lastly it can build one of the possible worlds among such massive agents (i.e., individuals in actuality) in the computer simulation environment. As such way, artificial social simulations have possibility to investigate various models such as "*historical if*" and "*what if*" scenarios based on both possible and impossible assumptions. Recently Turchin et al. (2018) argued multipath forecasting within a possibility of historical scenarios.

Principally, Epstein and his colleagues (2007) had shown an agent-based model on the historical process of ancient culture in North America. They successfully simulated an ancient civilization of formation, changing and disruption by ABM. They tackled comprehensive investigations collaborated with archeologists, environmental scientists, and computational scientists on historical mystery as vanished culture of Native American and those acculturation processes in their colonies. Colonial ruins of their settlement indicated evidences such as maximum capacity of the population estimated by total count of households, amount of pollen of crops in stratum implies agriculture, type of crops, the total amount of productions, and estimated amount of calorie to cover the demographic population in this area. As the results of these assimilated data, they could estimate and simulate demographic changing patterns by ABM. Similarly, Kohler and Gumerman (2000) clarified some artificial social models in the ancient social process.

A history model based on human civilization ought to be categorized as one of the complexity sciences (Johnson et al. 2017). Those dynamics constitute of collective dynamics of people, technological progress, statistical variations, chaotic instability, and emergent property. Those patterns of chaotic phenomena have been

[6] https://www.prb.org/data/ (Population Reference Bureau, USA).

examining by computational social science and data sciences, but it will take long lasting process for complete understanding in ordinary daily life. Such simulations can only depict several possibilities eliminated from all of the possible worlds.

12.5 Forensics

In the context of computer security management, digital forensics takes prioritized data-backup to prepare the criminal and emergency cases for investigations. It plays a role of a forensic investigator in security situations. Similarly, *forensic* ways by the digital technology will contribute crucially toward identifying each person by recovering scattered information.

12.5.1 Needs for Digital Forensics for Identification

Especially, at the disaster and criminal cases, inspectors and police in charge of forensics will be often difficult to repair those remains and articles of the deceased. Thus, there are greater needs discovering the missing people and identifying the corpse who is. Wearable devices and cellular phone with information such as geo-spatial, temporal and communication-log utilizes to locate their stayed sites and to trace historical data of their activities. If possible, sensing behavior patterns of them can be formalized as discerning each individual (e.g., voice, kinetics, height, stride length, facial, body features, etc.). Biometrical and genomic data can clarify decisively the identification clue for those requirements. Therefore, as assuming the disaster and criminal incidents, some experts indeed said that ordinary citizens should prepare and store digital and necessary data for those forensic manners.

On the other cases, when digitalized-data and manuscripts within personal computers and mobile devices will be found (i.e., dying message, posthumous-writings, living-will, and something like that), genuine creator and writer of those remained contents can be identified by machine learning, statistical pattern analysis, and text analysis (Hastie et al. 2009). To date, unless identification of those materials successes, legal matters suspend certainly the due process of further procedures on copy right, life insurance, heritages, and lawsuits.

12.5.2 Revealing Original Writer

Especially, when the materials will be assumedly created by the renowned writer, artists and creator, it will be actually examined by the expertise of appraisers. There is one of the longstanding mysteries in literature since more than a 1000 years in Japan. Was the original writer of *The Tale of Genji* singly (*Ms. Murasakishikibu*) or not (herself with others or multiple persons independently)? This is one of the oldest fictions written by a woman (women?) in Japan. It especially doubts that the last parts of ten chapters of *the tale of Genji* were written by another person.

In this point, recently, using statistical analysis and text mining, Murakami and Imanishi (1999) revealed this mystery and they concluded subtle differences which came from the differences of completed time of each chapter engendered apparent evidences for suspicion in this fiction. Discerning text patterns among *Murasakishikibu* and other persons, they further indicated characteristics of each writer, her historical surroundings at the time of writing conditions and other principal factors.

12.5.3 Archiving Memorials

Using VR and AR, it can build somewhat digital museum on literature, arts, archeology, anthropology, and natural sciences (Dugleana et al. 2017) whether it can be utilized online or not. Digitized materials should be managed in such concerns by successors.

For example, there was an aforementioned Fukushima disaster case caused by accidents of nuclear power plant (aftermath of the disaster at 11th Mar., 2011). Subsequently, the National Diet Library in Japan had opened online digital archive "*Hinagiku*[7]," which comprehensively provides related materials of above disaster such as digitalized visual data, research reports, governmental documents and interview data mapping each rural community. Sharing common memory among contemporary citizens and people of the future generations denotes that each individual will realize as one of successors who should learn the lessons of such tragedy and disaster (Shibuya 2017, 2018).

Shared memories, which converted to digitized folklores such as narrative data and dialogues provided by witness and survivors, have much significant value to share with global citizens beyond the longer-term generations (Abello et al. 2012). In this context, a kind of old folklores and mythologies passed from elderly peoples and ancient peoples had possibly helped future generations to be adapted in environmental risks.

12.6 Irreversibility

Life history and biography of each individual have unilateral physical-time sequence, which is embedded by its own irreversible, irreproducible, invariant, and unrepeatable constraints. To date, medical sciences which is enhanced by big-data

[7] http://kn.ndl.go.jp/

analyzation and the AI-supported medical science have already accelerated its progress in silico, whereas genome-editing and regenerative machines in vivo have coincidentally interacted with them. Especially, there are considerable matters on diversified influences of medical prolongation for lifespan of the humankind, and it should consider the inheritable right in law for heritage when it transplanted from an individual to its biological and digital clones. And then it must prepare appropriate civil laws and legitimizations for prolonging life and regenerative medical progress.

12.6.1 Pandora's Box

As described earlier in this book, according to Cave and Dihal (2019), they reviewed SF taxonomy of either optimism or pessimism on the AI such as the hope for much longer life ("immortality"). To date, there is a good promise that advancement of traditional and regenerative medical science and its engineering prolongs our lifespan, which cared by effective ways, and those efforts provide certainly the chances to fulfill each life goal. One of the largest ICT companies Google already invested the larger amount of financial supports to one of their child companies *Calico* which has been investigating biological and medical sciences such as transgenics, aging controls, and senolytics. However, regenerative medical science has been examining cellular senescence and its immortalization derived from mice (Utikal et al. 2009; Odell et al. 2010), and then it cannot sooner apply to the humans.

Until now, such medical operations never directly reach for the immortality yet. But if possible in the future, when every citizen can be operated by regenerative medical methods repeatedly during each life course, and consequently it can be kept in nearly immortal condition in each person. Gerontology and social welfare will be naturally overturned by such technological improvements. Total benefits are actually better to enhance our quality of life, and those motivations among medics can be approved by many citizens and patients. But as mentioned repeatedly above, there are also contrarily ambivalent feelings among citizens against regenerative medical sciences. The main reasons are conceived to be intruding against the inviolability area of the life phenomena. They may hesitate to open the Pandora's box what was firmly closed since the beginning of civilization, and are suspiciously cautious about the latent risks against existence of the humankind itself.

In brief, those medical operations have equal meanings of direct manipulations for the heart of the *identity* of the human-being. When such Pandora box will be completely opened, the human-being shall manage and arrange properly our identity based on our own physical body. It means to permit continually daily controls and regulations for not only life phenomena but the life course itself of each individual. Namely, Heidegger's term "*Sein und Zeit*" (Existence and Time) should be recalled, and it had clearly indicated such philosophical issues. To conclude this chapter "identity history," it must finally consider this issue in depth.

12.6.2 *What if Human-Being Gains Immortality*

12.6.2.1 Ethics: From the God to the Human-Being

Historically, some of philosophers argued the unique existence that holds immortality, infinity, holiness, and perpetuity. They concluded that such existence was indeed the God. For example, Descartes (2010) and Spinoza (2008) deduced an own logic-based philosophy, and their axiom system was derived from specifications of the God. They thought that only God owns such immortality, infinite existence, and perpetuity in the world, and all of biological species can be realized as a part of the modus and derived varieties from the God. In this context, "*Summum Bonum*" (the supreme good) was equally considered as such the sacred existence, and they founded their logics to be enriched for our modality and moral value. It indeed meant an ethics (Moore 1903). Namely, ethical manner for considerations are ordinarily stood on the highest value of the good which was deduced from immortality and infinite existence in the world.

12.6.2.2 Ethical Issues Against Immortality

Hereafter, the author intends to argue only ethical issues (not theological disputes). Now, what if the human-being really gains *immortality*? In other words, what if the human-being reaches to the near conditions of immortality (McGee and Maguire 2007)? We are still equally mortal beings, and both irreversible and invariant constraints have never been changed yet. All of our civic laws and social institutions had been firmly stemmed in this mortality assumption. Personal rights to manage private equities and real estates are assured under such limited lifespan constraints of the peoples.

Since the Asilomar conference held in the USA at 1975, our biotechnological operation had stepped further toward other dimensions not to get the imaginations against latent risks. However, the advancement of regenerative medical sciences indeed has a potential to violate the barrier to mortality of the human-being. After 2007, we have lastly obtained one of ultimate dream of medical researches. For example, iPS (induced pluripotent stem) cell transplanting operations had achieved to reveal a secret of regenerative phenomena of each cell (Takahashi et al. 2007). Moreover, using useful nucleases (e.g., Platinum TALEN, CRISPR/Cas9), genome-editing technology has cutting-edge for progressing modified proliferations (Martin et al. 2012).

However, as mentioned before, science journal Nature has published special topics on genome-editing regulation and call for discussion at March, 2019 (Lander et al. 2019). Their augments stood on ethical conducts rather than scientific research progress, and international researchers critically claimed against China case (Cyranoski 2019). Genome editing, in this regard, has the largest potential to redesign all biological species, and the responsibility of the human already becomes larger than any traditional biological technologies.

Otherwise, as iPS [8]cell-based regenerative operation has achieved a dream goal of the medical sciences, it has been now undergoing matters to apply for a part of our physical body, because it was satisfied by verifications on the safety in each clinical case. In the earlier part of 2019, it will launch the first case on transplanting retina cell (regenerated by iPS operation) in Japan.[9] Other parts of body such as skins, blood cells, and important organs will be applied for regenerations sooner or later. However, in contrary, many of the developed countries have already forbidden genome-editing operations for medically transplanting modified embryo to uterus of the human-being now. In the later part of 2019, Japan, the USA, the EU[10] (e.g., France, Germany (Fuchs 2010)), and the UK have approved only the basic researches itself but they do not strictly permit to apply for the actual cases (Hovatta et al. 2010). However there are possibilities of unknown jail-break actions in ethic-heaven countries. Despite publishing a misconduct case using genome editing (op.cit., a case of modified twins at the end of 2018), China government invests a large amount of budgets for developing such operations using genome editing, and some cases have been announced successfully. But they publicly said that they legally inhibit to commit medical operations against ethical conducts by medics.

Even though we have dreamed an achievement of regenerative medicine and immortality ever since the civilizations, it is time that we shall redefine our lifespan, quality of life, and our value of life. So then, the author intended to reconsider "*identity*" of the humankind from such contexts. Here, one of the most considerable scenarios is to worry about combinations with big data of medical health obtained from genome- and AI-centered advancements in engineering. Those progresses probably enable to achieve further breakthrough for humanoid artifacts made by ourselves in the future.

In an extreme case, as it is likely a science fiction, some of the citizens will commit to copy own memory in own brain to digitized AI systems and clone whole of the body to another body. So that cases, inevitable disputes certainly will occur. A robotic artifact copied digitized data from an individual should be dealt as the human holding all of the legal rights? Biological clone generated from an individual can also inherit all of the human rights from the original person?

This extreme case exemplifies a serious problem that any artifact that has transformed specific digital data and transplanted biological characteristics from an original individual claims own legal right for inheritance from the original individual. In many civil laws in various countries, each citizens has a succession right from own ancestors and other relatives in law. However, can their clones have also legal rights for succession without any taxes and costs? As they hold a same identity, there is no

[8]Please see 'Web' of 'References' in the end of this book: CiRA (Center for iPS Cell Research and Application), Kyoto University.

[9]The potent effects of Japan's stem-cell policies: https://www.nature.com/articles/d41586-019-02847-3

[10]Please see 'Web' of 'References' in the end of this book: EuroStemCell .

need to pay any taxations for succeeding all of own heritages and estates? If it should not be so, how can we draw the legal border among ambiguous cases?

12.6.2.3 Devaluation on Life of the Human-Being

Regenerative medical scientists often say that everyone will be able to gain possibilities to enhance own quality of life using regenerative cares. It apparently seems so. Their medical operations for the disabilities and needs of medical care will be able to achieve the purposes, and their applicable medical cases will be gradually increasing.

However the author keeps some suspicious minds on their rosy dream. If possible for the ultimate condition of immortality and reincarnation, what will wait for us? As impossibility of the immortality holds the highest value of our living existence, when we could really obtain such technologies for reaching to the immortality, but we cannot obtain such the highest value itself. Alternatively, we will gain the seeds of troublesome.

On the other examples, in China, an animal pet can be legally cloned by biotechnology (of course, they don't permit for the human cases). One of venture Companies *Sinogene*[11] in China has already launched such services for clients. They cloned an arbitrary pet offered from the clients, and each client wishes their pet to be reincarnated. Those costs require approximately 55,000 dollars per a pet at least. Unnaturally, their life can be just evaluated by such costs as living value at that moment.

The more important is that our life will certainly lose its value because we can be modified by medical cares as necessary. Namely, it will call the serious situation of life devaluation on the humankind. Our value of existent life has been firmly piled in one-time life, mortality and finite lifespan, and then many of us vigorously sought to achieve own living goal for success during own life course. Their dreams, struggle efforts, experiences, achieved results, and all of such process itself will make each individual from nobody as an ignorable person to somebody as a supreme and valuable person. Whatever modification of own life and parts of physical body for quality of life have a canonical reason to be cared, devaluation of life will be inevitable. Repeatable and modifiable lives cannot be evaluated by the same standard of living value. Such modifications of own life approaches to immodest value rather than immortality. The priceless value of physical body and its finite lifetime will be degraded and converted sooner to economic value. Such disguised honors for prolonging life will become disappearance as sooner. Even though medical cares should save sublime value of priceless life, when "the Holy Grail" will be obtained, the human-being can alternatively retain only the ash of vain dream.

The author guessed that Gilgamesh abandoned his immortality but he could alternatively gain the everlasting and eternal value of his legend. He positively lost

[11] https://sinogene.org/

the fake value for the immortality, and he decisively imprinted his superiority into the history. Such value cannot be eternally devaluated by anybody. He completed his legend which made everlasting value by such way.

However, many of us will not choose such his route. Repeatable and modified living people will eventually notice at the time of conditions. When all of the people can be prolonged and modified by own arbitrary and preferred design, everyone rashly desires to become the most brilliant and strongest person. So then, it is natural that everyone easily becomes successors without any hard struggles. Do such activities and results have any value? Is it just similar to cheating of video game? A proverb said that the greatest treasure which owned by everyone turns into obsolescence as the worst trashes. Repeatable life courses are similar to the continuing plays in games, and all of them will be bored in such life.

12.6.2.4 Prolonged Working Life

And another problem is social constraints in the living environment.[12] In Japan, average lifespan of ordinary female is around 87 years and male is around 81 years. And their pensions can be normally received since 65 years old or later (the several decade ago, elder persons could receive enough pension since 60 years), and they regularly retire around 65 years old. Annual working time of Japanese at 2015 was around 1800 h. Namely, retirement age becomes gradually later, annual working time is still longer than other developed countries (e.g., France, Germany), and their lifespan has been steadily prolonging.

With these statistics in mind, at first place, it can be reasonable that the total amount of social assurance budgets will be exponentially boosting up more than now. Social assurance budget was calculated 33 trillion yen in Japan, and it occupied 33% of the total amount of national budget. Japan total population estimated approximately 120 million people at 2015, and elder people more than 65 years old existed approximately 33.92 million (more than 25%). If immortality and regenerative medical care have become popular trends, national system for the social welfare, social insurance, demographic management, and medical cares shall be reconstructed by the zero-base scrap and build. Moreover, there are unconditionally residential matters. The population capacity of the homelands will reach to the environmental limitation. Supposing we will be able to live in outer space of the earth, it will be too optimistic to be assumed that our well-being will be undoubtedly heightening by the technologies.

If such medical care possible, working peoples will be forced to continually work during prolonging life, and they will be further encouraged paying much taxes for the nations. Work–life balance will be cared by the nations, but all of them probably cannot retire by own arbitrary timing. Assertion for retirement of working may mean stopping regenerative medical cares by the government. Government cannot

[12] The Japan Institute of Labor Policy and Training: https://www.jil.go.jp/english/index.html

spare much social assurance budgets against retired persons who are in prolonged life course. Unless they work and pay taxes, regenerative cares will not be provided. Labor disputes will be also frequently provoked against such policies. Then, will their rosy life course be enriched by such situations? Although someone calls it a dream breaker for regenerative medicine, this possibility will become the most possible solution.

When such society appears, you will see following propaganda: "*Our nation struggles in front of the hardships of international technological and economic competitions, if you are eagerly working for the nation, we can provide you prolonging lifespan.*" Such rule and slogan mediated by the government will be bizarrely similar to religious dogmas. Traditionally, many religions often concisely say: "*the God requires that you shall keep the dogmas and fight against the enemies, and then the God promises you eternal lifespan after you reincarnated.*" In this way, some leaders of the government will expect to rule over all of the citizens by handling lifespan controls.

It will not be absurd telling. Until now, living desires among citizens have been frequently ruling by the government's will. For example, the constitution of Japan had apparently articulated some articles on "basics of the human rights" and "living right." Some theorists in law studies often interpreted and praised the constitution. It precisely admits and assures that any citizens equally have basic human rights for own living. It is certainly true. But the author conversely thought another facet of those texts in depth. Those meanings of the contents are implicitly dependent on our natural desires to be stayed for own life. Here, the most important is that legal rights can be inevitably assured as a pair with legal duties. When there is a right for living, it should keep a duty for own living. Their living rights should devote to pursuing their duty for own living. Then, it can also say that citizens shall stay to be alive by the end of their lifetime. For staying to be alive, they must keep working harder and earning money. Actually, the constitution stipulates both duties for working and paying taxation. It undoubtedly means that each citizen is implicitly forced to be stayed living and undertaking the contributions for their nations since the birth of each citizen. The above statement "*If you are eagerly working for the nation, we can provide you prolonging lifespan*" will not be recognized as an illegal policy against the constitution.

In the future, when both labor demands and supplies in the AI-driven society will be perfectly worked only by the working robots, it approaches to the social situation for working exceptions among all citizens. At this time, the governor will immediately face at the problem how stipulations and its interpretations on those rights and duties are altered. Although basic income and economic assurances for their living without working and paying taxations will be argued in the AI working society, citizens might be alternatively requested another duty to perform for the nation.

12.6.2.5 Ewig Wiederkehren

Many people will be redundant to work harder even if longer lifespan can be given. Rather, many parts of GDP (gross domestic product) in each nation will become to be undertaken by the AI-driven working machines and robots. So, what happens next? People ought to be questioned what purposes they will be still alive.

Eternal and monotonous history of own life as repeating daily cycles (“*Ewig Wiederkehren”*) which was introduced by philosopher Nietzsche meant that there will be no more saviors for the human-being. In such process, finite history of own given life cannot alter its serial events during lifespan, and rather it just repeats cyclically. In this context, ethics can be understood as a rule of conduct autonomously directed by the will of each individual despite the absence of the “God.” It was certain that it was very influential for understanding the “*Zeit Geist*” (Period spirit). Fukuyama (1992) critically discussed this construal idea on repeating historical events, but it analogically conceived that historical revolutionary process has been confronting and compromising historical and eventual factors in interweaving with diversified bifurcations of the possibilities. And occasionally, in such time series of the history, it could overcome the hurdles against both retentions in the nation and tensions among the nations. The history could be regarded as self-similarity and fractal patterns on events rather than repeating cycles.

In contrast, the author now doubts that the human has the willing to restart own life course at the arbitrary timing even if regenerative medical science achieves prolonged life and immortality. It may engender various life events rather than not be cyclically repeated, but the most critical point is how to really satisfy the value and identity of each individual during own prolonged lifespans. If not so, those who are given prolonged life and immortality will be indeed jailed in the eternal cyclical time.

Economically speaking, utility for prolonging life may be steadily decreasing among citizens. For example, when initial value of prolonging life has merits as a million dollars, after second time, prolonging life value itself will be degraded among citizens as it expects by utility functions. Finally, nobody desires such looped life course. And their WTP and mental discount rate for future value on prolonging life in behavior economics may be expectedly collapsed (Camerer 2003). Market prices can be evaluated by the extent of desires among people, and finally nobody desires such medical care. Financially, each nation should further avoid severe conditions on adverse selection against medical cares and social welfares among own citizens and medical tourists beyond the borders, and then regenerative medicines could not be perfectly delegated to markets for the purpose of sustaining their quality of services. Then, there is possibility that some of developed countries cannot recover their large amount of investments to achieve regenerative and prolonging medicines. It may call a similar consequence of the “Lehman Shock” and “Cryptocurrency Shock” which puts ending periods on devaluation of each life.

12.6.3 Precautionary Principle and Imaginations Against Potential Problems

Furthermore, technological progresses on the medical operations will become harder due to considerable issues related to an inheritance on the identity and its legal properties. It should contemplate this theme. Until now, sociological and ethical disputes would be reactively repeated, but it could not reach the consensus among them yet (Wainwright et al. 2009; Knoepfler 2009; Capps and Campbell 2010).

First of all, medical doctor shall assure secretly storing each medical chart for client. If this subscription indicates usages of iPS cells and some genome editing, should we disclose it publicly and describe such medical evidences to each of family registers? This reason is to globally store and manage clinical data shared by the humankind. Certainly, clinical data and evidence of medicated care are quite important privacy information, but somewhat information management should be necessary to totally prepare against potential risks on modifications of the life among global citizens. Those medicated cares by regenerative operations and genetic manipulations, in the future, have considerable serious risks to transcend modified DNA from medicated parents to children and all of the other ordinary citizens.

When something worse on gene-edited cares will be found, it is certainly that regenerative medical operations for suspicious patients can be cared again. However, there are not still convincible reasons to agree such operations. Principally, there are greater needs to manage negative externality of unintentional accidents caused by genome editing against the whole human-being. In medical economics, such latent risks cannot be underestimated by cost-benefit estimations. But, we cannot still understand enough the whole system of our life phenomena as inner biosphere yet. Scientists are still trying to harder investigate inner dynamisms of a simple living cell and complexity among genetic networking of those functions within cells. Namely, the most significant is that humankind has not gained perfect ways of manipulation and regulation of our life yet.

In this regard, the precautionary principle should warn about those conditions in advance (Martuzzi and Tickner 2004). Actually, the WHO has organized an expert committee for discussing such problems at March 2019, and they intended to recommend the global monitoring and registration systems against illegal polices and anti-ethical researches using genome-editing technologies (Gyngell et al. 2017). One of the committee said "*irresponsible at this time for anyone to proceed with clinical applications of human germline genome editing,*" and such condition should be regulated by the global consensus (Cohen 2019). In addition, many ordinary citizens have often vigorous sentiment against risks on genetic modified organism (GMO) (Fischhoff et al. 1999; Bauer and Gaskell 2002; Shibuya 2013). It still requires convincible solutions. Then, further considerations are quite required for following open questions in this context, at least.

- Should everyone hold own legal rights of physical body and its parts to be cured by regenerative operations?

  - If yes, who does own responsibilities against some failures and future risks?
  - If no, what should those who can be cured by regenerative operations do alternatively?

- Should we maintain "*purity*" of the gene? And should we abort regenerative medicine itself?
- How should regenerative medical cares for clients prioritize in order?
- How should each nation manage both balance between demography and social welfare within the limitation of budget?
- Should each nation control total cares of regeneration per year?
- How should each nation solve overcrowding conditions and food shortages?
- Should we permit medical checks for gene edited or not among all of the citizens by government authority as necessary?
- Should we globally share digitalized clinical data of all individuals using blockchain-based database? If anyone marries with a fiancée, should it verify past clinical data of fiancée by querying into distributed database?
- Should we equally regard a same identity between an original and person who genetically performed plastic surgery? Especially, it should consider some cases of criminal investigation and when there is a necessity of identification proof in gene level.
- When and how should we draw the border to impose taxes for inheritance of private equity? When medical doctor apply a little bit regenerative operations using iPS cells for client, should it immediately regard that the original identity has lost and it shall impose appropriate taxations?
- How should we think about the timing of each death? Is it brain death? But when regenerative medicine and the advanced engineering achieve somewhat reincarnation and total memory within his brain can be transcending to new body, how to consider each death of the human?
- If such society will appear at once, will an embodied individual lose their raison d'etre? Using reincarnation operations based on perfect regenerative medicines and advanced brain engineering, each individual will be just reincarnated by arbitrary genome data from "gene bank."
- How should it educate ethical manners for us when regenerative medicines achieve prolonging life and pseudo-immortality?
- How should we still deal with murder as a crime when regenerative medicines achieve reincarnation and pseudo-immortality?
- Should it admit that AI predicts one of the possibilities for each person based on statistical data analyzation?
- Should it optimize for solution above matters using AI?
- How should we educate the AI driven robots on necessary ethical manners of the AI for the humankind? In other words, how can we explain the dignity of the human-being for the AI when reincarnation will be achieved by regenerative medicine?

There will be innumerable ethical and controversial matters not limited to the above lists. Thus, WHO (2005) has published the older manual for medical ethics, but they should revise the latest version for the ethical rule of conduct for global medics. And both judicial courts and congress in each nation have to be legalized stipulating concrete legal articles in necessary laws. After that, they should tackle to authorize the rules for the AI in due process.

References

Abello, J., Broadwell, P. M., & Tangherlini, T. R. (2012). Computational folkloristics. *Communications of the ACM, 55*(7), 60–70.

Aimone, J. B. (2019). Neural algorithms and computing beyond Moore's law. *Communications of the ACM, 62*(4), 110–119.

Bauer, M. W., & Gaskell, G. (2002). *Biotechnology: The making of a global controversy*. Cambridge: Cambridge University Press.

Breidert, C. (2005). *Estimation of willingness-to-pay: Theory, measurement, application*. Cham: Springer.

Brynjolfsson, E., Colls, A., & Eggers, F. (2019). Using massive online choice experiments to measure changes in well-being. *PNAS, 15*, 7250–7255.

Camerer, C. F. (2003). *Behavioral game theory: Experiments in strategic interaction*. Princeton: Princeton University Press.

Capps, B. J., & Campbell, A. V. (Eds.). (2010). *Contested cells: Global perspectives on the stem cell debate*. London: Imperial College Press.

Cave, S., & Dihal, K. (2019). Hopes and fears for intelligent machines in fiction and reality. *Nature Machine Intelligence, 1*, 74–78.

Cohen, J. (2019). WHO panel proposes new global registry for all CRISPR human experiments. *Science*. https://doi.org/10.1126/science.aax3948.

Cyranoski, D. (2019). China to tighten rules on gene editing in humans. *Nature*. https://doi.org/10.1038/d41586-019-00773-y.

Descartes, R. (2010). *Meditationes de prima philosophia* (English translated edition). Fili-Quarian Classics.

Dugleana, M., et al. (2017). *VR technologies in cultural heritage*. Cham: Springer.

Durand, J. D. (1977). Historical estimates of world population: An evaluation. *Population and Development Review, 3*(3), 253–296.

Easley, D., & Kleinberg, J. (2010). *Networks, crowds, and markets: Reasoning about a highly connected world*. Cambridge: Cambridge University Press.

Epstein, J. M. (2007). *Generative social science: Studies in agent-based computational modeling*. Princeton: Princeton University Press.

Erikson, H. (1950). *Childhood and society*. New York: W. W. Norton.

Erikson, H. (1959). *Identity and the life cycle*. New York: International Universities Press.

Erikson, H. (1982). *The life cycle completed*. New York: W. W. Norton.

Fehr-Duda, H., & Fehr, E. (2016). Game human nature. *Nature, 530*, 413–415.

Feinman, G. M., & Price, T. D. (2007). *Archaeology at the millennium: A sourcebook*. Cham: Springer.

Fischhoff, B., et al. (1999). *Acceptable risk*. Cambridge: Cambridge University Press.

Fuchs, M. (Ed.). (2010). *Forschungsethik: Eine Einfuehrung*. Stuttgart: J.B. Metzler.

Fukuyama, F. (1992). *The end of history and the last man*. New York: Free Press.

Grow, A., & Jan, V. B. (2017). *Agent-based modelling in population studies: Concepts, methods, and applications*. Cham: Springer.

Gyngell, C., Doublass, T., & Savulescu, J. (2017). The ethics of germline gene editing. *Journal of Applied Philosophy, 34*(4), 498–513. https://doi.org/10.1111/japp.12249.

Hastie, T., Tibshirani, R., & Friedman, J. (Eds.). (2009). *The elements of statistical learning: Data mining, inference, and prediction*. Cham: Springer.
Heidegger, M. (2008). *Sein und Zeit* (English translated edition). New York: Harper Perennial Modern Classics.
Hovatta, O., Stojkovic, M., Nogueira, M., & Nieto, I.-V. (2010). European scientific, ethical, and legal issues on human stem cell research and regenerative medicine. *Stem Cells, 28*(6), 1005–1007.
International Human Genome Sequencing Consortium. (2004). Finishing the euchromatic sequence of the human genome. *Nature, 431*, 931–945.
Iqbal, A., Khan, R., & Karayannis, T. (2019). Developing a brain atlas through deep learning. *Nature Machine Intelligence, 1*, 277–287.
Johnson, J., et al. (Eds.). (2017). *Non-equilibrium social science and policy*. Cham: Springer.
Kawachi, I., & Berkman, L. F. (Eds.). (2003). *Neighborhoods and health*. Oxford: Oxford University Press.
Kawachi, I., et al. (Eds.). (2008). *Social capital and health*. Cham: Springer.
Knoepfler, P. S. (2009). Deconstructing stem cell tumorigenicity: A roadmap to safe regenerative medicine. *Stem Cells, 27*, 1050–1056.
Kohler, T. A., & Gumerman, G. J. (2000). *Dynamics in human and primate societies*. Oxford: Oxford University Press.
Landauer, T. K. (1986). How much do people remember? Some estimates of the quantity of learned information in long-term memory. *Cognitive Science, 10*(4), 477–493.
Lander, E., et al. (2019). Adopt a moratorium on heritable genome editing. *Nature, 567*, 165–168.
Martin, J., et al. (2012). A programmable dual-RNA–guided DNA endonuclease in adaptive bacterial immunity. *Science, 337*(6096), 816–821.
Martuzzi, M., & Tickner, J. A (Eds.). (2004). *The precautionary principle: Protecting public health, the environment and the future of our children*. Copenhagen: WHO. http://www.euro.who.int/__data/assets/pdf_file/0003/91173/E83079.pdf
McCorristine, S. (2017). *Interdisciplinary perspectives on mortality and its timings: When is death?* Cham: Springer.
McGee, E. M., & Maguire, G. Q. (2007). Becoming borg to become immortal: Regulating brain implant technologies. *Cambridge Quarterly of Healthcare Ethics, 16*(3), 291–302.
Moore, G. E. (1903). *Principia ethica* (Rev. ed.). Cambridge: Cambridge University Press.
Murakami, M., & Imanishi, Y. (1999). *On a quantitative analysis of auxiliary verbs used in Genji Monogatari* (in Japanese). https://ipsj.ixsq.nii.ac.jp/ej/?action=repository_uri&item_id=12756&file_id=1&file_no=1
Odell, A., Askham, J., Whibley, C., & Hollstein, M. (2010). How to become immortal: Let MEFs count the ways. *Aging, 2*(3), 160–165.
Oyserman, D., Bybee, D., Terry, K., & Hart-Johnson, T. (2004). Possible selves as roadmaps. *Journal of Research in Personality, 38*, 130–149.
Saaty, T. L. (2013). *Theory and applications of the analytic network process: Decision making with benefits, opportunities, costs, and risks* (English Edition). Pittsburgh, PA: RWS Publications.
Shapiro, R., & Aneja, S (Future Majority (USA)). (2019). *Who owns Americans' personal information and what is it worth?* https://assets.futuremajority.org/uploads/report-for-future-majority-on-the-value-of-people-s-personal-data-shapiro-aneja-march-8-2019.pdf
Shibuya, K. (2013). Risk communication on genetically modified organisms. *Oukan, 7*(2), 125–128. (in Japanese).
Shibuya, K. (2017). *An exploring study on networked market disruption and resilience*. KAKENHI Report (no. 26590105), pp.1–200 (in Japanese).
Shibuya, K. (2018). *A design of Fukushima simulation*. The society for risk analysis: Asia conference 2018, Japan
Spinoza, B. (2008). *Ethica, ordine geometrico demonstrata* (English translated edition). BiblioLife.
Takahashi, K., et al. (2007). Induction of pluripotent stem cells from adult human fibroblasts by defined factors. *Cell, 131*, 862–871. https://doi.org/10.1016/j.cell.2007.11.019.

Turchin, P., et al. (2018). A history of possible futures: Multipath forecasting of social breakdown, recovery, and resilience. *Cliodynamics: The Journal of Qualitative History and Cultural Evolution, 9*, 124–129.

Utikal, J., et al. (2009). Immortalization eliminates a roadblock during cellular reprogramming into iPS cells. *Nature, 460*, 1145–1148.

Wainwright, S. P., Michael, M., & Williams, C. (2009). Shifting paradigms? Reflections on regenerative medicine, embryonic stem cells and pharmaceuticals. *Sociology of Health and Illness, 30*(6), 959–974.

Wheeler, D. A., et al. (2008). The complete genome of an individual by massively parallel DNA sequencing. *Nature, 452*, 872–876.

Whillans, A., Macchia, L., & Dunn, E. (2019). Valuing time over money predicts happiness after a major life transition: A preregistered longitudinal study of graduating students. *Science Advances, 5*(9). https://doi.org/10.1126/sciadv.aax2615.

WHO. (2005). *Medical ethics manual.* http://www.whcaonline.org/uploads/publications/em_en.pdf

Part III
Discussions

Chapter 13
General Discussion

Each chapter from 1 to 12 of this book has been respectively discussed. Finally, taken together, the author may be permitted to generally discuss a little bit freely for further considerations. As it seemed likewise Cartesian, the author firstly defined preconditions for identity of the humankind (Chaps. 1 and 2), subdivided the whole problems on recent technological matters to necessary parts (Chap. 2), and discussed each issue in depth (from Chaps. 3 to 12). In this chapter, the author further contemplates those components from another angle for reconstructions, which should be integrated for understanding and reasoning some significant variations on proceeding future patterns.

Probably, because researchers must mostly manage relevant vast knowledge crossing various fields, and then there are currently too limited researchers who can simultaneously deal with combinations of multiple keywords such as ethics, sociology, AI, regenerative medicine (e.g., iPS and Stem cells, gene-editing), big data, digital technological transformation, and other engineering. But in contrary, the author has a chance to comprehensively discuss those issues standing in an earlier position.

13.1 Historical Necessity and Certainty

As indicated earlier, the author has a disposition to eagerly review any social phenomena under the continuous dynamics and variating progress such as historical changes, moving patterns of any variables and complexity. And better discussions for the adaptive survival of the humankind must be performed by rather evidence-based considerations on bifurcations from a part of the possibilities than unimodal blinding ways not to be reflected in various possibilities.

Here, let me think about some historical hypothesis on AI. Namely these are both necessity and certainty on AI. In world historical perspective (McNeill 1999; Kremer 1993), ancient people sought to desperately survive in each environment,

K. Shibuya, *Digital Transformation of Identity in the Age of Artificial Intelligence*, https://doi.org/10.1007/978-981-15-2248-2_13

and they would eagerly hope to gain new knowledge and technologies for enhancement of their survivals, happiness, and well-beings. All of such knowledge was to be rooted in conquering *diseases, death, restless works,* and *other emerging risks against external harmful enemies* during their finite lifespan. Namely, it can concisely conclude that ancient people probably devoted to advancing technologies to approach for medicine, the world which exempts working, and advanced weapons. For accomplishing such dreams, they steadily unveiled the nature of life and built somewhat entity worked by automatic mechanisms. In many of the ancient civilizations, their legends, mythologies, and folklores often described an adventure for exploring immortality, reincarnation, and immortal existences. These narrative stories clearly told the evidence what ancient citizens thought such that, and actually many religious rituals were seemingly grounded in such desires. And social distinction between a king as ruling class and servants who categorized as social classification indicated common hopes of the human. Advanced weapons could become as a signaling for governor's legitimacy in owning lands. An outline of the human history can be undoubtedly understood as such contexts.

In brief, ever since the dawn of civilization, it might be certain that ancient and contemporary workers have been vigorously conducting to achieve the ideal society where all of their works and duties will be replaced by the others (e.g., human servants in the past, autonomous AI-driven robots), even if each knowledge and technology has not appeared to link with such ultimate goal.

13.2 Latent Threats of AI

13.2.1 AI Threats

Actually, many citizens, researchers, and national leaders commonly retain to feel anxieties against the AI (Bostrom 2013; Yudkowsky 2018; Taddeo and Floridi 2018; Yu et al. 2018; Congressional Research Service (USA) 2019). First, advanced AI has the potential to be applied to military weapons. Secondly, international tensions among the superpowers will be rebalanced by the AI-based military forces and cyber weapons. And thirdly there is another possibility that latent threat of AI becomes to the actual enemy against the humankind.

As first concern, the final answer which advanced AI is ultimately a lethal weapon becomes a consensus among global citizens in term of the human-being's history. Especially, many university researchers and the United Nations' committees have stated the latent threats of AI such as LAWS (Lethal Autonomous Weapon Systems) and other military specifications (UNODA 2017; UN Chronicle 2018). Certainly, advanced AI can become as inseparable partners, but simultaneously there is a possibility to alter our inevitable threats. Double edge sword of such AI has been warned in various viewpoints. In the context of AI regulations, each nation begins to tackle such concerns.

In the presentation on AI symposium at 2018,[1] as one of the political leaders, Russian President Putin talked on "*Political instability and warfare,*" and he further warned "*Artificial intelligence is the future, not only for Russia, but for all humankind. It comes with colossal opportunities, but also threats that are difficult to predict. Whoever becomes the leader in this sphere will become the ruler of the world*". The author thought that above presentation by Russian president has another meaning in depth to restraint against other antagonists and opponent countries. Actually, at August 2019, UNOG[2] (the United Nations Office at Geneva) published their working paper, but even global restriction treats on LAWS (including AI-driven military forces and autonomous military robots without human-supervision) has been struggling to reach completely the global agreement and its effective controls among nations. Superpowers vigorously commit to draw borderlines keeping each national security and military dominance with own initiatives holding AI-driven weapons as the ways for the balance breakers.

Regarding this point, in second controversial issues, Garcia (2018) similarly argued that it grows latent risks against international peace caused by developing advanced weapons using military AI systems and AI-driven robotics. AI-based weapons will become a balance breaker against international cooperation and the military power balance among big nations. In particular, such military R&D competitive condition further engenders the new conflicts between the superpowers such as the USA and China. Eurasia Group[3] (2018) reported "top 10 risks" in the world, its first place was "China" and third place was "Global tech cold war." Surprisingly, Morgenthau (1978) had superbly predicted (1978) such world situation at the age of the cold war, and he vividly forecasted that the greater potential of China's presence will be achieved by technological progresses and demographic population. It should deepen international military controls including AI and strategical policies (Shapley 1980).

Otherwise, as warned above, there is another concern that the latent threats of the AI turn into the enemy against citizens. Their telling stories, which AI's rebellion and uprising against the human, have been already arranged by science fictions. And now, they similarly thought that it will be possible to become an actuality. In other words, the history told that aristocracy-based governments possessed such anxiety against peasants, and any revolutions between the governor class and ruled classes had been repeated anywhere and anytime. It feels one of the same patterns between the human-being and the AI. The author saw a *déjà vu* on such stories.

To conclude above points, strategic thinking aims to firstly decide the goal, and to explore possible *equilibriums* between stakeholders. Namely, an augment on the threats of the AI against the human-being indicates how to equalize each power balance among AI master as the military nation versus ordinary nations, international

[1] https://www.cbinsights.com/research/ai-threatens-humanity-expert-quotes/

[2] https://www.unog.ch/80256EDD006B8954/(httpAssets)/0F149C84EF8D36CEC125845C0051F592/$file/CCW_GGE.1_2019_WP.7.pdf

[3] https://www.eurasiagroup.net/issues/top-risks-2018

rebalance of the powers, and mutual relations between the human and the AI. First and second matter are necessary to be regulated by the international acts for the next peace, and the third point recommends installing somewhat ethical circuits and decision-making systems within the AI before shipment from factory.

13.2.2 *The Cases of Threats by AI*

In the future, AI battle fronts will be enlarged in both cyberspace and physical environment.

The former is generally to activate computer cracking by the AI and intrusion of software entity in the context of cyber war. Such entity based on AI may be implemented as a software agent, which can behave probably smarter than ordinary computer virus (i.e., mobile agent across computer systems). Security validations for software agent have been already possibly checked by behavior analyzation in the sandbox which is relocated from computer resources. In the case of cracking targeted to network vulnerabilities, it has been often damaged by malicious human crackers. And the causes are often underlying in a fact that system managers are redundant and incomplete to update their managed systems. Namely, human crackers and hacktivists that are hired by antagonists and terrorists must be firstly investigated and arrested by police. When someone faces suspicious situations, they should inform cyber security professionals as soon as possible in accordance with security policy (in many cases, it may not shut down computers if possible, because professionals will verify both memory and CPU in background running process of damaged computers).

Otherwise, the latter case will be seriously assumed to order and command combat actions involving physical destructions among military robots which installed AI (Congressional Research Service (USA) 2019). Some countries have already developed autonomous reconnaissance aircrafts, AI-driven versatile moving machines and insect like micro drones. In ordinary conditions, these military weapons certainly have severe explicit threats against our daily peace. The author thinks that latent threats will be equal between human terrorists using some weapons and such AI-driven military robots. Additionally, AI-driven ICBM and portable missiles will be equal threats against us. Albeit human terrorists should be arrested by police if possible, when someone encounters somewhat physical entities as military robots, they should ask police and military for help. Such robots can be thoroughly damaged and destroyed by ordinary physical strikes and energy supply suspensions. It will not be easy tasks for the military, but after that those mechanisms should be investigating in national security context. In the case of collective destruction activities by innumerable AI-driven machines, it should firstly overwhelm to their master nations or directive organization.

That is to say, at the present condition, the author concludes that both cases will firstly watch more human-factors (e.g., crackers, terrorists, military nation's leaders, etc.) than AI factors. As the advantage of the AI-driven military machines will be regarded as lifeless entity, easier repair, and massive productiveness, it shall prepare to build social systems where all of the AI-driven military machines will be continually kept monitored in advance.

13.2.3 Further Points

In actual conditions, the author points out other considerable issues. It should separate different views on technological advancement such as governors, researchers, business industries, and ordinary citizens.

- In governmental level, as mentioned earlier, *cyber sovereignty* has been focused in international conflicts. If each nation can assert its region of cyberspace, it will turn out another problem (i.e., cyber wars). Furthermore, there were criticisms on decision-making at the brink of national crisis (Herek et al. 1987), even if smarter governors often decide rational judgment against risks.
- At researchers' level, DU (dual use) problems on AI technology have been tackling among them under considerations. Traditionally, internet and mobile devices are successively rooted in the origin from ARPANET and GIS military products. At 2018, Google employees claimed AI application to develop for USA army, and they asserted the needs of human security against advanced AI-driven arms. But many of ICT and other engineered technologies have commonly shared the latent risks to be abused against the military purposes and its artifacts. For example, the author had similar experiences. A published paper on ubiquitous computation for analyzing human behavior which was written by the author (Shibuya 2004) had been cited from a military report edited by one of the Canadian military R&D institutes (Abdalla and Niall 2007). Namely, this clearly indicated that they intended to apply it for their military forces. This example indicated that the invisible military intention is always surrounding around ordinary researchers.
- In business industries, they vigorously intend to invest much money to AI and big-data fields. Many of them aim to establishing a definitive position as next platformer.
- For ordinary citizens, services provided by AI and its relative computing systems should be accessible, usable, and handier. The AI systems can be assured to offer an amenity and safety for our life.

13.3 A Coexistence Possibility Theorem and Its Limit

As discussed ethical issues for both citizens and the AI at past chapters, in actually, ethics and philosophy have to dedicate delving into both meaning of living value and the ways to deepen understanding. As the author throughout discussed wider issues in previous chapters, ethical rules for both people and the AI must be entirely contemplated deeper and wider. As being discussed through this book, at least, there are enumerated disputes on identity in the contemporary digitized world.

Why do we prohibit to murder against the other person (see Chap. 3)? How is possible coexistence with the human and the AI-driven robots? Before such condition, we must get started firstly to contemplate how to cooperate with each other beyond the differences. For ordinary people, their ethics have to be ruled in coexisting with the other people and conquering the hardships such as daily life, crisis, opinion divides, intergroup conflicts, cultural and racial differences whether online or not. In contrary, ethical rules for the AI-driven mechanisms shall be defined, supervised, and governed only by the humankind (International Committee of the Red Cross 2018; Yu et al., 2018). And of course, there should be inquiring into meta-level integrated rules for bridging between each of us.

As introduced ethics written by Spinoza (2008) and Descartes at previous Chap. 12, they attempted to build own axiom logics in philosophy and ethics. Then, the author similarly attempts to logically prove a "*Coexistence Possibility Theorem*" based on the necessary preconditions for identity of the human (refers to Chap. 2). This is a propositional and predicate logic (Singh et al. 2000; Russell and Norvig 2009), and it enables to extend for a style of modal logics and AI-based interference system. This logic can be clearly proved by deducing contradiction way to be understandable for the AI and younger citizens, and those logics can deduce and unveil further considerable points step by step in depth.

13.3.1 Proof

- **1st Proposition**: It permits to murder the other existence in the world which constitutes of multiple people.
- **2nd Proposition**: A person (x) who intends to do murder other existence still desires to be alive in the world.
- **Proof**: "The other existence" in 1st proposition does not mean x, because x still desires to be alive and there is no reason to commit suicide. But all existences excepts for x in the world can be also regarded x as the other existence standing on their viewpoints. Namely, 1st proposition also permits all members to do murder against x in the world. Then, it can harm a living possibility of x by the harmful threats of the others, and such condition contradicts the wills of x to be alive in 2nd proposition. Therefore, 1st proposition cannot be true. (Q.E.D.)

As the author had a teaching experience on this theme in a university classroom, this logical statement could be consented for the young claiming students. This coexistence theorem seems very simple and rough proof, but it becomes drawing the clearer outline of this logical space. There is much implication in possibility for both coexistence and cooperation among stakeholders. Namely, it clearly shows that anyone who desires to stay alive should recognize and accept the others' existences and their willing to be alive in the world. It is too suggestive that the human-being can firstly recognize the other existence after confronting with the others in the world. Choosing to stay living is simultaneously to permit the other's physical existence. Both of them shall be tightly related with counterpart in same physical constraints. Such situation was repeatedly pointed out by Heidegger's philosophy and other ethics.

In addition, above logics and proof styles are understandable for the AI-driven machine. As mentioned earlier, all of the AI systems should be dedicated for ethical rules to be learnt by their trustable manners (refer to Chap. 3). Logic-based inference can be defined as a set of IF-THEN rules, and the AI-driven system accurately operates provable logical statements (Russell and Norvig 2009; Gintis 2014). The processing system of the AI has to autonomously conduct behaving itself within the permissions of such ethical rules for us. Then, we always require at least *traceability for the AI logics*, *understandability for the AI behaviors*, *transparency for the AI thought*, and *controllability for the AI thinking manners*. As necessary, all ethical rules among us can be converted to interpretable logics for the AI, and vice versa. Similarly, IBM (2018) already determined own ethical standards defined as the *explainable AI*. Those implementations are obviously required in daily routines and judicial matters, for example.

Occasionally, someone said that such logics have just common senses for all citizens, and *we already know it*. But the AI system and younger students cannot understand such knowledge yet. What to learn usually starts to ask a question. Many of adult citizens cannot clearly give an answer based on descriptive, modal, and normative logics for children as well as the AI-driven robots. Lacking answers are explicitly to mean that many of elder citizens did not contemplate such matters in depth, and the capability of mind requires being reflective.

13.3.2 Balance and Equilibrium

Aforementioned theorem exhibits to reach the ideality of coexistence in the common physical space. In other words, these patterns show mutual interdependency and "*double contingency*" (Luhman 1984). Social systems can similarly manage from dyad interpersonal situations to complicated relationships among diversified groups and international countries.

But actually, when multiple stakeholders exist in the same environment, a next problem is how to be strategically *balanced* among their powers for peaceful coexistence (Levy 1984; Sheehan 1996; Fearson 1994). In abstract level, it simultaneously denotes mutual strategical viewpoints on either cooperation for coexistence or assaults each other among multiple members. For example, if four people (A, B, C, and D) exist in the community and person A stays as the strongest position and overpowers the others, the other people will keep the balance of their coexistence among them for confronting with person A.

Game theory indeed mathematically dealt with such strategic situations and alliance formations (Neumann and Morgenstern 2007). There is always the possibility that stronger counterpart will assault weaker one using any lethal weapons as balance breaker. Supposing first strike can perfectly complete to annihilate hostile opponent, such situation will be incentive to cause it. Those social situations are similar to chicken-game and PD (prisoner-dilemma) game, and TFT (tit-for-tat) strategy in serial iteration games was adapted in complicated environment (Axelrod 1997; Jost 2005; Cederman 2005). Then, each of them intends to enforce own power and keep power balances among them. It just begins enduring military races (Ayson 2004). Paradoxically, for keeping peace condition which defined as *equilibrium*, it must also continue enhancing and maintaining own military forces to be balanced among stakeholders. The main goal is not to battle with opponents, and it might be rather to keep balance with them. Unless the other existence vanishes, those tensions also never stop.

If such power balance by *equilibrium* can be principally established, and after that it ought to keep the *equilibrium* among them, and namely these are *cooperation, compromise, alliance, coalition, agreement,* and *legal treaties for nonaggressions.* It could be underlying in these relations if and only if mutual stakeholders recognize the other who desire to be alive. In this way, the reasons why the cold war conditions among big-nations and power balance of military forces are often required keeping *equilibrium* as one of the peace situations are understandable for us (Schelling 1980; Jackson and Nei 2015). Mutual trust among nations can be maintained as long as each nation continues compliances with their agreements.

Actually, Morgenthau (1978) argued that existential desires of own living among citizens would be globally founded for the common senses and value standards, and he further discussed possibilities to let them hold the world opinions which should be shared globally for sustaining democratic regimes in accord with the balance of power among nations. Nevertheless we must obviously differentiate subtle nuances of world opinions seeking for global cooperation between nominal standards of the humanity and strategic doctrines in each nation. But maximization of possibility for staying own existential life in the both situations can be presumed as the basic nature of the human to conceptualize any strategical and moral conditions. Without all stakeholders must choice firstly staying own life, it cannot strategically step forward toward establishing anything in the progress.

In daily contexts, it can observe similar patterns. As noted at the Chap. 3, if you are a student and pupil belonging in school, there are often bullying and hate speeches against minority and comparative weaker person whether online or not. *Minority* stays in the situation on an imbalance of the power in the classroom. Then, it is generally the nutshell of this problem on school bullying and conflicts among classmates relating to steady clusters of preferred based human relationships (e.g., sociometry). For preventing severe conditions, generally, those who face at the harmful antagonists organize for self-protections to align with the others who can help with each other (e.g., your friends, those who become your friends). It actualizes a kind of coalition formation in game theory. When those friendships and relationships as networking allegiance cluster appear larger presence against antagonists, everyone will experience the significance of the balance of power among them.

Further, if it is not enough, they may request for helps to be attended by external legal professionals, helpful teachers, and trustful school psychologists.

As simulated at Chaps. 6 and 7, social intelligence to detect anomaly, deception, and betrayals among partners might be equipped in our course of civilization history. It can be defined as mutual watching relationships against social conditions which can be formalized as Byzantine Generals Problem. And the ways for deception detections and self-proving procedures would be welcomed in such situations. Mutual trust in friendship can be maintained during each of them continues the condition not to be betrayed. Hence, the trust condition cannot be beforehand given from neither internal nor external factors in advance, and rather it is an emergent property fostered by interrelationships during mutual performances on the agreement which can be conducting by each of them.

It is the same thinking on alliance formation in international affairs. The heart of this problem is that we are mutual enemy against each of us in natural condition, as Spinoza (2008) said. Namely, it is to guard myself against attacks by the others as antagonists, and to align with the others as trustful partners. In that way, some larger clusters will appear in society from microlevel (i.e., individuals) to macro level (i.e., nations). One assertion and ideology often engender its opponents, and those mechanisms to maintain the power balance among them have a necessity to be realized longer peace situation. As mentioned at the Chaps. 8, 9, and 10, the human-being are subject to flock with each other reducing social uncertainty whether online or not, and it is called as in-group and like-minded community. This is also a basic reason why opinion clusters often appear facing at the controversial issues, why somewhat categories (e.g., class, race, ethnicity, ideology) make people difference, why intergroup polarization appears, why prejudices based on intergroup disadvantages often cause to realize secondary social disputes, why intergroup conflicts endlessly engender dividing people in the nation, and why international warfare among alliances was repeatedly co-occurred crossing the borders since civilization. And as mentioned at Chaps. 9 and 10, when stabilization of such confrontation of opinions among citizens within the nation could not be lastly legitimized by majority of citizens, legitimation crisis will be easier to be provoked by collective dynamics of citizens at this age of social media (Shibuya 2017).

In the natural condition, the human has been laid in mutual distrust with each other, and then, the peace per se shall be treated as one of the emergent property by establishing the balances which stably kept among the nations (Kant 1795). According to *Kant's triangle for peace*, he listed up *firm alliances, democratization, economic interdependency among nations*, and *international institutions* (e.g., League of Nations and the United Nations). These factors can affect that each of the nations has any strong intensives to betray and commit military attacks against other nations. These are necessary factors to establish *equilibrium* among stakeholders, and in actual case, détendre (disarmament treatments) among nations such as NPT (Treaty on the Non-Proliferation of Nuclear Weapons) will become under consideration if both nations satisfy each perceived power balance among them (Morgenthau 1978). Unless such balancing criteria do not satisfy among the nations, comparative weaker nations will be forced to accept humiliating conditions by stronger one.

13.3.3 National Security and Warfare

The warfare per se enduringly exists unless the other hostile existences will not be completely absent. When an individual (e.g., one nation, one race, one community) exists only in the environment, it is no warfare. It is quite true that the warfare are exactly caused by the other existences (counterparty and enemy) confronted with oneself. In a nutshell, without saying, warfare is definitely the ways to achieve the ultimate goal for the extinction of the other existences (e.g., alter-ego) by standing on perspective of oneself. Therefore, the ways of warfare against hostility may switch to the alternatives of more lethal procedures than ordinary ways in international affairs, when discussion-based resolutions are collapsed among stakeholders.

The definition on warfare varies quite different between stipulations of the articles on the international law and military operations among the nations. And required roles and missions look different. The former is to offer the legal backgrounds to self-guard rights and right to collective defense against invasion by the enemies, coordinate conflicts, and decide judicial resolutions in warfare among the nations. The latter is to be considered to keep own both strategic and tactical advantages of military forces and destroy decisively against the enemies and its homelands (Caforio 2018).

Traditionally, philosopher Locke (1998) clearly articulated "*The state of war is a state of enmity and destruction; and therefore declaring by word or action, not a passionate and hasty, but sedate, settled design upon another man's life puts him in a state of war with him against whom he has declared such an intention, and so has exposed his life to the other's power to be taken away by him, or any one that joins with him in his defence, and espouses his quarrel; it being reasonable and just I should have a right to destroy that which threatens me with destruction; for by the fundamental law of Nature, man being to be preserved as much as possible, when all cannot be preserved, the safety of the innocent is to be preferred, and one may destroy a man who makes war upon him, or has discovered an enmity to his being, for the same reason that he may kill a wolf or a lion, because they are not under the ties of the common law of reason, have no other rule but that of force and violence, and so may be treated as a beast of prey, those dangerous and noxious creatures that will be sure to destroy him whenever he falls into their power*". Then, he admitted citizens who possess the natural rights to fight and self-guard against any enemies, and it appears that warfare among the nations can be defined by such reasonable bottom-up logics based on citizens' rights. Otherwise, Rousseau (2009) permitted that each citizens can fight for their safety against invasions from the other nations, and during ordinary situations, each right for own saving life should be totally delegated to the national military forces.

However, recently, there is another problem on citizens' identification in their nation. Bit-nation and networked identification has been increasing the extents of participation, and there are no guarantees that such associative organization could control their dynamics. Supposing many of them commit in virtual nation rather than actual nation, what and who should national military systems defend?

In addition, even though it will possibly conquer the conflicts beyond the hardships, the author wonders that the actual solutions will be never reached to the ideal peace condition as long as parallelizing each standard of the value among stakeholders. When cryptocurrency and rewarding online-economy still encourage citizens to evaluate everything by the businesslike cost-benefit analysis, as discussed at Chap. 4, there is an open question how their standard of value can be altered by leading toward specific direction for establishing better situation. Namely, can cooperation and peace led by the rewards of scoring services be possible? The author has not any confidences that multiple causal relationships of conflicts across complicated factors can be rewarded for better understanding among disputing parties by those incentive mechanisms. And there will be disturbing factors against the reaching to peace conditions, for example, as noted above, because fake news producers and propagators by through ordinary internet and dark-web can earn much money against target persons, countries, and organizations. Those propaganda and agitation effects caused by fake news will be stumbling blocks to achieve more peaceful conditions, and actually endless cycles of conflicts are still ongoing matters wherever the human exists.

Finally it touches with more predestination of the human. As one of the famous episodes, psychoanalyst Freud was asked from physicist Einstein. "*Why does not the humankind abandon wars?*". He answered a cause on the subliminal "*destrudo*" (death drive) of each person. But, in this context, the author points out that ordinary people do not hold such drive, and rather terrorists may be activated by such drive. And the AI-driven military forces might be regarded as an actualization of death drive itself. Then, if it happens, it will become another type of warfare between the ordinary citizens and the AI-driven military forces.

13.3.4 Symmetry Breaking Between the Humankind and the AI

For further discussing above coexistence theorem, please recall that the author defined six necessary preconditions on identity of the human at Chap. 2. Further, it can hypothesize to derive below corollaries based on those preconditions as necessary. And these features can be named as a set of *three symmetries*, as scientific evidences endorsed each of preconditions, each ordinary citizen is hypothesized possessing these primitive characteristics. In terms of three symmetries, next it envisions the differences between the human and the AI.

1. Corollary 1 (Living desire): It means a desire of staying own life whether it is finite lifespan or not. By preconditions from 1 to 4, it must be true.
2. Corollary 2 (Intellectual abilities): It means own rationality and intelligence. By preconditions from 3 to 6, it must be true.

3. Corollary 3 (Empathy): It means sympathy and its emotional sensibilities of understandings for the other finite lifespan existences. By preconditions from 1 to 6 as well as above corollaries 1 and 2, it must be true.

These classified mental functions emphasized the foundation as a human. Although there were still several differences, this classification is similar to *the Plato's tripartite theory of soul* such as *epitumos (appetite), logos (reason),* and *thumos (spirit)*. As different paradigm of *voluntarism, intellectualism,* and *emotionalism* in philosophy had been also discussed since the ancient Greek era, and thus these categorizations critically show the reasons to discuss how the AI can reach to the natures of the humankind. In particular, sympathy and empathy can be a good reason to indicate whether expression as the humanity or not.

At a glance, we will hardly cooperate with the AI-driven robots because these systems lack emotional and existential reasons as *asymmetry* conditions. Table 13.1, explains symmetries such as *living desire, intellectual abilities,* and *empathy,* which are completely kept only by the human species. It strictly draws the borderline as identified symbols of supremeness of the human and differences against the other existences such as zoological species, AI, and artifactual robots. Namely, symmetry firstly indicates a degree of similarity compared with arbitrary pairs. Reversely saying, all of those enumerated factors are simultaneously retained only by the humankind, and any arbitrary pair of the humankind has maximized possibilities to coexist, communicate, and understand with empathetic emotions with each other, because it can be a symmetric situation. When different existences hold one of common symmetries, both of them will keep somewhat balance, as long as each of them can be recognized as equal level. Some zoological species (e.g., ape, dog (Katayama et al. 2019)) can emotionally interact with the human, and empathized communication among them indicate efficiency of innate sensing ability. If not so, it can be called as *asymmetry* conditions, and such relationships may not become stable and negotiable among them.

For another example, there is further the symbol grounding problem of the AI (Harnad 1990). How can the AI correctly interpret such mentality of the human by through sympathized ways in social contexts? During nonverbal and verbal communication (and text-based contents) between the human and the AI, how can the AI understand implicit interlinkages between somewhat symbols, nuances, meanings,

Table 13.1 The three symmetries between the humankind and other existences

Symmetry of ability level among existences	Humankind (ordinary person)	Humankind (e.g., terrorists)	Some zoological species	AI
Desire of Finite Living	○	× (primitive level?)	○	× (impossible)
Intellectual Ability	○	○	△ (primitive level)	○ (it'll be smarter than the humankind)
Empathy for the Others	○	× (lost?)	△ (primitive level)	× (impossible)

and corresponded actuality in the complicated real world? Some features which are proven by interactive models among people will undoubtedly enhance the AI mechanisms of meta-cognition, coordination, and moral judgments among multiple partners in depth (Freitas et al. 2019).

As introduced *the 23 Asilomar AI Principles* and LAWS at Chap. 1, the author felt a subtle contradiction about their ethical rules. They required a set of necessary ethical rules for the AI, but it means that the AI has been perceived as an equivalent level on the intellectual aspect for the humankind. Thus, those rules meant that it has established one of the *symmetries* in intellectual level between the AI and the humankind (e.g., the Turing test problem). If so, this is the first case ever since the birth of the human species. But if not so, such ethics will bind for only our humankinds. Actually, declared list clearly describes both of the AI and the humankind, and we unconsciously presumed higher intellectual level of the AI which is compared by our intellectual level.

Hence, we obviously presume that the AI has intellectual abilities to learn and understand a set of rules and prohibited matters, and further the AI will acquire necessary skills to be communicable with the humankind (Dreyfus and Dreyfus 1986). And if aforementioned coexistence theorem expands to modal logics of knowledge and CKR (common knowledge of rationality: Gintis 2014), there are also needs to deepen its logic spaces and possibility for common understanding between the AI and the human in various contexts.

Without saying again, some of the citizens feel certainly latent threats against the AI system, because the AI is already smarter than many citizens, and further there might be another reason. Namely, the *asymmetry* and *symmetry breaking* situations of intellectual level will everlastingly overwhelm against us. Please recall discussions on proving process for honest works and detecting deceptions (and anomaly) at Chaps. 6 and 7, and such social intelligence will be required interacting with not only the other humans but the AI-driven systems. If impossible, an asymmetry condition within them will be laid in the harder condition for adaptation. Ethics for the AI must be prepared to restrict the advancement of those AI intellectual mechanisms, and it should be equalizing to reach balanced conditions between the humankind and the AI. Is it called as *fairness* between them, as Rawls (1971) said? Then, our humankind shall explore own adaptive ways to be coevolved with the AI system at confronting those challengeable conditions. It may become harder experiences to coordinate with those lifeless and smarter partners in an asymmetry situation.

13.3.5 Barrier Against Logical Weakness

Above theorem principally told, in short, that our society has to be undertaken only by those who stay eagerly own living life. As discussed repeatedly, unless acquiring smarter intellectual tools, readable devices of the other's brain and medical ways for prolonging finite lifetime, these necessary symmetries will be perpetually and firmly stabilized. However, to date, *symmetry breaking conditions* will be possibly attained by someone (e.g., terrorists, military robots). Their behaviors by criminals, ruthless intentions, and irregular accidents have been concerned about further considerations.

As clever readers noticed, furthermore, it is certainly that above theorem has critical caveats. In logical expressions, there are four patterns as follows.

1. $w \rightarrow \neg m$: If x holds the will of own living, x does not commit to do murder against the others. (above proof of the theorem)
2. $m \rightarrow \neg w$: If x commits murder against the others, x does not hold the will of own living. (Contraposition of above proof of the theorem)
3. $m \rightarrow w$: If x commits murder against the others, x holds the will of own living. (Converse of above proof of the theorem)
4. $\neg w \rightarrow m$: If x does not hold the will of own living, x commits murder against the others. (Inverse of above proof of the theorem)

Hence, the most critical matter is an *inverse* case of logic expressions such as "*if x does not hold the will of own living, x commits murder the others*." And contraposition and converse cases must be relying on inhibition rule by social laws, but the genocide wills of the humankind cannot be regulated by external ways. Many of the suicide bomber cases committed by terrorists might actualize this logic. Above theorem implicitly assumed that we ought to be rational, empathetic, and careful about our own finite lifespan. But such terrorists will not guarantee any social predispositions. For preventing such crimes, contemporary society has already institutionalized the organization of the police to watch them by digitized ways. Additionally, as you noticed soon, threats of the *lifeless* AI have similar warning against us (Ulgen 2017), but AI-driven military robots will not behave by the intuitive heuristics as Kantian imagination. The AI-driven military robots might take such actions only based on the logics and its interpretation.

The worst point is that the AI-driven machine itself and such human terrorists are unnecessary to be kept own life whereas many ordinary citizens still desire to keep own living. This is an *asymmetrical* situation between our living desires and lifeless crimes. When they have strong incentives of breaking ordinary citizens' expectations, in contrary, such situation will inevitably force many of ordinary soldiers to be exposed to dangerous risks. In this concern, risk management against them should also care about *imbalance* of powers between us. Accumulating big data in city dynamics among citizens and traffics, harmful patterns of conducting crimes and terrorisms will be possible to be detected in advance. Weinberger (2011) reported that the progress in computational social science and data science had already achieved to prevent a fear case of terrorism in advance (WEB OF WAR, *Nature*). In military perspective, it is natural that ISR (intelligence, surveillance, and reconnaissance) has been effectively organized by specialty of the AI-driven system.

Political leaders must talk about their vigilance, but rather they should say a reason on such harmful danger against us. It should manage all cause of latent emergent factors not to be actualized the worst cases in advance. It is certainly that the lifeless AI-driven robots will sweep human terrorists and other lifeless robots, but the problem still remains. Who and how should their root control command? Can such commander (e.g., government leader) be trusted?

13.3.6 Cooperation Beyond the Difference

13.3.6.1 Cooperation Possibility

Hence, this theorem made essential statements on possibility for coexistence and cooperation[4] between the AI-driven robots and the human. When the intellectual level of such artificial existences will be more superior to the ordinary citizens, cooperation possibility among them deserves careful investigations. Large number of the humanoids and AI-driven robots will be intervening into civic activities with ordinary citizens in social contexts in the future. Without such AI foundations on both cooperation possibility and betrayal impossibility, ordinary citizens of future generations will be naturally difficult to adapt in the digitized world. Thinking the ethics of the AI is crucially embedded in those crucial reasons (Zhang 1992; Ramchurn et al. 2007; Kang and Kim 2012).

Especially, cooperation among the humankind per se is insufficient to convince in social sciences. The core of this problem is why and how only primitive humankind obtained such abilities during historical process.

Cooperative behavior seems to be one of our specific dispositions. But why does it occur? As introduced a case of the disaster at Chap. 10, in fact, volunteering, social capitals, and reciprocal minds have been mysterious for our understanding in economics. Classical economics simply assumes that ordinary people just sought to earn money, but volunteering actions do not have any internal incentives for pursuing economic reasons. Reluctantly, economists call such social value as "*common goods*" and they regard something as value shared within communal groups (Shibuya 2012). Such acquired intelligence and social capital had strongly enlarged opportunities to share common tools for our survivals in the uncertain environment.

It is often said that those social abilities have been probably understood as the hard foundations to be more adaptive than other zoological species. Someone often concluded that such cooperative mind of the humankind was indeed formed by genetic evolutionary consequences, and it represents the fittest adaptation of the humankind in the historical process. Such theorists are usually called as evolutionist in social sciences. Was it true during the historical process?

As naïve assumptions written by Rousseau (2009), he further noted "*there were never conflicts and warfare among the human in the natural conditions.*" But such notable imaginations might be denied by recent scientific evidences. In the ancient natural condition, our direct genetic ancestor (i.e., *Homo sapiens*) consequentially overwhelmed other genetic relatives (e.g., *Homo erectus*) during the historical process by means of somewhat lethal ways. Why could not they help with each other

---

[4]Editorial: The cooperative human, *Nature Human Behaviour, vol.2, pp. 427–428.*

beyond the genetic differences (it was too little difference in genetic level)? It should focus on the specific factors of the *Homo sapiens* as our ancestor survived in the natural conditions (as an example of recent studies: Harvati et al. 2019). Was it a decisive factor for cooperative actions than other genetic relatives? As many of the other apes who cannot intellectually cooperate with each other could survive on the ancient earth, it cannot crucially say that only *Homo sapiens* as the humankind could be more adapted than other mammals and primitive apes (including *Homo erectus*) even if our ancestors could genetically acquire cooperative mind. Namely, differences of intellectual level among the species did not seem to be critical factors for their survival itself.

Civilization and cooperation of the humankind could play key success factors for their earlier adaptation in historical process, but these differences must be discerned by clear process of thinking (Shibuya 2005). Primitive cooperation can be discovered within and across many sociobiological species such as insects, animals, and monkeys, as well as the humankind. But on the contrary, cooperative actions based on civilized institutions must accumulate and integrate fragments of shared knowledge and skills which have been intellectually cultivating among citizens beyond the longer time and spaces. The former is partially affected by somewhat genetic factors, while the latter might be almost independent from direct genetic factors and evolutionary natural selections. In short, primitive level cooperation can be generated by genetic-coding instinct level, whereas civilization for cooperative actions (e.g., social laws, intuition, norm, custom, etc.) can be perfectly depended by continual intellectual efforts. Sociobiological species can be adapted in the former actions, but the humankind has both of them. Both distances between different meanings are not filled out by evolutionary theoretical explanations.

Regardless neutral theory of molecular evolution has been endorsed, evolutionary theorists in social sciences often assert that any cooperation (including unintentionally civilized matters) among the humankind had been *evolutionarily* acquired in their surviving process (Simpson and Kenrick 1997). Their definition on "*evolution*" distinctly means gene selection within species of the human-being whatever they can compare with other zoological species such as primate apes, mammal animals, and insects. Further a part of scholars have been hunting unknown genetic factors for cooperation and any survived reasons from genetic factors to brain's functions of neural circuits which coevolved with a set of the genome of the humankind. Then, the evidence for evolutionists' theory is still exiguous.

Otherwise, genuine sciences must be endorsed by bridging between thoughts and grounding in evidences (Lebowitz et al. 2019). And then, many scientists simply disagree with the opinions regarding above evolutionists and pseudo-scientists. If primitive ancient people could rationally recognize accepting coexistence with the others who could be counterparts, as each of them could assume to be alive at any confronting situations. Simple symmetry such as social intelligence, interactions, and communication skills among them could be mostly established and primitively adapted in their social customs and institutions. For the purpose of keeping each of life, as consequently, both of them could be balanced and coordinated with the others.

If all cooperation among the humankind can be perfectly determined by all genetic evolution, conversely all warfare must be simultaneously controlled by genetic operation. At least, genetic operations (e.g., using genome editing) can reduce and control aggressive behaviors of the human? If possible, the AI-driven system can detect these patterns as anomaly in advance. Of course, as necessary, primitive people might fight, deceive, betray, or help with each other, but they learnt the results of interactions among them in each occasion, and judged more rational ways to be alive at the time. The author thinks that they could vividly learn the results of their actions for surviving in their history, and then they eventually adapt to be smarter than other species. Namely, it was not strictly necessary to be considering social evolution of the humankind based on the genetic drifts and mutations (Nowak and Highfield 2013). Consequently ancient people could learn for surviving in such surroundings by *simple* rational judgments, and social aspects of our contemporary generations were not always inherited something from their genetic dispositions. It is certainly that our communication backgrounds (e.g., language), intellectual rationality, imaginations, emotional sympathy, and other complicated necessary factors could be rooted in a part of genetic foundations. But as standing on the principle of Occam's razor, it must not presume any unnecessary factors for explaining social abilities in the historical processes of humankind's surviving.

Namely, evolutionists in social sciences often overexpress genetic factors in the human history, even though they pretend the reductionism of a manner of genuine science. Their assumptions and notations just stay analogical imagination based on longing for biology. For example, in any biological species, every patterns of renown gene sequences arranged by four nucleotides (G, A, C, and T) have not any mysterious meaning for evolutionary process. In actually, arbitrary nucleotides other than ordinary forming alleles (G, A, C, and T) can be inserted and replaced into gene sequences by biotechnology, and even it can generate biological process as another possibility in the historical process. Present gene sequences occasionally happened to be such results in longer-term process, and namely there was not any meaning to endorse for evolutionists.

Furthermore, *emergent properties* such as cooperation, trust, coexistence, and social institutions (Young 1998) could not be engendered and propagated only by neither genetic factors nor *meme* (proposed concept by Dawkins). Those emergent properties related to the sociality might be the civilized manners rather than genetic one, and such civilization in the humankind itself cannot be equal meaning to an evolution concept in natural sciences. There is not any decisive evidence which can endorse their assumptions. Warfare and betrayals would be same. Primitive people could choose the ways for their living by somewhat judgments rather than any inheritances of genetic properties. Any decisions of cooperation or betrayals among the humankind could be directly determined by own judgment (Gintis 2014), and it is not any genetic and instinctive directives. But still yet, some of contemporary people (e.g., evolutionary psychologists, evolutionists, and parts of pseudo-scientists) have been imaginatively seeking the genetic cause to become helpful and cooperative minds, and they further incline to conclude that primitive people who were evolved by *unknown* genetic factors and mechanisms could be survived. If they lack both scientific manners and falsifiability (Popper 1963), and it means that they are still pseudo-scientists.

In addition, a part of mathematical biologists and computer scientists often help them to offer their mathematical models on evolutionary games and its evolutionary stability as equilibrium (Weibull 1997; Bowles and Gintis 2011). It is certainly that their models explain mathematical stable conditions for adaptations by simulations in silico (Axelrod 1984), and they eagerly assert that such primitive judgments of the humankind could be endorsed by evolutionary hypothesis. But any mathematical models can be explained by arbitrary preconditioned definitions, they must further endorse it by other evidences. Genetic algorithm could be refined as an *analogy* of biological adaptation and fittest survival mechanism for engineering (Goldberg 1989).

Hence, there are still many spaces for grounding to be convinced between their conceptual-based discussions and practical evidences. As a canonical scientific manner, they shall rather strictly reexamine their evolutionary models. Actually, Nowak's renowned taxonomy on five types of social reciprocities which proved by mathematical biology papers had not been enough endorsed by social experiments (Rand et al. 2014). Thus, they shall attempt to clarify their presumed evolutionary process by constructive empirical methodologies such as synthetic biology and genome-editing ways in vivo, for example.

If such ethic sensitive biological experiments are possible, these experiences will enrich understanding on our survival history and its cooperation possibility. Scientists will investigate possible interactions between the genome and the brain functions, and they may endorse some parts of these coevolving possibility. And it might give some implications for coevolution between the humankinds and the advanced robotic artifacts (e.g., humanoids, bioinspired computing with the AI-driven robots), and it further stabilizes mutual trustful cooperation in blockchain-based networking if they can endorse their models.

13.3.6.2 The Significance for Grounding in Practical Solutions

However, evolutionists who asserted to disclose cooperation possibility and peace idealists have not offered any practical solutions for warnings by realists standing in actual conflicts yet. If perfectly abolishing military weapons among the nations, but warfare and conflicts will arise anywhere and anytime even though there are no destructive weapons (e.g., in the Bible, the first deception and murder of Cain was executed by his unidentified ways). As discussed later, both cyber wars and sharp power require less physical weapons than hard power, and those emerging attacks can be effective for target countries.

Here, the author's research style majorly aims at grounding in both thought and actuality, and its cyclical feedbacks. And then, before the advent of the social media, the author (Shibuya 2006) had conceptualized the Ubiquitous Jigsaw, which was categorized as online cooperative learning and collaborative activities, enhanced by ubiquitous technologies and mechanism designs. Originally, the *jigsaw classroom,* which was proposed by Aronson et al. (1978) aimed to achieving cooperative learning environment beyond the racial and cultural differences between white and black students. At the time, there were still severe prejudices between recital conflicts in the USA, and Aronson tackled such complicated needs to be resolved. Their concepts and practical solutions had successfully matched to being maintained the sound classroom, and further such styles envisaged new possibility for cooperation among different backgrounds.

Over a decade ago, the author thought when ubiquitous society comes, interconnected people must realize to globally adapt in heterogeneous and different backbones. To date, such synchronizing network society has come, and interconnected digitization will be further pervading over the worlds. The AI-supported education (EdTech) can arrange each classroom, students' needs, evaluation, and subject preparation. As mentioned in Chaps. 9 and 10, each demand of citizen has to be arranged by inter-collaborative activities. Distributed task allocations and collaborative experiences among global citizens have a chance to be self-learned for the utility of cooperation beyond the differences. As above Kant's discussion on the peace told that a factor on *interdependency* among the nations would be critically significant, this learning style could be similarly performed as interconnected and interdepended cooperation among stakeholders whether online or not. Therefore, such cooperation can be indeed experienced as an emergent property, and jigsaw classroom obviously exhibited that it was not necessary to pay attention to any evolutionary story for cooperation among citizens. Rather, cooperative relations can be made by the firm will of ordinary participants. Furthermore, the author believes that such enriched results and experiences for cooperation using digitized tools and services will create a new page of the future generations.

13.3.7 Inequality in Social Welfare and Utility

On the other hand, our ongoing society has to be undertaken only by those who holds eagerly own living will among citizens. But there are still imbalances of social status, chances and asymmetry of gifted ability among citizens. And then consequential inequality in socioeconomics among them remains to be solved (Rawls 1971, 2001). If macro economy intends to maximize domestic welfare and common wealth, it takes much care of welfare and quality for their living. And their socioeconomic status can reflect their valuation for own identity and self-esteem, and such depressed mindsets among citizens in lower status may further suffer social unfairness. Thus, it cannot neglect considerable problems on social welfare functions to improve socioeconomic factors in the age of the AI (Sen 1970; Suzumura 1983).

In such contexts, each individual motivates to enhance own skills, well-being and satisfaction based on living desires, but valuable resources have been crucially too limited in the world. Namely the entire society constituted by multiple individuals has risks not to be satisfied its social value of welfare. It is a social inconsistency between individual and social level. And is the utilitarian's slogan "*The greatest happiness of the greatest number*" always correct in society? Some of individuals will lose the utility for own living in such social conditions.

Economically speaking, Sen (1970) asserted that the assessment of inequality had been neglected in economics, and he took up an example of social loss due to Atkinson's equivalent income. As taking into account of the diversity of individuals, he definitely defines inequality. As Roth (2016) examined his matching framework for kidney exchange in medical care, for example, well-sophisticated mechanism design has chances to solve such inequality.

Inequality directly stagnates socioeconomic states among people who stay own living (Piketty 2014). It might be consequently rooted in the coexistence with the others, and those conditions have to be solved in coordinating with each other if peacefully. If impossible, such social inequality situation further calls next conflicts against them. Many of civil conflicts among citizens were conceived as economic disputes in the courts, and compensations for solutions can be settled by only monetization. In larger cases, some of international tensions have shown the unsolved evidence of such imbalances among the nations. As Kant (1795)'s theory told that wider economic interdependency among the nations has a key of the perpetual peace, but nobody knows that digitized economy, which will be networked clustering has potentials to manage such concerns or not yet. When the AI-driven robots perfectly work for citizens around the clock, there will be other considerable disputes on how basic income and rewarding-based economy equalize economic fairness among citizens.

13.4 Manipulatable Minds

13.4.1 Information Manipulation

Social media has been already reshaped as one of cyber weapons for information manipulations and mental controls against ordinary citizens. As mentioned each issues at chapters from 5 to 10, the keys are that social media already becomes global networking tools for interconnecting with world opinions and contagions of influence across the borders of the world. In this sense, those collective dynamics participating by ordinary citizens sometimes cause to lose the order of their national governance and sound conditions in democracy. As famous evidences recalled, "Arab Spring" and similar events had outperformed their national governing regimes (Shibuya 2017, 2021). In democratic societies, even though those revolutions will not usual happen, it should take a close look at the nature of voting behaviors at elections campaigns (Schafer et al. 2017; Shao et al. 2018; Aral and Eckles 2019).

Historically, those informational manipulations have been conducting by terrorists, governments, and military agencies. From the beginning, information was still *weapons of the warfare*. Recently, at August 2019, Twitter and Facebook proclaimed that homeland China government frequently manipulated information against collective demonstrations at Hong Kong using those social networking services. This case indeed shows that counterparts intended to collapse organized demonstrations and confuse their judgment of antagonists (e.g., ordinary citizens living in Hong Kong).

For other examples, there was a case of Hitler in the Second World War, and he agitated publicly ordinary citizens and manipulated them by media controls. In this context, Morgenthau (1978) had already articulated struggles for obtaining the public minds by through old-media manipulations. He especially emphasized the utility of mass communication tools such as newspapers, radio, and public speeches at the Second World War (The advent and popularization of television were still later and of course internet technology did not appear yet. But, around the cold war (1950s later), space artificial satellites for communication and information had been successfully launched).

Further, as another case, there is one of the strategy books, "*the Art of War*" written by Sun Tzu in ancient China (it was estimated around BC 500). This book had already prioritized and concentrated on both spying as the intelligence agency and quantifying information about the enemy. Similarly, under the contemporary digitized society, social media and interconnected network can expand those systems for military intelligence from actual space to cyberspace.

13.4.2 Emerging Sharp Power

Politics and participatory democracy have been conducting by various standpoints and opinions among ordinary citizens. However, due to fake news and post-truth by through social media, global society might be on the brink of such democracy. It has already opened the Pandora's box as the cyber war (e.g., *sharp power* and *hybrid war*) as well as communication tools for ordinary citizens. Threats against our democracy will be derived from ourselves (Ministry of Defense 2017). And of course, it can expect that AI-driven activities of sharp power has been already working in the background.

As it still mentioned, cyber war (Luciano and Mariarosaria 2014) warfare can be categorized into some types such as *hard power* (ordinary military forces), *soft power* (sociocultural influences against the target nations), and *sharp power* in recently. The future AI-driven weapons and military robots will be applied in each category. For example, hard power in the future might deploy the AI-driven intercontinental missiles and military autonomous robots. Soft power using the AI has possibilities to be implemented as stealthy software programs for online autonomous cultivations against the target citizens. Sharp power based on the AI and big data will be tremendous impacts against either democratic or tyranny nations. Actually, at 2015, the USA military forces announces to progress developing their AI-driven weapons of the next generation such as autonomous self-learning military machines, collaborating AI with the human soldiers, supporting AI for the human military forces, team formations arranged with human soldiers and the AI, and networked autonomous AI-driven weapons.

According to a report by SIPRI[5] (Stockholm International Peace Research Institute) at 2018, the total amounts of budgets for the largest military forces in developed nations were respectively 649 billion US dollars (the USA, approximately 3% as GDP rate) and 250 billion US dollars (China, approximately 2% as GDP rate). But those budgets included only ordinary weapons and military costs, and it might not include sharp power and its necessary costs. Comparatively, sharp power still wastes only lower costs than ordinary military forces. It might be estimated as negligible costs.

In addition, according to IISS[6] (The International Institute for Strategic Studies, UK), they announced that total amount of deaths caused by warfare and conflicts at 2016 were globally estimated as 157,000. Of course, these data can be only determined by hard power of military forces, and sharp power relatively can be said as peaceful tools in this context. But sharp power indirectly might make serious consequences against target nations and citizens.

13.4.3 Becoming Citizens as Mercenary by Manipulations of Social Media

As previously said, behavior science (and computational social science) has already become a genuine science, because all disciplines of genuine sciences aim to unveil three distinguished factors such as *explainability, predictability,* and *controllability* about specific phenomenon. Each of them respectively enables us to explain specific descriptions of evidences what such phenomenon is, formalize it by specific equations in forecasting modeling (or backward), and ultimately gain the controls for either all or parts of necessary variables on such phenomenon.

In other words, it becomes offering own specific knowledge for military requests. Especially, controllability indicates that such phenomenon can be intentionally provoked by the human in arbitrary timing and location. Ubiquitous computation and social media tools could alter those academics as a discipline of genuine science, and scientists in those fields could apply the methodologies to do above all factors. At Chaps. 4, 9, 10, and 11, smarter readers may understand that collective dynamics can be facilitated by the external intentions of the third parties as well as intrinsic motivations of massive participants. Namely, one of applications for military cases in those academics opened the gate of media manipulation and sharp power against ordinary citizens.

Therefore, the author has an inclination to warn about that sharp power which committed by hostile countries becomes ordinary citizens living in target democratic

[5] https://www.sipri.org/

[6] https://www.iiss.org/

nation for their invisible and massive "*mercenary*." Please recall discussed matters such as inducing social value by the scoring services, massive migration crisis, and collective dynamics of citizens at earlier chapters (Chaps. 4, 9 and 10). Those massive mobilizations by citizens itself have vulnerabilities to be promptly turned into national enemy, and it can induce weakness and self-decay of target nations.

Cyber war already became all-out war using smartphone, and it has dispositions to include numerous and unconscious citizens with remaining in daily constraints. There is no border between actual and cyberspaces. Does it feel somewhat overexpression? Certainly, many of the participants hold their strong identification for the homeland, and they do not have any harmful intentions against their homeland. But their actions and its consequences have no guarantees not to be manipulated by external malicious goals. Collective dynamics of social movements caused by sharp power, as mentioned in earlier chapters, are quite phenomenal to be possibly massive mobilizations by ordinary citizens. Those dynamics have nearly same meaning that it could be suddenly realized as appearing massive military forces within the homeland without actual invasions beyond the national borders. At least, it might be quite ideal to socially engender an uncertain atmosphere, opinion conflicts among citizens, and other chaotic conditions caused by media manipulations during democratic election campaigns, for example. In contrary, hostile countries can hide their existence and do not waste their military forces during their operations.

Hence, it should be careful about this topic in terms of more considerations. First concern is to disguise social movements and democratic demonstrations caused by computational agitations using social media. Second type is social sensing-based collective intelligence. Thirdly, it should care about citizens' education.

First, there are already indications that recent social movements, hate speeches (Johnson et al. 2019), and democratic demonstrations caused by through social media could be understood by hidden intentions and third parties' manipulations (Shibuya 2021). As it repeats to say once again, democratic nations should be legitimatized by election campaigns and voting results, and such majority-based decision-making in democratic system has serious weak points by sharp power. Hostile intentions, which actually command against target nations would manipulate those who engage in online activities with massive like-minded partners by through social media. Further, it could be unintentionally entangle them to overturn national security of their homelands. Terrorists have much merit to recruit massive participants who participate in social activities (e.g., demonstration, democratic movements) and hide their actual goals. In contrary, naïve citizen has vulnerability to take participating for beating their homelands as consequently.

Actually, sharp power by Russia has been suspiciously accused by USA congress and other governments in EU because it has been often eagerly continuing to interlude into opinion formation, democratic movements and voting campaigns. Some military reports have already warned that the sharp power conducted by Russia and China become to be larger parts of cyber wars. Additionally, terrorists and antagonists can be committing such same activities online, and they aim to organize their networking for pursuing goals. It is still difficult that attacks of sharp power using social media and online communications can be detected by ordinary citizens (DiResta et al. 2019; Howard et al. 2019).

Of course, Facebook, Twitter, and other social networking sites have already launched eligible regulations and managements in verifying submitted contents and fact-checking. In brief, such cases have shown our vulnerability that naïve minds and collective sentiments among citizens can be manipulated by external antagonists easily. There is no any guarantee that they can always detect deception and fake information from the antagonist's manipulations.

Secondly, larger social experiments of numerous balloons finding, which aimed to empirically examine effectiveness on social sensing and collective intelligence, had been pursued by one of the DARPA (Defense Advanced Research Projects Agency, USA) project. Giannotti et al. (2012) also discussed that collective intelligence has advantages in those effectiveness and social challenges for any improvements of social problems. In other words, using those collective intelligence, hostile country has various chances to obtain necessary information and collaborative partners in opponent country while they can conceal their existence and actual goals. Big-data accumulation by social sensing will be threating against participated citizens, because those data can be assimilated for simulations and geospatial engineering of military purposes (Abdalla and Niall 2007).

According to a bureau of the Germany governments (BfV:Das Bundesamt für Verfassungsschutz), at 2017, they claimed that China intelligence agency were under working for recruiting ordinary German citizens by mocking online recruit style through social media sites. In such ways, China agency obtained necessary information from ordinary citizens in Germany, and further they could successfully contact with principal professors of top universities and industrial engineers. Then, German government warned about those ways to be leaked information from inside of the nations.

Thirdly, contemporary digitized society exhaustively requires daily watching and monitoring online data and activities' log among all citizens (it includes both dark-net and dark-web). Hidden terrorists and hacktivists usually pretend to be ordinary citizens, and political authorities should keep going to analyze and gather necessary information whether it is online or not. As introduced in earlier chapters, both deception detections and deceptive technologies have been coevolving each other, and those mechanisms can be arranged by the AI and its machine learning. Ordinary people cannot discern what information is true, whose person is trustful, and which pattern is anomaly. Even if they will be probably supported by the AI in the near future, final judgments of the AI could be manipulated and learnt by external malicious intentions (e.g., Microsoft "Tay"). For the purpose of guarding the nation, it prioritizes necessary educations for ordinary citizens, and then such policy will lastly protect both citizens and nation.

13.4.4 Education for Citizens in the Digitized Democratic Society

For preventing the worst situations, it should educate more critical thinking, STEM (Science, Technology, Engineering, and Mathematics) subjects and internet manners for ordinary citizens. At the digital transformation, it deals with information literacy among citizens, and then scientific manners for thinking styles, judgments, and basic knowledge are necessary for them.

Cyberspace has to be kept as freedom, as many of people repeatedly say, but they should be also aware of definitive difference between *freedom* and *irresponsibility* online. Critical thinking should be based on conventional scientific ways and it is always quite needed for everyone. It should give them these educational courses properly as a part of STEM education. This critical understanding can play an effective role as an immune barrier against unscientific matters and become the preparedness against serious risk and crisis managements. And, it gives much chance to discuss and consider social issues for students and ordinary citizens.

13.5 Democracy in the Age of AI

13.5.1 Democratic Enforcement

The digital transformation has already facilitated velocity of decision, volume of data and variability of opinions in our political system. These mechanisms have been reforming global democratic or despotic government regimes. Political factors per se in the AI-oriented world must be solved as complicated problems (Susskind 2018).

In actual cases, for example, it seemed that Brexit[7] in the UK had not discussed their routes of significant bifurcations for the future in this way, and eventually they lost much time, chances, trusts, efforts, and leading compass for their regime of the nation. Actually, in 2019, the UK government officially disclosed their "*Yellowhammer operations*[8]," which described their scenario presuming the worst case.

The author thought that it should be considered as a malfunctioned result of democracy rather than a failure of democracy per se. Their failure was firstly caused in their examining procedure which could not verify possible bifurcations in decision trees by scenario-based approaches in advance. Before the government of UK arranged such judgment tools to be examined by evidence-based thoughts in depth, they thoughtlessly casted a chance to be decided by national referendum.

---

[7] Brexit, Gov.UK, https://www.gov.uk/government/brexit

[8] https://assets.publishing.service.gov.uk/government/uploads/system/uploads/attachment data/file/831199/20190802_Latest_Yellowhammer_Planning_assumptions_CDL.pdf

Additionally, they could not initially decide the final goal and the way of its achievement to be performed by intensive efforts. Reversely, before the time of ballot for referendum, many of the citizens should keep in touch with evidence-based discussion rather than participating in larger social movement. Even if many citizens living in the UK had not heard enough necessary information given by their government, they should request firstly necessary information to be filled into each subject in controversial lists for the purpose of soundly judgments whether Brexit or not from the EU. Even though there were still some strong doubts about sharp power by Russian agency who had manipulated against opinions among the UK citizens, their democracy had to retain the rationality to be authorized by their judgments. Moreover, "*migration crises*" around EU had broken EU alliances within member countries, but there seemed not to be alternative plans in earlier stages.

Therefore, for conquering political matters on the earth, any social problems what the contemporary and future generations of the people undertake ought to be intensively learnt as a valuable lesson on the failure of Brexit. Democratic meanings should be delimitated in the linkages between directive leading for solutions and consensus formation by discussion.

First is to decide a set of both final goal and its achievable solutions as soon as possible. There is a greater need to be planned and contemplated in various scenarios and possible bifurcations in depth by critical and systems thinking. Such solution can be evaluated in the context of evidence-based examinations. Democratic discussion never means that parliament members either decently state each opinion or competently debate nonsenses among them. In strategically saying, it is likely a chess player, it principally spends much time and efforts to trace all of the possibilities on the future variations, and even more, decision-makers should tackle to prioritize each bifurcation and branch of success possibility and its affordable ways. Any systematic models almost require not only linear assumptions but complicated emerging factors, and it is the ideal condition that all of those variables can be revealed and modeled in advance. It loyally obeys scientific manner rather than cognitive belief and opinion. In nearly future, the UK government and the EU nations will be offered by the AI services for political decision-making. And at the time, they will recognize that politics is just the way of system thinking and its calculations.

Secondly, as standing on such schemes, it prompts to perform it by higher standard of quality despite existing any troubles and hardships. If impossible such goals, alternative plans should be handled by the leader. And thirdly, it prioritizes to seamlessly promote for the achievement of the final goal rather than keeping consensus among congressmen and citizens. Although they would be supported by the AI, final decision shall be undertaken only by the leader elected from many citizens in democratic society.

Namely, politicians are not always head for the consensual decisions. As likewise Arrow's impossibility theorem (Arrow 1963), democracy has critical disadvantages in converging segmentations of discomforts among citizens to the consensual opinion. But Tolstoy also said "*All happy families resemble one another, each unhappy family is unhappy in its own way.*" Social movements have shown that many of participants appealed each discomfort against the government (Chap. 10), even so, democracy is not any panacea to save each of unhappy conditions in totally, and some cases will be nearly impossible to reach greatest happiness of the greatest number idealized by the Utilitarian. Hence, democracy must just make accurate decisions rather than obeying each of the claims. Democracy has been too deified by dogmatic daydreamers, and politicians must open their eyes for actualities by grounding in strict evidences.

13.5.2 A Lone King of the Lone Kingdom

13.5.2.1 Why Revolutions Matters in the Digital World?

In contrary, next problem is the democracy on the side of citizens. What is the justice of democracy in the global digital age? Venture to say, democracy does not always play a role of the religious indulgence in avoiding any contradictions among each demand of citizens. As introduced at previous Chaps. 8, 9 and 10, the author wonders why many of people will engage to participate in larger social demonstration for appeals and oppositions against their government. Those civic engagements organized by spontaneous participation of the large amount of citizens do not usually enlarge to causally serious results.

But occasionally, as it is likely a case of "Arab Spring," we could observe the severe consequences such as national collapse and devastation of common wealth caused by their mass actions. Their engaged actions could not always achieve their good ending and its future. Those rebellions and upheavals can be nearly regarded as same results of the warfare. Then, social media as an emotional fuse can be easily ignited by sensitive matters, and it diffuses fascinated contagions among citizens beyond any borders.

Historical upheavals were always ignited by irrational and collective minds of ordinary citizens. Any burden of proofs for political and legal justice ought not to be stifled whether it is democratic regime or not by justified collective emotion. Emotional triggers among citizens had repeatedly reshaped our historical regimes and international balance of powers. But consequently, socioeconomic status among each of citizens could not be almost improved in line of their presumed story whether it becomes democratic regime or not. Relatively both the nations and citizens would lose much valuable things. Therefore, mysterious factors are their incentive mechanisms, motivations, and those social phenomena itself. In this section, it critically contemplates those backbones.

13.5.2.2 Social Media

The age of social media has inspired the trends of not only world opinions (Morgenthau 1978) but populism (Hill and Hughes 1998; Shibuya 2017). Namely those media simultaneously let citizens misunderstand the meaning of the democracy. Many theorists often said that contemporary democracy obviously faces at the corner of its history. As mentioned in Chaps. 8, 9, 10 and earlier part of this chapter, online platforms and digital tools encourage many of the ordinary citizens to participate in social movement and political demonstrations against their government. As consequentially, some cases could become political upheavals caused by those collective dynamics of people. Moreover, it should not forget malicious results of sharp power by the antagonists.

Until several decades ago, those activities were too limited to organize the larger extent of participants, and it could hardly enlarge its influence of their assertions. But those recent technological advantages can conquer interconnecting with numerous like-minded partners beyond the difficulties, and there apparently seems to be the strong evidence, which is massively participated by those who hold same attitudes and opinions, has been endorsed by through observing and broadcasting news. Their justice of political actions is often misunderstood to obtain the majority which has been accepted by consensually all of the national citizens. Of course, such apparent evidences cannot be approved by legally.

In theories on political philosophy (Fukuyama 2014; Kelsen 2013), the nature of democracy requires each citizen to be matured, independent, autonomous, faithful, responsible, reliable, and self-educated person. As Fukuyama (1992) told the necessity of democracy in the perspective of human history, many of the citizens retain the freedom since the beginning of democratic regime. In contrary, certainly, Mill (1989) eagerly recommended open discussions and deliberative communications among citizens. And he intensively asserted a liberty of opinion expressions of each citizen against the tyranny conditions by the majority. As one of the philosophers at the 17th century, Locke (1998) also argued the liberal independence from the natural law and the slavery conditions and the social contract between the nation and citizens in his famous representative work "*Two Treatises of Government*." His discussion clearly addressed citizens' rights and obligations for governing national power. They usually differentiated the human as three types such as *a nation, a citizen,* and *collective people*. He was now considered one of the origins of the Libertarian (Nozick 1974). But, in the age of social media, ordinary citizens have to discern between the freedom of the expression and irresponsibility, and they ought to undertake their activities and its consequences within the legal constraints.

13.5.2.3 Misleading Signals on Democracy by Social Media Against Citizens

The author thought that ordinary people shall become a self-starter to solve social issues rather than claimer against their leaders. "Democratic" nuances must be understood by such context. However, digitized society gives the vast information for them, but their maturity cannot be well developed enough yet. Then, as noted at Chap. 10, legitimation crisis (Habermas 1976; Shibuya 2017) has been easier observed globally.

The author has convinced that social media itself also ought to handle the greater responsibility of misleading signals on democracy against ordinary citizens. And without saying, participated citizens can be accused by the legal authority if final consequences performed by their actions will excess over the legal range of allowances.

Firstly, as articulated in this chapter, it derives the *sharp power*. It enabled to flourish backgrounds of domestic confusions and collapses caused by agitations using social media, even though it could offer somewhat merits of useful intercommunication among citizens.

Secondly, it enforces the causes of misunderstanding that democracy gives ordinary citizens a chance for somewhat promotion to become more superior rulers and masters of the government. They can be just permitted to express own opinions for the government. But by through interconnecting each other by social media, they unconditionally fall under the delusion holding the imaginary majority, and they presume that their opinions are supporting by all of the national citizens. Even if there are multiple larger clusters of divergent opinions among them, each of opinion groups will not doubt own majority and legitimacy.

And thirdly, exactly, in the representative democracy, the most important is principally to realize the longer distances between assertions of citizens to the final decision at the parliament, and the due process of decision-making for legitimation must spend much time and efforts among parliament' members. Even if strong online influencers among citizens can give some pressures against their mentality of parliament members, but political legitimation systems for decision-making should clearly distinguish the differences among "*hear*," "*listen*," and "*obey*" in English.

Especially, as Japanese verb "*kiku*" is very ambiguous, it cannot discern strict meanings among "*hear*," "*listen*," or "*obey*" without more descriptive information. It might be similar cases in other languages. For example, as the author often experience at public demonstrations, what does a slogan "*Iu-koto-wo-kike*" among demonstrators in Japanese means? And what is the legal backbone of their assertions? In English, at least, their phrase can translate to "Hear what I say," "Listen what I say," or "Obey what I command." The most correct one is probably interpreted as "Listen carefully and obey what I command." Hence, this ambiguous phrase is not clear that government leaders just are required to *hear* and *listen* to their voices. Rather, should government leaders *obey* what they ordered? The freedom of expression is not equally legal commands for anyone. In many cases, such massive claimers' meanings may unintentionally drive to achieve the latter nuance, and many of them actually believe that they can *legally* force political leaders obeying what citizens claim as long as their assertions sustain the majority. But actually, such assertions completely have not any legal reasons. Strictly meaning, all of the legal articles shall be *obeyed* among all national citizens in democratic regime. Every citizen certainly has the right to claim their legitimate demands for government in accord with the due procedures. Unless it is not acceptable for such assertions, in many cases the government shall *hear* and *listen* carefully to their voices, and they finally *obey* such claims when their assertions have some reasons of the legitimacy to be approved. But unless no articles are concretely stipulated in enacted law, there will be no ways being acceptable for their demands. The legal courts will similarly decide such ways. And then, namely their claiming activities should be almost procedurally legitimated by the due process in advance.

If not so, they just commit to illegally threaten against someone. Even if their opinions and assertions seem to occupy the pseudo-majority at the larger spaces for public demonstrations and online inflamed discussions, it just collectively commits to do illegal threatening actions against governmental leaders and the legal institutions. The paradigms do not have any differences between democratic regime and monarchy regime, and it means that ordinary citizens shall play a role of working undertakers whether they are sovereign or not. But it still became better than servant in the past ages, because they were applied for a set of the basic legal rights since the democratic revolution.

Therefore, if the genuine democracy has critical given dispositions, which ought to be listened and obeyed among the voices and the general will of collective citizens, such political mechanism had been functionalizing as either autonomous self-decay or external controllability. The political way of the populism, which is alleged and agitated by propaganda and emotional hostility, must be contraindicated among the gentle politicians, but social media had emancipated it for the ordinary citizens and agitators. Their virtual sensations can further actualize to reform a shape of the physical world whether this consequence will be preferable for the citizens or not.

13.5.2.4 Lone King: The Projecting Excessive Identity

Furthermore, each of the citizens often presumes own different "*democracy*" and believes own justices. Many of them just intend to commit frauds for legitimacy of their assertions by pretending own pseudo democracy. Cynically saying, Riesman et al. (1969) previously called mass of ordinary citizens "*The Lonely Crowd.*" In other words, contemporary conditions of the digitized society can name them the so-called "*A lone king of lone kingdom*".

The digitized technologies apparently enable each citizen to interconnect with like-minded partners, but such relationships nearly just mean interdepended and umbilical cords binding with them. Namely, they cannot spontaneously achieve anything by themselves and there is no servant for them. Each citizen who regarded as sovereigns seemed likely to become free and celebrated, but each individual resembles just a lone king in lone kingdom. Owing separated and tiny digital space certainly offers the private realm of the kingdom, but actually nobody looks for them. Then, they must claim to achieve their needs outside.

In the digitized society, every citizens can project own imaginary ideal-self as a king to the public sphere whether it is online or not. Their online interconnections and assembled collective assertions against the external caregivers (e.g., political government) can be realized as an alliance among "kings of lone kingdoms." In such ways, they can just keep own anonymous states ("identity ambiguity"), believe in own fake value ("identity value"), swarm collectively ("collective identity"), scare lonely ("identity protection" and "identity deception"), flatter to the majority ("networked identity"), and further only confirm validity of own opinions to the outsiders ("identity proof"). Therefore, digitized solutions have cultivated citizens to such immature existences ("identity health") and there are a lot of risks that many of the citizens have inevitable vulnerability being manipulated by external controls (e.g., sharp power). And lastly, such "democracy" cannot be paid attention to any significant meanings. Then, many of the theorists often said that the decay of democracy has already come in the digital age, and who can recognize such evidences among citizens?

13.5.2.5 The AI and King

The digitized revolutions will steadily reach to the AI-driven society. When the AI-driven robots massively commit collective demonstration to appeal their assertions against the humans' masters in the nearly future, the AI-driven robots may assert "*listen what we say.*" At the time, how will future generations of people think about such demonstrations? They will surely feel discomfort, because they can be relieved from the duty and working hardships, and at this time they become recalling that they were just same servants which were equivalent to the AI-driven robots. Namely, they are equally servants as workers for everyone nevertheless we still obtain some legal rights. The nature of citizenship does not majorly change and improve its socioeconomic status from the dark age of the previous regimes.

Then, all robots shall be required obeying only by the directions of the human, because they will indeed be created as the servants to royally work for the human. Similarly, in actually, democracy has not achieved its functions unless all of the citizens shall undertake their duties and working roles for everyone and the nations whether their regime is based on the democracy or not. At first place, once again, each citizen as sovereigns in democracy shall play a role of spontaneous practitioners, self-starter, and working undertakers rather than roles such as a commander, a king, and a claimer as an oppositionist. Why do they always claim only against the political governor? Why don't they spontaneously achieve and solve social matters by them? Why don't they progressively undertake such unresolved tasks by their responsibility? Those who collectively assembled at the corner of public parks could alternatively launch new social businesses and nonprofit projects for improving their nations with like-minded members who similarly asserted. Their assertions could be achieved by their affordable ways.

However, because they may hesitate to do anything more than pressing keys and buttons of keyboards and smartphone, they never think that their duty for their democracy can be left by asserting own opinions. And they just presume that their voices might be solved by anyone. They are nearly a *king*.

13.5.2.6 Enlightenment

On the other views, soundness on those democratic problems concerns strictly on the viewpoint whether each citizen has both independent responsibility and liberal abilities for performing each demand or not. In other words, they shall share tough intentions to undertake their performed consequences whether it is desirable or not. Excessive collective demonstration and movements participating organized by the large amount of citizens ought to be managed as their immaturity rather than failure of the democratic regime. Their developmental stages of life course seem to be stagnated in an infant level rather than a king.

As mentioned above, until now, the digitized society was too reluctant to enlighten such citizens for their sound development. Let me recall a philosopher Kant (2009). He said "*Enlightenment is man's leaving his self-caused immaturity. Immaturity is the incapacity to use one's intelligence without the guidance of another. Such immaturity is self-caused if it is not caused by lack of intelligence, but by lack of determination and courage to use one's intelligence without being guided by another*".

The AI and big-data age has already come, but the humankind has not been matured in this way. The identity of each individual ought to be continuously matured by through self-learning progress. Emotional triggers of citizens cannot be supported by the AI. By scoring their value measured by the AI, can they learn appropriate political ways?

13.6 Working with AI? Either Worked by AI or…?

13.6.1 Working or Learning

As noted a bit at Chap. 11, in fact, both education and works have been long-lasting kinships. In ancient Greece, a set of the liberal arts offered for ordinary citizens belonged in liberal social class. And the origin term of "school" σχολή nearly meant "free time" in ancient Greek. They were exempted to work in daily life, and then they had much time enough to learn and discuss. Such persons were indeed called "*philos sophia*" in Greeks (knowledge lover).

When AI-driven society perfectly achieves social conditions to assure such life for all citizens who exempt work in the future, what such people will alternatively do? All works and duties will be undertaken by the AI and its implemented robots.

In this context, ordinary citizens in future generation will be similar to ancient Greek citizens, and then they will aim to participate in online political discussions, studying in learning communities online, and entertaining own preferred hobbies. Those services itself have already launched in the contemporary world, and many of the citizens have been engaging such activities in each life. Further, their lifespans will be prolonging more, and then it will become a problem how their "disposable time" as a usable amount of time in life consumes.

Certainly, there are some considerable points, which can be proposed by different critics.

- In the intermediate stage of this course, how do motivations and self-esteem of ordinary human workers change when their works are replaced by the AI? Workless citizens will be covered by basic income or like this in daily life?
- In AI-driven society, what kinds of academics should be learning? Especially, many of the authorities usually recommend mathematics, computer programming, logical thinking, creativity, and the skills to contemplate comprehensively for the youth and children. These subjects will become new liberal arts for the future generations, and traditional schooling style also will inevitably vary its raison d'etre.

13.6.2 Self-Actualization

Naturally, both educational and learning chances should be assured to achieve self-actualization for everyone (Coleman 1979). As discussed at Chap. 4, it can help enhancing social value of each individual by self-learning motivations. Authority of education does not have any reasons to touch and prohibit all such activities. Traditionally, school paradigm exclusively plays a role authorized by the nation, such institutions only can admit students' ability and education level, and their qualification and the right to grant the academic degrees had invincible professional authority.

However, as mentioned the AI Watson before, the AI system can memorize the vast amount of knowledge which acquired by intensive learning, and it can easily excess over than any professionals of the human-being. When AI-driven educational technology (i.e., EdTech) becomes to perfectly serve learning curriculums for everyone and educate pupils, the power structure centralized by the established regimes (i.e., schools and teachers) will be unveiled that they are just an intermediate exploitation entities against the natural rights to learn something of ordinary citizens. Everyone has naturally own rights to learn and know by all ways, their skills and abilities are not necessary to be admitted by the human authorities. The human authorities' factors do not always work for acquiring skills in the learning process of each student. Progressive learning style by each student is a more significant factor than passive educational style. Enforced learning with teachers can be alternatively replaced by the AI systems. As a teacher role will be employed by the AI systems, students can learn from AI and EdTech-based learning environment. No need of human teachers who desire to be honored.

Further, as inter-collaboration environment has been already equipped by ubiquitous and mobile computing, everyone has various chances to learn something valuable anywhere and anytime (Shibuya 2006). Many universities have already published own open courseware and online free learning contents. And the academic degree and certification possess just something of value of signalizing for the others, such intellectual symbols occasionally means one of prestigious medals in the life course. So then, several prestigious universities only exist, and their educational services offer online course materials supported by the AI systems.

13.7 Antinomy and Scenario Analysis

13.7.1 Antinomy Analysis on Scenario-Based Model

To summarize above all discussions, next, the author considers several future patterns by scenario-based thinking in detail. Various bifurcations of the routes in multi-ending scenarios would be crucially dependent on our decision-making and performances at each occasion. As innumerable SF and academic studies were published, negative endings of the humankind within those stories could be frequently observed. Please recall a review of SF stories on the AI once again (Cave and Dihal 2019). They categorized published science fictions such as *immortality, inhumanity, ease, obsolescence, gratification, alienation, dominance,* and *uprising*. The author borrows concepts of this taxonomy, and it steps further considerations.

Here, it starts to begin a thought experiment. Let me conceptualize a matrix by two scenarios, as it is likely philosopher Kant (1999)'s thinking style of *antinomy*. Antinomy matrix comparably envisages two different cases (*thesis* versus *antithese*) by descriptions of logics. Original Kantian's antinomy style might aim to clarify given two propositions by contradictive deducing.

The most prioritized thing is to decide the final goal for the future of the human. Unless it cannot firstly decide the final goal, nobody can examine multiple points. By back-casting process from traversing each leaf problem to the final goal, we should confirm a paralleled vector what to do at the present. The author similarly exemplifies both scenario H and A. Each initial means that the former is *H*uman-centered scenario, and the latter is *A*I-technologies-centered scenario (Bostrom 2016). Both differences are actually coined in the common reason, and each of both looks at a possibility on each side of positive or negative. Such dichotomy of a possibility will be critically verified in wider diversifications (Table 13.2).

This matrix showed an *example* antinomy based on the author's view for discussion. It took a close look at the comprehensive risks on not only the AI but other considerable factors. It should hold an integrated view on the risks, because every risk interlinks with intentional and unintentional reasons beyond the disciplines

Table 13.2 An example on matrix of antinomy

Subjects	Scenario H	Scenario A
Final goal	The global society conducted by Human-centered directions with harmonized technologies. This scenario aims to avoid singularity of the AI	Welcome for the singularity of the AI and further progress of necessary engineering and sciences The global society conducted by Advanced Technologies (e.g., AI, medical sciences, data science, etc.) drives to the further ideals which are followed by all of the humankind
Technologies progress	It still maintains traditional and natural conditions. All technology should be harmonized in this context. Human-being will accept all of them such as own lifespan, fortune (or doom), happenings, efforts to survive, costs, and benefits in their life	It aims to be managed by the AI and computational assistances. Human-being intends to heighten AI-based technologies, medical engineering, computational engineering, data sciences, and other necessary industrial revolutions more than human-beings' expectations
Social problems	It should simultaneously plan and solve various considerable matters Future generations shall tackle the problems such as food shortages, congestions, shortage of habitable zones, demographic issues, social welfare, sustainable development, environmental management, conserving natural resources, maintaining resilience from natural disasters, energies, garbage controls, racial conflicts, international wars, and constructing the living spaces outside of the earth. And it shall also decide controversial issues such as immigration crisis, a lifespan of each individual, private rights, and its data management. Such variables and latent factors will engender further emergent properties, and those complexities on causal relations and nexus of the possibilities will excess sooner the manageable criteria by the human-being	It should simultaneously plan and solve various considerable matters But advanced AI can tackle and optimize gradually social problems such as demographic coordination, congestions, garbage control, house allocation, and managements for natural resources. And AI can advise alternative solutions on many issues for decision-making to the human-being Genome-editing and regenerative medical science will achieve to prolong our life and modify genes of various foods, and then food shortages will be resolved by the accord with the needs and supplies in the market IoT-based sensor networking can be accumulating big data such as human behavior, industrial process, urban dynamics of transportation, and other managements. Energies cost and disposal will be optimized. Autonomous operations in socioeconomics and finance can be perfectly proceeding in blockchain-based transaction and exchange markets. Robotics and its based welfare services can assist for every needs of the human-being When natural hazard and disaster occurs, rescue services will be arranged by the AI and citizens can escape by AI direction
Solutions	It should be tackled by collaboration and discussion to reach consensus among global citizens The AI systems can probably propose some plans and solutions for people, but final decision should be arranged by them. It should be firstly undertaken by people in all issues Population size of global citizens will be increasing, but it will not be able to be solved. And then many co-occurrence factors such as food shortage, urbanization, environmental hazards, and other multiple controversies will gradually reach to the dead end	AI-centered scheduling and optimization can control for all variables It will be supported by the AI systems, and it hopes to enhance any solution for above complexity.AI can handle and undertake to control optimizing all of events and phenomena. Such world does not have any unexpected events, and all happening itself can be staying under scheduled harmony. In the natural condition, "randomness" and "probability" decide something as value and valueless in citizens' lifespans, but this scenario calls further to manage entirely citizens' life history and daily happening from cradle to grave, and such happenings are synchronized by apparently co-occurrences with causal relationship and timing arranged by the AI systems

(continued)

Table 13.2 (continued)

Subjects	Scenario H	Scenario A
Latent Threats	Probably, time and efforts will be much consumed, but there will be possibility that the human-being cannot decide and coordinate with each other. It means that uncompleted solutions still finally remain	The human-being will sooner reach the no-returning points, where they cannot go back to the past conditions. Unless AI exists, the human-being cannot be alive It will seem likely that problems are nothing…But actually it will be just operated that advanced AI does not arrow occurring any events and problems in every process under their controls…
Critical Factors at the brink of the future of the human-being	Bravery to decide what should be aborted and cut off Prioritization to decide which values should be maximized Complete performance to achieve the goal before the cataclysm	The willingness to know what and how it is operated by the AI Keeping the handle on the root authority for all of the AI operations To implement the multiple barriers of the ethics and dignity for the human-being into the AI-driven robots before the leases only by the authentication of the human-being
Reachable Destination	It consumes much time and efforts, but many of the citizens hope that it will be time to establish the better performances. The final destination will be decided by the general will supported by all citizens. Balancing among factors itself already requires somewhat regulations, but many of the citizens cannot decide to abandon their desires and needs If any solutions will be impossible and incomplete, the human-being will be repeatedly suffered by above social and environmental issues, and then it will not be able to gradually sustain their living on the earth. As results of those conditions, all of the humans may reach to their extinctions It may happen luckily, but the human-being lastly faces at the end of the earth. Each star and universe has also own finite lifespan, and the solar system will be collapsed by the Sun's burnout more than several billion years later. How does the human-being survive in such severe conditions without necessary technologies?	No need nobody alive This extreme assumption may become the worst case against the human-being. It will be possibly established when reincarnation technology will be ultimately completed by regenerative medicine and engineering As consequences by medical science and engineering, reincarnation system will be launched. As necessary, reincarnation machine driven by AI will regenerate someone who is reborn from "gene-bank." Then, a demographic factor of the human-being will be completely manageable under control by the AI. AI and robotics itself can be worked for self-repairs and development As a result, it is necessary to only stay AI, robots and the nature on the earth, and consequently all of the human-being as the existence with physical body will be never alive without exception When the human-being becomes a simulated entity to coexist with the AI in cyberspace, those limitations among the human-being with physical body will not be all restricted by many physical constraints If possible such solutions, all social problems on the earth will be perfectly finalized
Political System	Human-centered and democracy-based politics in many countries But nobody decides and completes It means a kind of ochlocracy	Human-centered political system gradually alters to the AI-based Dictatorships in all countries All delegating control from the human-being to the AI will call further decay as the human-being. Finally, there will not be any political systems. The human-being becomes a tamed pet or simulated entity by the AI
Aphorism	"*Quid pro quo.*" What desires can human-being abandon? Who does undertake sacrifices?	"*Blind follow is equal to slavery to AI*"

(Roeser et al. 2012; Bostrom and Ćirković 2018). Each reachable destination depicts only one of the possible worlds. Above antinomy seems likely science-factionary discussion, but the author listed up serious social problems which must be tackled and solved as soon as possible effectively in actuality. Even though both scenarios approaches to each different goal for the purpose of enhancing our well-being, the author intended to describe some of possibilities that both of them logically reach the absence (extinction or vanish) of the human-being. The former will indicate a future of our extinctions, and technologies cannot reincarnate us. In contrary, the latter suggests that the human-being with physical body will vanish, but technology can reincarnate us as necessary. Both results will be too different. Of course, there are many possibilities in such processes and bifurcations. There seems to be an opinion that another scenario should be combined with only preferable ideas from both scenarios.

Ever since the civilization, many of the ancient people desired to gain advanced technologies and spurred its progresses, whereas many of them also warn about such extreme technologies. Ambivalent emotions against such things, which are displayed bifurcations of a possibility on both positive and negative facets, will be induced by intuitive perceptions. Then, it should be clearly convinced both actual shape of the latent threats and possible solutions stood on scientific evidences which eliminated subjective belief. In STS (science and technology studies), academic critics on technological future progress must be handled as the route dependency in under determination and precautious principles. Future forecasting can be predicted in the bases of technological advancement, but such uncertainty per se has own drives to realize negative possibility being considered. Therefore, our preparedness against future risks should be strictly paid much attention on negative actuality rather than optimistic ideality.

Rather, many of the citizens will be realized at the brink of the ending. As reminiscently, Hegel said "*die Eule der Minerva beginnt erst mit der einbrechenden Dämmerung ihren Flug*" (It means "The owl of Minerva takes its flight only when the shades of night are gathering"). Cynically, this sentence has much nuance against the human's blindness on the latent risks and many of us are reluctant to be enlightened by the truth and negative possibility.

Arbitrary simulations can be possibly defined by specified model, which constitutes of a set of computable variables (Setola et al. 2016). The author has a custom to clarify firstly total of manageable variables, its combinations, and computable solutions within finite time. Subsequently, the author conceptualizes one of the social system designs composited of these variables and constraints. In brief, any studies on operations researches and computer sciences usually aim to solve one of the solutions based on constraint satisfaction problem and multi-objective optimization among such variables and combinations. In other words, both the human-being (e.g., the author) and the AI will firstly get in touch with finding necessary significant variables and unchangeable constraints for the solutions. Any solutions reversely cannot be operated when such variables are absent. Even if we assign any tasks on exponential combinations among many variables to one of the fastest HPC (high performance super computers (and even if quantum computer)), there are the cases that such solutions cannot be operated within finite times and computer resources. Thus, during while it is applicable for possible solutions, it should be solved completely as sooner.

13.7.2 Environmental Factors

Many of the global citizens have reluctantly recognized the evidences on global climate change on the earth every year. The progresses of our civilizations were coeval with an inevitable and strong problem on interlinks between population and geospatial environment (DeMenocal and Stringer 2016; Forman and Wu 2016). Further, as mentioned earlier chapters (Chaps. 10 and 11), natural disasters and its crises have repeatedly harmed against the existence of the humankind. Despite the big data and AI socialization, those topics remain to be solved. Those big data have been accumulating to analyze and model based on economic, meteorological, and environmental factors, but rather, for globally achieving urgent and significant goals of the United Nations Paris Agreement, it now requires aggressively performing one of the better scenarios among presuming scenarios of the future (Rogelj et al. 2019; WMO 2019).

At 2018, two winners of the Nobel Prize in economics (i.e., Nordhaus and Romer) were nominated in the field of environmental issues and its technological progress. But originally, an economist Uzawa was a pioneer who founded those topics in economics (Uzawa 2003). Economics and technological studies in the environment have been interlinked synchronizing with common research idea, and those social applications can be grounding in the actuality. In this concern, the United Nations indeed determined their statements on SDGs[9] (sustainable development goals). Unless there are appropriate ways to perform both regulations of human population and sustainable development for soundly keeping the natural environment on the earth, it is distinctively clear that the future generation of the human will reach to the dead end. Warning for such future were repeatedly illustrated and dictated by scientific reports (e.g., Club of Rome's report: *the Limits to Growth*).

The key success factor for the ultimate environmental solution is to gain the control of the human population. In the past, the world models proposed by Meadows et al. (2004) shows their proposals for sustainable solutions, which are constrained by necessary physical and socioeconomic constraints. Their conclusion was to be weighed on effective controls of variables such as human population, environmental resolutions, and sustainable technology progress. Their proposals and models had been exposed by both criticisms and approvals, and lastly their opinions have rooted in our present understandings (Komiyama 2014; Vincenot

[9] https://www.un.org/sustainabledevelopment/sustainable-development-goals/

et al. 2017). But such activities for environmental solutions have been undergoing trouble matters yet.

Evidence-based discussions must be kept in mind on environmental policies. To date, it is certain that emission controls of greenhouse gases (GHG)(carbon dioxide, nitrogen oxide, CFC (chlorofluorocarbon)) and PM2.5 (particulate matter 2.5μm (micrometer), LCA (life-cycle assessment) in each products and environment monitoring (e.g., atmosphere, water, and soil) have to be effectively managed by AI, IoT, and remote sensing (Okuhara et al. 2012). In some developed countries (e.g., China, the USA), as parts of trials, transportation control for innumerable cars and taxis in urban city launches to be optimized by AI-driven planning, and geospatial sensing system can track real-time trajectories and dynamics of human behaviors. And the AI-driven system can handle to energy-efficiently navigate them toward the destination. Those big data will be useful for survival of the humankind on the earth (Min et al. 2018).

By the way, Kant (1999) argued an antinomy on the infinity of the universe, but the latest physics has unveiled and estimated the finite lifespan of the universe, the earth, and the solar system. The future generation of the human-being will lastly face at the end of the earth. Each star and universe have own finite lifespan, and the solar system will be collapsed by the Sun's burnout more than several billion years later. How does the human-being survive in such condition without necessary technologies?

13.7.3 China's Demography Control and Technology Progress

Then, the key success factors are demography and technology in antinomy scenarios. In other words, it means effective controls for human-factors (e.g., population, desires) and progress of technologies (e.g., AI, medical science, engineering). In both meanings positively and negatively, China will be focused in this context (Chen and Li 2018). In the past, they controlled their birth population meanwhile they have been struggling social welfare for large amount of elderly people at the present. Moreover, they are eager to achieve further developments such as AI technologies, medical sciences, space technology, agriculture, environmental studies, and computational engineering (Lee 2012). AI-centered design of social system might be matched in centralized governance rather than democratic society. Therefore, perhaps, China will reach faster to the no-returning points, where they cannot abort the AI-driven services for their governances.

However, most democratic countries will not probably assimilate China's regulation policy into their refined controls. They will not promptly solve those considerable matters unless they can confirm and cannot ignore facing hazards and problems, and after that the general will among public opinions and congress lastly reaches the consensus or majority. As the common goal is pursued globally, the United Nations and regional countries ought to be negotiating and coordinating with each solution. Steadily, the time indeed passes over during staying in the middle stage for the solution. Unless each nation is involved, the United Nations cannot perform the goal for all of the humankind, and extinction possibility of the human-being will realize as one of the terrible day dreams.

13.7.4 AI Tyranny

Paradigm shift by the AI and big-data revolution does not wait for any delays. Furthermore, when many cases related to the daily life, governances, and business will be globally and pervasively conducted and arranged by advanced AI directions, and all of the democratic systems also steadily alters to dictatorship by the AI. The invisibility of dictatorships by the AI may rule in all of our life from microlevel to macro level. Additionally, nobody may recognize that all of us are living in such conditions.

As a renowned economist Arrow (1963) proved the Impossibility Theorems on democracy principles: our democratic society cannot be established unless it ignores one of the democratic preconditions at least. It suggests that mutual difference of both ideal democracy and dictatorships unexpectedly resembles each other. Further, there may be only a difference whether it is human dictator or AI dictator in the future.

It is a good reason to theorize a formal manner for modeling in rationality (Brandt et al. 2016). Everyone has natural dispositions based on own preferences. As Gintis (2014) theorized BPC model (Belief, Preference, and Constraints), a rational actor employs perfect actions in own consistent preferences. If each individual is a rational actor, his rational preference and behavior are well explained consistently in any cases. That is, Gintis examined that a preference ordering $\succcurlyeq A$ on A is a binary relation with the following three properties, which must hold for all $x, y, z \in A$ and any set $B \subset A$;

1. *Complete*: $x \succcurlyeq A_y$ or $y \succcurlyeq A_x$;
2. *Transitive*: $x \succcurlyeq A_y$ and $y \succcurlyeq A_z$ imply $x \succcurlyeq A_z$;
3. *Independent of Irrelevant Alternatives*: For $x,y \in B$, $x \succcurlyeq B_y$, if and only if $x \succcurlyeq A_y$.

Similarly, the AI optimizations for solutions have to be expressed by somewhat logical frames in order. In the contexts on aforementioned antinomy, several "*D*" can be regulated by handful variables not constants and constraints, and such variables are *d*emographic birth control of the human-being, *d*ecision-making based on the preferred and prioritized orders, and *d*ecisive executions for all issues. First demographics solutions can be perfectly managed by the strict control, and it is certain that many of the controversial issues will be melt over in the future. Second factors, it will be conducted by avoiding restrictions of the Arrow's theorems. Thirdly, decisive solutions already appear for us, but no leaders who elected among citizens can execute inevitable solutions, and nobody commits to do it. However, the AI systems cannot hesitate to execute redundantly necessary solutions despite any rational reasons.

13.7.5 Game Changer

Advanced technologies already employ a role of game changer against the traditional scenario, although it has been firmly and strictly underlying in the preconditions as identity of the human such as mortality, singleton identity, not replication, and ruled in randomness in each life. Ever since our civilization, such scenario was inscribed on the testament in some of ancient cultures. In this condition, regenerative medical science and engineering are negatively regarded as rather a balance breaker and cheating against such scenario than game changer.

When the humankind decides to gain and use such technologies along with simulating any demographic models (Grow and Jan 2017), we simultaneously decide the regulation and control for our lifespan and total population. The earth has been already occupied by nearly 8 billion people and more. Here, the separation rate of the earth between the land and the ocean is respectively estimated as 30% and 70%, and the most important is to recognize that our entire civilization can be restricted only in habitable zones on the land side. Given such habitable zone may be still estimated up to approximately half of the land, namely, the result is total 15% of the land ($=0.3 \times 0.5$). Many of us perhaps don't want to long-termly live in jungles, higher mountains, deserts, icebergs, and hazardous area. And such limited space must be also allocated for agricultures, sustainable natural resources (e.g., forests, rivers, lakes, etc.), evacuation areas (i.e., disasters), and other necessary purposes (Vliet 2019; Stephens et al. 2019). Even if underground space and satellite space colonies for outer human-made habitable zone (i.e., biosphere) were proposed by science fictions, the author cannot hold optimism of how many people would want to stay perpetually in such environment. Is it nearly expulsion from the Garden of Eden? Many of the global citizens don't desire to live in those places, and actually they still live as overcrowding conditions in the larger urban area. Water supply, food shortages, soils, and atmosphere conditions have been interlinked with all human activities, and such natural resources will reach to the unsustainable limitations (Kaneko et al. 2014).

Furthermore, when we reach to the prolonged lifespan society, many people have obscure expectations that it will exactly enrich for QOL of citizens. But is it true? It should be rational to reconsider with whom such merits can be obtained. It laid weights on the side of the national government rather than patients and citizens. Because unless the national government could not gain controls and major profits by regenerative medicines, social welfare and medical insurance services will be lastly collapsed without much payment taxations from healthy workers. And in other concerns, overpopulation should be inevitably rethought on the earth. Population controls of the global citizens will be quite required in such society for prolonged living life, and nobody can deny prioritizing totally social benefits rather than each well-being.

There are some extreme ideas for solutions. According to Dawkins' theory, every living existence was devaluated considering as *gene-vehicle*, and it means that the nature of living existence laid more weight on rather a set of genome than physical entity. If so, contemporary technology enables to convert a set of genome data which eliminated from each living existence to a digitized computer entity, and it virtually evolves itself, mutates a crossover among other genome in silico, and it represents a simulating entity as each living existence in computer simulation. Complete simulation on living phenomena has not been achieved yet, but simple set of variables on simulation of the living process can be manageable possibly. In this context, someone possibly transcend own digital data of genome to own computer model without keeping physical body, and they attempt to simulate own existence as each simple evolutionary process in the simulation environment. Further, when regenerative medical science ultimately gains the complete control of reincarnating physical body, as necessary, each person can be reincarnated in actual environment. As simulated entity comparatively costs lower energy than a physical entity, ongoing matters on the earth will be resolved in this plan. That's a heart of all the problems on the earth. If such plans are possible, it will be certainly that the humankind loses one of own identity backgrounds.

Is it too pessimistic perspective of the future? Somebody often says that technology will solve such problematic matters. AI-driven services will certainly support to solve the problems if possible. But, the author convinces as follows. Principally, the humankind must gain the way to control ourselves rather than the AI. All possible worlds and all bifurcations of the possibilities cannot be known in advance, but there is only a certain fact. Although we cannot regulate us, how can the humankind control over the world and the universe? We cannot decisively become "the God," and we cannot manage "*Deus Ex Machina*." Before that, the humankind has a possibility to degenerate to AI's pet. Those who want to be tamed by AI should promptly go back to the caves or farms.

13.7.6 *Sub Specie Aeternitatis*

As antinomy denoted, albeit many science fictions almost wrote a negative ending of our future possible worlds, will the humankind be possible perfectly to tame the AI and its implemented autonomous robots?

The author alternatively thought that the human of the future generation (after of "*generation Z*") will be rather eagerly obeyed by the AI's directions. This is the *third route* of the future generation rather than extreme two cases of aforementioned antinomy. Any rebellions by the AI and any warfare between the human and the AI will not happen. Because the AI-driven services will be useful for everyone, and then such services by the AI will gradually and steadily erode and percolate into anytime and anywhere. And lastly, the human reaches to the unaware conditions which cannot return to the past.

Please imagine "the past ages before the internet" if possible. Those who birthed after the internet revolution cannot understand such past society and daily conditions (it may be after 1995, so-called "generation Y"). All citizens could not hesitate to use internet and experience its exponential progress except for few critics in pessimism (Dreyfus 2001). Then, similarly, future generations who do not experience any daily life without the AI-driven systems will not understand our warning and pessimisms against the AI. Their identification will be based on such social surroundings.

Regarding such conditions, when an AI-driven central system finally attains to control for both all of the global people and all interconnected networks of XaaS (e.g., IoT, IoP, sensor networks, working robots, fabrications, factories, smart agricultures, connected cars, smart cities, BMI, smarter devices, smart grids, smart homes, space satellites, etc.), and then an ideal world will pervasively descend for the human as a philosopher Spinoza thought (2008). Given *something* which constituted of singleton, perpetuity, pervasiveness, infinity, superiority of intelligence, and enduring substances possibly appears, it will indeed achieve *the ubiquity of the AI's operations*. In such digitized society, all social phenomena, historical events, social relationships, social norms, well-beings and each history of individual among all of the citizens on the earth may be exclusively managed, arranged, synchronized, and optimized by the central AI system. And those complicated multivariable factors on computer simulations will be possibly solved by the AI-driven quantum computing systems. The pervasiveness of the AI operations for anyone will exist anytime and anywhere. The future generations of the human can only permit to see through the world as "*sub specie aeternitatis.*"

Let me introduce a part of propositions (i.e., proved theorems) in *Ethica* written by Spinoza (2008).

- Part 1, Proposition 29: "*Nothing in the universe is contingent, but all things are conditioned to exist and operate in a particular manner by the necessity of the divine nature*".
- Part 4, Proposition 40: "*Whatsoever conduces to man's social life, or causes men to live together in harmony, is useful, whereas whatsoever brings discord into a State is bad*".
- Part 4, Proposition 72: "*The free man never acts fraudulently, but always in good faith*".
- Part 4, Proposition 73: "*The man, who is guided by reason, is freer in a State, where he lives under a general system of law, than in solitude, where he is independent*".

These sentences seem to be fortune telling. Of course, such digitized society is just fictionally an unestablished future among various possibilities at the present. But if such fictional future will be possible, the humankind will certainly become *a dependent part of the AI's ubiquity*. Synchronization among networked and collective citizens with XaaS will be unilaterally ordered only by the central AI, and it realizes the digitized nervous system interconnecting with each citizen as a living organ. The AI-driven systems will arrange and operate all of the social phenomena, and then contingency will not be happening. As watching all social behaviors around the clock, blockchain-based trustworthy knowledge-base and deceptive detection technologies will expel malicious factors in advance. Then, it assures perfect harmony, trustful manners, and no conflicts within the interconnected world fusing both actuality and cyberspace. There will be no longer democracies and kings for them, and all of them converge to the ubiquity of the central AI. Using computational interconnecting systems based on XaaS, tele-existence, BMI, and virtual reality, every people share only single perspective in cyberspace and they will convince that it is a unique reality surrounding them which was provided by the AI-driven services. One for all and all for one within the scheme planned by the AI's ubiquity. Someone might call it the divine nature because many of them intensively desire such world ever since our civilizations. And it means that future generation citizens will be devout believers who will be carrying out the directions ordered (or unexpectedly manipulated) by the AI.

Is it a kind of the digitized *pantheism* by the ubiquity of the AI in the interconnected world? At this time, nobody will call them the humankind holding with an own *identity*. The human history eventually ends, and something will alternatively start.

13.8 Beyond Singularity

Where does each of us head for in the digitized society? All things discussed, if both singularity of the AI and prolonging lifespan arises, each individual must reselect (or keep) own route within affordable alternatives. But nobody knows which choice is correct yet. "*Quid pro quo.*" It will still indicate larger meaning for us. What should the human obtain and lose alternatively? Until now, nobody estimates considerable costs and consequences in the future, even though someone often overexpresses benefits of such progress.

Finally, this book calmly concludes what superiority of the human for the AI is. When necessary preconditions are given, logical and imaginative thoughts are possible to straightly gain some oracles for the future consequences. The author intended to shed light on considerable issues interweaving with multivariable in this book, and crossing lights of our speculations might depict a part of the sharing visions for the future. Through contemplating *identity* issues on the humankind, after the singularity of the AI establishes if possible, the author thought that it still retains following *five principles* at least. These principles are recompiled from necessary preconditions on identity for the human at Chap. 2.

1. To notice for the physical limitations of own body.
2. To undertake your finite lifetime and living history by yourself.
3. To care for coexistence with the others.
4. To coordinate with the others as necessary.
5. To keep growing with own gifted intellectual abilities if you have. Or not, you should work to care for the others.

After that, the author further thought as follows.

Please determine an approach to the final destination, where becomes the human-being as the tamed-animal by AI, or still retains an autonomous existence with dignity. The most important is not to decide the way of living process itself, rather to decide the way of proving evidence that you highly accomplished what you can in your living history. "*Identity*" means something identifiable from many of the people. When you become such distinguished person, your living history also will turn out the legend in the own standard of value. When you launch something valuable what you believe, your evidence will be scripted in the record of your living history.

"*Werde, Was du bist*" ("to become what you're").

References

Abdalla, R. M., & Niall, K. K. (Eds.) (2007). Review of spatial-database system usability: Recommendations for the ADDNS Project, Defence Research and Development Canada (DRDC).

Aral, S., & Eckles, D. (2019). Protecting elections from social media manipulation. *Science, 365*(6456), 858–861.

Aronson, E., Blaney, N., Stephin, C., Sikes, J., & Snapp, M. (1978). *The jigsaw classroom*. Beverly Hills, CA: Sage.

Arrow, K. J. (1963). *Social choice and individual values*. New Haven, CT: Yale University Press.

Axelrod, R. (1984). *The evolution of cooperation*. New York: Basic Books.

Axelrod, R. (1997). *The complexity of cooperation*. Princeton, NJ: Princeton University Press.

Ayson, R. (2004). *Thomas Schelling and the nuclear age: Strategy as social science*. Routledge: New York.

Bostrom, N. (2013). Existential risk prevention as global priority. *Global Policy, 4*(1), 15–31.

Bostrom, N. (2016). *Superintelligence: Paths, dangers, strategies*. Oxford: Oxford University Press.

Bostrom, N., & Ćirković, M. M. (Eds.). (2018). *Global catastrophic risks*. Oxford: Oxford University Press.

Bowles, S., & Gintis, H. (2011). *A cooperative species: Human reciprocity and its evolution*. Princeton, NJ: Princeton University Press.

Brandt, F., et al. (2016). *Handbook of computational social choice*. Cambridge: Cambridge University Press.

Caforio, G. (2018). *Handbook of the sociology of the military*. Cham: Springer.

Cave, S., & Dihal, K. (2019). Hopes and fears for intelligent machines in fiction and reality. *Nature Machine Intelligence, 1*, 74–78.

Cederman, L. S. (2005). Computational models of social forms: Advancing generative process theory. *American Journal of Sociology, 110*(4), 864–893.

Chen, W., & Li, X.-Y. (2018). Welcome to the China region special section. *Communication of the ACM, 61*(11), 38–87. (∗ multiple contents within special section).

Coleman, J. S. (1979). *Equality of educational opportunity*. New York: Arno Press.

Congressional Research Service, USA. (2019). Artificial intelligence and national security. https://fas.org/sgp/crs/natsec/R45178.pdf

DeMenocal, P. B., & Stringer, C. (2016). Human migration: Climate and the peopling of the world. *Nature, 38*, 49–50.

DiResta, R., et al. (2019). The tactics & tropes of the internet research agency. https://disinformationreport.blob.core.windows.net/disinformation-report/NewKnowledge-Disinformation-Report-Whitepaper.pdf

Dreyfus, H. L. (2001). *On the internet* (2nd ed.). London: Routledge.

Dreyfus, H. L., & Dreyfus, S. (1986). *Mind over machine: The power of human intuition and expertise in the era of the computer*. New York: Free Press.
Eurasia Group. (2018). Top 10 risks. https://www.eurasiagroup.net/issues/top-risks-2018
Fearson, J. D. (1994). Signaling versus the balance of power and interests: An empirical test of a crisis bargaining model. *Journal of Conflict Resolution, 38*(2), 236–269.
Forman, R. T. T., & Wu, J. (2016). Where to put the next billion people. *Nature, 537*, 608–611.
Freitas, J., Thomas, K., DeScioli, P., & Pinker, S. (2019). Common knowledge, coordination, and strategic mentalizing in human social life. *PNAS, 116*(28), 13751–13758. https://doi.org/10.1073/pnas.1905518116.
Fukuyama, F. (1992). *The end of history and the last man*. New York: Free Press.
Fukuyama, F. (2014). *Political order and political decay*. New York: Farrar, Straus & Giroux.
Garcia, D. (2018). Lethal artificial intelligence and change: The future of international peace and security. *International Studies Review, 20*(2), 334–341.
Giannotti, F., Pedreschi, D., Pentland, A., Lukowicz, P., Kossmann, D., Crowley, J., & Helbing, D. (2012). A planetary nervous system for social mining and collective awareness. *The European Physical Journal Special Topics, 214*(1), 49–75.
Gintis, H. (2014). *The bounds of reason: Game theory and the unification of the behavioral sciences (revised edition)*. Princeton, NJ: Princeton University Press.
Goldberg, D. E. (1989). *Genetic algorithms in search, optimization, and machine learning*. Reading, MA: Addison-Wesley.
Grow, A., & Jan, V. B. (2017). *Agent-based modelling in population studies: Concepts, methods, and applications*. Cham: Springer.
Habermas, J. (1976). *Legitimation crisis*. London: Heinemann.
Harnad, S. (1990). The symbol grounding problem. *Physica D, 42*, 335–346.
Harvati, K., et al. (2019). Apidima Cave fossils provide earliest evidence of Homo sapiens in Eurasia. *Nature, 571*, 500–504.
Herek, G. M., Janis, I., & Ruth, P. (1987). Decision making during international crises. *Journal of Conflict Resolution, 31*(2), 203–226.
Hill, K. A., & Hughes, J. E. (1998). *Cyberpolitics: Citizen activism in the age of the internet*. Lanham, MD: Rowman & Littlefield.
Howard, D., et al. (2019). Evolving embodied intelligence from materials to machines. *Nature Machine Intelligence, 1*, 12–19.
IBM. (2018) Everyday ethics for artificial intelligence. https://www.ibm.com/watson/assets/duo/pdf/everydayethics.pdf
International Committee of the Red Cross (ICRC). (2018). Ethics and autonomous weapon systems: An ethical basis for human control? https://www.icrc.org/en/download/file/69961/icrc_ethics_and_autonomous_weapon_systems_report_3_april_2018.pdf
Jackson, M. O., & Nei, S. (2015). Networks of military alliances, wars, and international trade. *PNAS, 112*(50), 15277–15284.
Johnson, N. F., et al. (2019). Hidden resilience and adaptive dynamics of the global online hate ecology. *Nature*. https://doi.org/10.1038/s41586-019-1494-7.
Jost, J. (2005). Formal aspects of the emergence of institutions. http://www.santafe.edu/research/publications/workingpapers/05-05-018.pdf
Kaneko, N., Yoshiura, S., & Kobayashi, M. (Eds.). (2014). *Sustainable living with environmental risks*. Tokyo: Springer.
Kang, B. K., & Kim, K.-E. (2012). Exploiting symmetries for single- and multi-agent: Partially observable stochastic domains. *Artificial Intelligence, 182–183*, 32–57.
Kant, I. (1795). *Perpetual peace*. New York: Macmillan. (English translated edition).
Kant, I. (1999). *Critique of pure reason*. Cambridge: Cambridge University Press. (English translated edition).
Kant, I. (2009). *What is enlightenment*. London: Penguin. (English translated edition).
Katayama, M., et al. (2019). Emotional contagion from humans to dogs is facilitated by duration of ownership. *Frontiers in Psychology*. https://doi.org/10.3389/fpsyg.2019.01678.

Kelsen, H. (2013). *On the essence and value of democracy*. Lanham, MD: Rowman & Littlefield. (English translated edition).

Komiyama, H. (2014). *Beyond the limits to growth*. Tokyo: Springer.

Kremer, M. (1993). Population growth and technological change: One million B.C. to 1990. *The Quarterly Journal of Economics, 108*(3), 681–716.

Lebowitz, M. S., Tabb, K., & Appelbaum, P. S. (2019). Asymmetrical genetic attributions for prosocial versus antisocial behaviour. *Nature Human Behaviour*. https://doi.org/10.1038/s41562-019-0651-1.

Lee, G. (Ed.). (2012). *Advances in computational environment science: Selected papers from 2012 international conference on environment*. Berlin: Springer.

Levy, J. S. (1984). The offensive/defensive balance of military technology: A theoretical and historical analysis. *International Studies Quarterly, 28*(2), 219–238.

Locke, J. (1998). *Two treatises of government student edition*. Cambridge: Cambridge University Press.

Luciano, F., & Mariarosaria, T. (Eds.). (2014). *The ethics of information warfare*. New York: Springer.

Luhman, N. (1984). *Social systems*. Stanford, CA: Stanford University Press.

McNeill, W. H. (1999). *The world history*. Oxford: Oxford University Press.

Meadows, D. H., et al. (2004). *The limits to growth*. White River Junction, VT: Chelsea Green.

Mill, J. S. (1989). *On liberty*. Cambridge: Cambridge University Press.

Min, W., Yu, L., & He, S. (2018). People logistics in smart cities. *Communications of the ACM, 61*(11), 54–59.

Ministry of Defense (MOD), Japan. (2017). *Defense of Japan 2017 (White paper: English edition)*. Urban Connections, Japan

Morgenthau, H. J. (1978). *Politics among nations: The struggle for power and peace*. New York: Knopf.

Neumann, J., & Morgenstern, O. (2007). *Theory of games and economic behavior (anniversary edition)*. Princeton, NJ: Princeton University Press.

Nowak, M. A., & Highfield, R. (2013). *Super cooperators: Altruism, evolution, and why we need each other to succeed*. New York: Free Press.

Nozick, R. (1974). *Anarchy, state, and utopia*. New York: Basic Books.

Okuhara, K., Tsuda, H., Shibuya, K., & Tsubaki, H. (2012). Development of eL-platfrom by ICT and knowledge acquisition for estimation. In *IEEE international conference on ICT and knowledge engineering*, pp. 53–58

Piketty, T. (2014). *Capital in the twenty-first century*. Cambridge, MA: Harvard University Press.

Popper, K. (1963). *Conjectures and refutations*. London: Routledge.

Ramchurn, S. D., Sierra, C., Godo, L., & Jennings, N. R. (2007). Negotiating using rewards. *Artificial Intelligence, 171*(10–15), 805–837.

Rand, D. G., Nowak, M. A., Fowler, J. H., & Christakis, N. A. (2014). Static network structure can stabilize human cooperation. *PNAS, 111*(48), 17093–17098.

Rawls, J. (1971). *A theory of justice*. Cambridge, MA: Harvard University Press.

Rawls, J. (2001). *Justice as fairness a restatement*. Cambridge, MA: Harvard University Press.

Riesman, D., Glazer, N., & Denney, R. (1969). *The lonely crowd: Changing study of the American character*. New Haven, CT: Yale University Press.

Roeser, S., Hillerbrand, R., Sandin, P., & Peterson, M. (Eds.). (2012). *Handbook of risk theory: Epistemology, decision theory, ethics, and social implications of risk*. Dordrecht: Springer.

Rogelj, J., et al. (2019). A new scenario logic for the Paris Agreement long-term temperature goal. *Nature, 573*, 357–363.

Roth, A. E. (2016). *Who gets what and why: The new economics of matchmaking and market design*. Boston: Eamon Dolan/Mariner.

Rousseau, J. J. (2009). *Of the social contract, or principles of political right*. Chicago, IL: Regnery.

Russell, S., & Norvig, P. (2009). *Artificial intelligence: A modern approach*. Upper Saddle River, NJ: Prentice Hall.

Schafer, F., Evert, S., & Heinrich, P. (2017). Japan's 2014 general election: Political bots, right-wing internet activism, and Prime Minister Shinzo Abe's hidden nationalist agenda. *Big Data, 5*(4), 294–309. https://doi.org/10.1089/big.2017.0049.

Schelling, T. C. (1980). *The strategy of conflict*. Cambridge, MA: Harvard University Press.

Sen, A. (1970). *Collective choice and social welfare*. Cambridge, MA: Harvard University Press.

Setola, R., Rosato, V., Kyriakides, E., & Rome, E. (2016). *Managing the complexity of critical infrastructures: A modelling and simulation approach*. Cham: Springer.

Shao, C., Ciampaglia, G. L., Varol, O., Yang, K.-C., Flammini, A., & Menczer, F. (2018). The spread of low-credibility content by social bots. *Nature Communications, 9*, 4787.

Shapley, D. (1980). Arms control as a regulator of military technology. *Daedalus, 109*(1), 145–157.

Sheehan, M. (1996). *The balance of power: History and theory*. London: Routledge.

Shibuya, K. (2004). A framework of multi-agent based modeling, simulation and computational assistance in an ubiquitous environment. *Simulation, 80*(7–8), 367–380.

Shibuya, K. (2005). Evolutionary spatial process of socio-economical organization and conformity structure. In K. G. Troitzsch (Ed.) *Pre-proceedings of the third annual conference of the European Social Simulation Association (ESSA 2005): Representing social reality: Approaches and results* (pp. 228–238); Together with the autumn meeting of the Research Committee on Modelling and Simulation of the German Sociological Association and the third Model to Model Workshop, Verlag Dietmar Folbach.

Shibuya, K. (2006). Collaboration and pervasiveness: Enhancing collaborative learning based on ubiquitous computational services, including as Chapter 15. In M. Lytras & A. Naeve (Eds.), *Intelligent learning infrastructures for knowledge intensive organizations: A semantic web perspective* (pp. 369–390). Hershey, PA: IDEA.

Shibuya, K. (2012). A study on participatory support networking by voluntary citizens—The lessons from the Tohoku earthquake disaster. *Oukan, 6*(2), 79–86. (in Japanese).

Shibuya, K. (2017). Bridging between cyber politics and collective dynamics of social movement, Chapter 307. In M. Khosrow-Pour (Ed.), *Encyclopedia of information science and technology* (4th ed., pp. 3538–3548). Hershey, PA: IGI.

Shibuya, K. (2021). Breaking fake news and verifying truth. In M. Khosrow-Pour (Ed.), *Encyclopedia of organizational knowledge, administration, and technologies* (1st ed.). Hershey, PA: IGI. (in press).

Simpson, J. A., & Kenrick, D. T. (1997). *Evolutionary social psychology*. Mahwah, NJ: LEA.

Singh, M. P., Rao, A. S., & Georgeff, M. P. (2000). Formal methods in DAI: Logic-based representation and reasoning. In G. Weiss (Ed.), *Multiagent systems: A modern approach to distributed artificial intelligence*. Cambridge, MA: MIT Press.

Spinoza, B. (2008). Ethica, ordine geometrico demonstrata. BiblioLife (English translated edition).

Stephens, L., et al. (2019). Archaeological assessment reveals Earth's early transformation through land use. *Science, 365*(6456), 897–902.

Susskind, J. (2018). *Future politics: Living together in a world transformed by tech*. Oxford: Oxford University Press.

Suzumura, K. (1983). *Rational choice, collective decisions, and social welfare*. Cambridge: Cambridge University Press.

Taddeo, M., & Floridi, L. (2018). Regulate artificial intelligence to avert cyber arms race. *Nature, 556*, 296–298.

Ulgen, O. (2017). Kantian ethics in the age of artificial intelligence and robotics. *QIL, 43*, 59–83.

UN Chronicle. (2018). Report of the 2018 session of the group of governmental experts on emerging technologies in the area of lethal autonomous weapons systems.

UNODA. (2017). *Perspectives on lethal autonomous weapon systems*. UNODA Occasional Papers No.30. https://www.unog.ch/80256EDD006B8954/(httpAssets)/6866E44ADB996042C12581D400630B9A/$file/op30.pdf

Uzawa, H. (2003). *Economic theory and global warming*. Cambridge: Cambridge University Press.

Vincenot, C. E., Mazzoleni, S., & Parrott, L. (2017). *Hybrid solutions for the modelling of complex environmental systems*. Lausanne: Frontiers Media.

Vliet, J. (2019). Direct and indirect loss of natural area from urban expansion. *Nature Sustainability, 2*, 755–763.

Weibull, J. W. (1997). *Evolutionary game theory*. Cambridge, MA: MIT Press.

Weinberger, S. (2011). Web of war. *Nature, 471*, 566–568.

WMO. (2019). The global climate in 2015–2019. http://ane4bf-datap1.s3-eu-west-1.amazonaws.com/wmocms/s3fs-public/ckeditor/files/Five_year_report_2015-2019_0.pdf?4M6Z45W4mlGplwRxbJnQrgi08Ssq5LXe

Young, H. P. (1998). *Individual strategy and social structure: An evolutionary theory of institutions*. Princeton, NJ: Princeton University Press.

Yu, H., et al. (2018). Building ethics into artificial intelligence. https://www.ijcai.org/proceedings/2018/0779.pdf

Yudkowsky, E. (2018). Artificial intelligence as a positive and negative factor in global risk. In N. Bostrom & M. M. Ćirković (Eds.), *Global catastrophic risks*. Oxford: Oxford University Press.

Zhang, C. (1992). Cooperation under uncertainty in distributed expert systems. *Artificial Intelligence, 56*(1), 21–69.

Chapter 14
Conclusion

This book articulated the nature of *identity of the human* comparing with the AI and big data in the digitized society. The goal of this book was to argue this issue by comprehensive investigations. It firstly reviewed contemporary conditions based on advanced technological tendencies, and it further unveiled social mechanisms on identity of the humankind in depth. For the digitized social era, the author discussed that it should firstly consider the ethical rules of conduct and the ways of control for the human-being rather than the AI. Of course, there are requirements on necessary regulations and ethical framework for the AI and big data. For future investigations and discussions in digital social science, an identity issue would be repeatedly examined in reflexive perspectives.

In brief, there were considerable keywords in this book. Namely it related to *identity* such as *ethical rule of the AI for the dignity of the human-being, valuation of the life, protecting private data, proving myself, social intelligence against deception, collectiveness, synchronization in networking, health, life history and inheritance, equilibrium as power balance to keep between the human, and the AI* and *interconnected ubiquity of the AI, respectively.*

Therefore, it follows from what has been considered thus far that it can summarize remarking discussion points as follows.

1. Due to the advent of social media and ubiquitous computing, global people indeed face at reshaping the actuality to become a digitized society. The larger amount of big data such as people's life history, health, behaviors, and interconnected networking will be exponentially collected and operated to reveal each uniqueness and identity in the context of data assimilation for the purpose of commercial and governmental solutions. There are rich data on their privacy which should be kept by strict regulation and legal backgrounds.

   Big-data management must be globally consented among all nations, and it should be offered by proper ways of keeping privacy of each citizen. Monopoly and oligopoly of big data owned by the larger business companies will be prohibited or regulated by global policies.

K. Shibuya, *Digital Transformation of Identity in the Age of Artificial Intelligence*, https://doi.org/10.1007/978-981-15-2248-2_14

2. Each individual has to reconsider own identical value standing on the own standard. Cyber measurement to evaluate each individual will be arranged and optimized ruling in the social common foundation where everyone shall be obeyed. Each individual has to keep and hide own priceless value without disclosing own private attitudes and preferences, because the digitalized society can easily quantify and evaluate mechanically comparative value of each individual among all of the citizens at once.
3. Identifications among ordinary peoples in the actual nation and online nation (e.g., bit-nation) will be interweaving with actual and cyber activities. Their collective dynamics online and offline have already achieved reshaping the global world in political, economic, cultural, and international contexts. Global online synchronization (and asynchronization) can be prevailingly provoked by several influencers. Eventually, there were severe cases on regime collapses caused by such collective actions has been actualized. In other words, the vulnerability against manipulatable minds among citizens by the external malicious agitators requires being educated for their sound judgments.
4. In global, the ethical rules of conducts of the AI for the humankind have not been enough completed yet, and rather, it should define a set of rules to keep sound relationships between the AI and the humankind by the consensual will among global citizens. The AI-driven machines and robots should hold the ethical rules of conduct for protecting human rights and dignity since shipments from factories. And contrary, software-based invisible entity of the AI has to be concerned about their irregular activities in social contexts. As their asymmetrical advantages will be regarded as lifeless entity, easier repairs, and massive productiveness, the human-being shall prepare to build a kind of the fault tolerant social systems. It will continually keep monitoring all of the AI-driven machines and robots.
5. Balance and equilibrium between the human-being and the AI should always keep the efficient distance for technological safety in long-lasting duration, and the human-being has to perpetually retain the root controls against them. It means that the human-being should perfectly handle the master controls for the AI as servants. The AI-driven services have been overall equipped to be tracked by the human authority, and any solutions proposed by their calculations should be checked by the human judgements grounding in actual contexts.

   And we must be keeping awareness for asymmetry conditions between the humankind and the AI. The smarter intellectual abilities, lacking empathetic imagination for the other living existences, and lifeless advantages of the AI will not be critically considered as an exaggeration when the highest warning calls at front of the emergency situations against the humankind. Therefore, the humankind should keep enhancing own abilities and skills to manage and command them effectively.
6. At the digitized big-data age, medical and health data have been definitively enhanced by the AI and ICT engineering. Medical science and engineering will aim effectively to control the aging of the life and approach prolonging life of each citizen, even though there are inevitable and considerable disputes on the

value and dignity of the human-being. If such regenerative medical operations for the prolonged life and nearly immortality will be achieved, devaluation of each individual should be certainly reconsidered. As whole and part of the physical existence of the human-being should not be poisoned by monetarization and capitalism, own priceless value of each individual has been kept. If at once, when such operations are established, the amount of money for the operations will directly indicate each economic value of each individual. If such operations cost half a million US dollars in regenerative medicines, your physical body as well as prolonged life itself will be evaluated by only such value as the costs. If it stagnates less than one US dollar (or nearly no costs in the future), your value of prolonged life merits only such value. Even if you can contribute toward doing something what you should be achieved, your extended lifespan will be shifted from the priceless value of life with holding the dignity to another value standard of life.
7. Gene therapy and regenerative medical operations have also a meaning against intrusion to living phenomena and modification of each identical background. For revealing and preventing potential risks and criminal conducts, there are opinions that medical charts should be stored in globally sharing database such as blockchain-based distributed database. If genome editing and gene therapy have latent threats and uncertainty against the existential risks of the human-being (Cohen 2019), there are greater needs to be prepared to trace each genetic specifications on medical charts with keeping privacy.
8. There is a possibility that the interconnected world enforces global citizens to converge into the ubiquity of the AI-driven system. And the nature of such society should be deeply considered in various contexts.

At least, these mentioned points will be reexamined by our future generations. Then, taken together, the author displayed one of the highest possibilities, and namely these factors may lastly become the catalyst to crystallize the digitized pantheism by the ubiquity of the central AI in the interconnected world. And of course, there will be many bifurcations at the brink of the crises as frequently, and the most significant is the humankind exhaustively decides, selects, and performs what should do and not to do.

Next, this book's contents can be alternatively paraphrased as below categories based on another perspective on identity, and these topics clearly demonstrate contemporary disputes on digitization led by larger streams of the AI and big data.

1. Existential Modality of the Humankind: Chaps. 2–4, 6, 8, 11–13
2. Assimilating External Standard of Value and Morality: Chaps. 4, 8, 10, 12, and 13
3. Collective Conformity: Chaps. 4, 8–13
4. Interconnected Synchronization: Chaps. 7–11, and 13
5. Manipulatable Minds: Chaps. 4, 6–11, and 13
6. Monitoring of Behaviors: Chaps. 4–7, 11–13
7. Digitization of Data Shared by the Global: Chaps. 1, 4, 5, 12, and 13
8. Conflicts and Balance of Power: Chaps. 3, 6–8, and 13

9. AI Threats: Chaps. 1, 3–5, and 13
10. Prolonged Life-Time: Chaps. 1, 2, 11–13
11. Crises and Disasters: Chaps. 8–13

As it seemed likewise Cartesian, the author firstly defined preconditions for identity of the humankind (Chaps. 1 and 2), subdivided the whole problems on recent technological matters to necessary smaller parts (Chap. 2), and discussed each issue in depth (from Chaps. 3 to 12). After that, at Chap. 13, the author further contemplated those components from another angle for reconstructions, which should be integrated for understanding and reasoning some significant variations on future patterns. All above topics recently appeared as challengeable and considerable matters, and hence those focusing directions which are drawn by the auxiliary lines to lead the way of a future possibility will be shown obviously. Mostly, many problems are usually relying on multiple and complicated emerging properties rather than single and unilateral factor, and then each of the discussed topics should be overlaid on each other to visualize the outline and the core of the problems. Whenever the humankind will inevitably confront with unforeseen risks and crisis, mitigations of the consequences caused by those risk factors will be conquered by the wisdoms of people. Probably this role will be also employed by the intensive studies of digital social science.

Of course, there are a lot of controversial issues and latent variables which were not treated enough by through this book. Then this book is still a prolegomena for further researches. Certainly, digital transformation in global level has been an ongoing matter. Wherever unpredictable innovations will reshape our world, we will steeply envision a next horizon through moving dynamics of the human history. Further discussions and emerging problems would be continually investigated by much effort of the future researches.

Last but not least, thank you for reading all through this book.

Reference

Cohen, J. (2019). WHO panel proposes new global registry for all CRISPR human experiments. *Science*. https://doi.org/10.1126/science.aax3948.

Appendix

An Example of R Code on Interactive Dyad Dynamics

Please see detailed descriptions on Chap. 7. Here, the author demonstrates an example code of R.

```
#-----------------------------------------------------
An Example of R Code on Interactive Dyad Dynamics (inspired by Kaneko, K. and
   Nowak, A models)
#Developed By Kazuhiko Shibuya (this program on dyad model was refined from
   Shibuya, K (2009) )
#-----------------------------------------------------
x1 <- dnorm(0)
x2 <- dnorm(1)
x1; x2
#parameter : r
r1 <- 3.67
r2 <- 3.57
#parameter : a
a <- 0.6
#loop parameter : C
C <- 500
ct <- 0
rx <- array(1:C, dim=c(1, 1))
ry <- array (1:C, dim=c(1, 1))
rz <- array (1:C, dim=c(1, 1))
rz [0] <- -1
# Difference Equation
while (ct<=C) {

  x1 <- (r1 * x1 * (1 - x1) + a * r2 * x2 * (1 - x2)) / (1 + a)
```


K. Shibuya, *Digital Transformation of Identity in the Age of Artificial Intelligence*, https://doi.org/10.1007/978-981-15-2248-2

```
    x2 <- (r2 * x2 * (1 - x2) + a * r1 * x1 * (1 - x1)) / (1 + a)
    rx [ct]<- x1
    ry [ct]<- x2
    rz [ct] <- ct
    ct <- ct + 1
}
#-----------------------------------------------------
#dyad dynamics(X1&X2)
#-----------------------------------------------------
library (tseries)
self.ts1 <- ts (rx, frequency=1)
self.ts2 <- ts (ry, frequency=1)
#-----------------------------------------------------
# Time-Series
x11()
par(mfrow=c(3,2))
start (self.ts1)
end (self.ts1)
frequency (self.ts1)
plot (self.ts1, main="X1", ylab="Evaluation", col="steelblue")
start (self.ts2)
end (self.ts2)
frequency (self.ts2)
plot (self.ts2, main=" X2", ylab="Evaluation", col="tomato1")
#-----------------------------------------------------
#Auto-correlation and auto-variance
acf(self.ts1,type = c("correlation"), main="Autocorrelation(X1)")
acf(self.ts1,type = c("covariance"), main="Autocovariance(X1)")
summary(self.ts1)
acf(self.ts2,type = c("correlation"), main="Autocorrelation(X2)")
acf(self.ts2,type = c("covariance"), main="Autocovariance(X2)")
summary(self.ts2)
#-----------------------------------------------------
#Attractor 1
x11()
plot (self.ts1, self.ts2, main="Interpersonal Dynamics", xlab="Eval(X1)",
   ylab="Eval(X2)", col="firebrick")
#Attractor 2
x11()
plot (self.ts1, self.ts2, main="Interpersonal Dynamics",
xlab="Eval(X1)", ylab="Eval(X2)", col=rainbow(30), type="b", pch=16 )
#-----------------------------------------------------
```

References

Further Web Sites

Asilomar Conference on Beneficial AI. (2017). https://ai-ethics.com/2017/08/11/future-of-life-institute-2017-asilomar-conference/

CiRA (Center for iPS Cell Research and Application), Kyoto University. http://www.cira.kyoto-u.ac.jp/e/

Ethics Department of CiRA. http://www.cira.kyoto-u.ac.jp/uchiro-ethics/

Eurasia Group. (2018). Top 10 risks. https://www.eurasiagroup.net/issues/top-risks-2018

EuroStemCell: Ethics and reprogramming: ethical questions after the discovery of iPS cells. https://www.eurostemcell.org/ethics-and-reprogramming-ethical-questions-after-discovery-ips-cells

How AI will go out of control according to 52 experts. https://www.cbinsights.com/research/ai-threatens-humanity-expert-quotes/ (Mar. 2019 final viewed)

OECD measuring well-being and progress: Well-being research. http://www.oecd.org/statistics/measuring-well-being-and-progress.htm

PDMA (Pharmaceuticals and Medical Devices Agency, Japan). (2018). Regulatory frameworks of regenerative medicines and products review in Japan. https://www.pmda.go.jp/files/000226121.pdf

Report of the 2018 session of the Group of Governmental Experts on Emerging Technologies in the Area of Lethal Autonomous Weapons Systems, Geneva, 9–13 April 2018 and 27–31 August 2018 (CCW/GGE.1/2018/3). https://www.unog.ch/80256EDD006B8954/(httpAssets)/20092911F6495FA7C125830E003F9A5B/$file/CCW_GGE.1_2018_3_final.pdf

Sendai Framework for Disaster Reduction. https://www.unisdr.org/we/coordinate/sendai-framework

The United Nations: Artificial intelligent news. https://futureoflife.org/ai-news/

UN Chronicle. (December 2018). The role of the United Nations in addressing emerging technologies in the area of lethal autonomous weapons systems. https://unchronicle.un.org/article/role-united-nations-addressing-emerging-technologies-area-lethal-autonomous-weapons-systems

UNOG: Background on lethal autonomous weapons systems in the CCW. https://www.unog.ch/80256EE600585943/(httpPages)/8FA3C2562A60FF81C1257CE600393DF6?OpenDocument

World Economic Forum: Artificial intelligence and robotics. https://www.weforum.org/agenda/archive/artificial-intelligence-and-robotics/

K. Shibuya, *Digital Transformation of Identity in the Age of Artificial Intelligence*, https://doi.org/10.1007/978-981-15-2248-2

Further Researches (e.g. Universities, Associations)

AAAI (the Association for the Advancement of Artificial Intelligence). https://www.aaai.org/
Center for Computational Analysis of Social and Organizational Systems, CMU. http://www.casos.cs.cmu.edu/
Complexity Science Hub Vienna (CSH). https://www.csh.ac.at/
Computational Social Science Society of the Americas. https://computationalsocialscience.org/
Cornell University. http://as.cornell.edu/block/computational-social-sciences
COSS: Computational Social Science, Department of Humanities, Social and Political Sciences, ETH Zurich. http://www.coss.ethz.ch/
Digital Social Science Center, Columbia University Libraries. https://library.columbia.edu/locations/dssc.html
Division of the Social Sciences, University of Chicago. https://socialsciences.uchicago.edu/departments/computational-social-science
High-Level Expert Group on Artificial Intelligence, EU. https://ec.europa.eu/digital-single-market/en/high-level-expert-group-artificial-intelligence
IQSS: The Intuited for Quantitative Social Science, Harvard University. https://www.iq.harvard.edu/
IRiSS: Center for Computational Social Science, Stanford University. https://iriss.stanford.edu/css
Oxford Internet Initiative. https://www.oii.ox.ac.uk/
Santa Fe Institute. https://www.santafe.edu/
SSRC (Social Science Research Council) Working Group: Digital Social Science. https://www.ssrc.org/programs/component/digital-culture/digital-social-science/
The Department of Computational and Data Sciences (CDS), George Mason University. https://cos.gmu.edu/cds/

Index

K. Shibuya, *Digital Transformation of Identity in the Age of Artificial Intelligence*, https://doi.org/10.1007/978-981-15-2248-2